RADIO STARS

ASTROPHYSICS AND SPACE SCIENCE LIBRARY

A SERIES OF BOOKS ON THE RECENT DEVELOPMENTS
OF SPACE SCIENCE AND OF GENERAL GEOPHYSICS AND ASTROPHYSICS
PUBLISHED IN CONNECTION WITH THE JOURNAL
SPACE SCIENCE REVIEWS

VOLUME 116
PROCEEDINGS

RADIO STARS

PROCEEDINGS OF A WORKSHOP ON
STELLAR CONTINUUM RADIO ASTRONOMY
HELD IN BOULDER, COLORADO, U.S.A., 8–10 AUGUST 1984

Edited by

ROBERT M. HJELLMING
National Radio Astronomy Observatory,
Socorro, New Mexico, U.S.A.

and

DAVID M. GIBSON
New Mexico Institute of Mining and Technology,
Socorro, New Mexico, U.S.A.

D. REIDEL PUBLISHING COMPANY

A MEMBER OF THE KLUWER ACADEMIC PUBLISHERS GROUP

DORDRECHT / BOSTON / LANCASTER

Library of Congress Cataloging in Publication Data

Main entry under title:

Radio stars.

(Astrophysics and space science library; v. 116)
Workshop sponsored by the Joint Institute of Laboratory Astrophysics, University of Colorado.
Includes indexes.
1. Radio sources (Astronomy)–Congresses. 2. Stars–Radiation–Congresses. 3. Radio astrophysics–Congresses. I. Hjellming, Robert M. II. Gibson, David M. III. Joint Institute for Laboratory Astrophysics. IV. Series.
QB475.A1R35 1985 523.8 85-11766

ISBN-13: 978-94-010-8892-3 e-ISBN-13: 978-94-009-5420-5
DOI: 10.1007/978-94-009-5420-5

Published by D. Reidel Publishing Company,
P.O. Box 17, 3300 AA Dordrecht, Holland.

Sold and distributed in the U.S.A. and Canada
by Kluwer Academic Publishers,
190 Old Derby Street, Hingham, MA 02043, U.S.A.

In all other countries, sold and distributed
by Kluwer Academic Publishers Group,
P.O. Box 322, 3300 AH Dordrecht, Holland.

TABLE OF CONTENTS

PREFACE

This book is the proceedings of a workshop on stellar continuum radio astronomy that was held in Boulder, Colorado on August 8-10, 1984. Although it was originally intended to be a small workshop with participants mainly from North America, it evolved to a workshop with 72 participants from twelve countries (U.S.A. 52, Canada 3, the Netherlands 3, United Kingdom 3, Australia 2, Ireland 2, Italy 2, France 1, Mexico 1, Switzerland 1, West Germany 1, and U.S.S.R. 1). This workshop was sponsored by the Joint Institute of Laboratory Astrophysics (JILA) and the University of Colorado.

In order to preserve a workshop atmosphere, while still presenting both extensive reviews and contributed papers, an experimental format was adopted. All contributed papers related to the topics of the day were presented in poster form in the early morning and were accessible all day. During each morning (or afternoon) session review papers were presented, followed by a coffee break in the poster area adjacent to the conference room. Then the review papers and contributed papers were discussed for roughly one and a half hours. The last session was devoted to invited panel papers and discussion of current and future problems in the field of stellar radio astronomy.

Because of the format of the meeting, the discussions did not necessarily occur after each paper. For this reason they are presented in this volume after the papers appropriate to each session. We have edited the discussions using the hand-written sheets provided by participants and the tapes of the sessions kindly provided to us by Peter D. Jackson. Thanks are due to Jeff Linsky and Dave Abbott for organizing and carrying out the distribution of the discussion sheets that allowed major parts of the discussion to be carefully recorded by the participants.

This book represents an August 1984 overview of the observational and theoretical work on stars that emit continuum radio emission. Many sub-fields of radio stars are covered extensively: the physics of stellar radio emission, stellar winds, interacting stellar winds, bipolar flows, radio-emitting pre-main-sequence stars, dwarf (dMe) flare stars, active binary stars (particularly RS CVn and Algol systems), jets from stars, and x-ray stars that produce both unresolved and extended radio emission.

Financial support for the Workshop was provided by the National Science Foundation, the Research Office of the University of Colorado, and the International Union of Radio Science (U.R.S.I.). We also thank Dr. David W. Norcross, Chairman of JILA, for allowing in-kind contributions of staff time, the JILA auditorium, and mailing expenses in support of the workshop. The Local Organizing Committee would like to acknowledge the exceptional assistance it received in planning, advertising, and carrying-out the Workshop from Debbie Cook (UC-Conference Services), Olivia Briggs (JILA, Sec.-at-large), Lorraine Volsky (JILA Publ. Office), and two student workers at JILA, Lori Dewender and Dominick Dirksen. We also acknowledge Doug Johnson's (JILA) assistance on financial arrangements.

We thank Betty Trujillo (National Radio Astronomy Observatory) and Debbie Brook (New Mexico Tech) for secretarial assistance in the production of this book, and Carol Hjellming for providing the occasional help and advice of a professional editor.

Robert M. Hjellming
David M. Gibson

SCIENTIFIC ORGANIZING COMMITTEE

D.C. Abbott (JILA)
G.A. Dulk (U. Colorado)
D.M. Gibson (New Mexico Tech)
R.M. Hjellming (NRAO), Chairman
S. Kwok (U. Calgary)
J.L. Linsky (JILA)
D.B. Melrose (U. Sydney)
R.L. Mutel (U. Iowa)

LOCAL ORGANIZING COMMITTEE

G.A. Dulk (U. Colorado)
D.M. Gibson (New Mexico Tech), Chairman
J.L. Linsky (JILA)

MEETING PARTICIPANTS

D.C. Abbott (JILA)
R.C. Altrock (NSO)
T.R. Ayres (LASP)
T.S. Bastian (U. Colorado)
M.C. Begelman (U.Colorado)
J.H. Bieging (UC-Berkeley)
N.G. Bochkarev (Sternberg Inst.)
T.J. Bogdan (HAO)
J.A. Bookbinder (Harvard)
P.L. Bornmann (NSO)
C. Boyle (U. Glasgow)
A. Brown (JILA)
J.C. Brown (U. Glasgow)
C.J. Butler (Armagh Obs.)
I. Cairns (U. Sydney)
B.G. Campbell (Caltech)
P.J. Cargill (U. Maryland)
K.G. Carpenter (JILA)
G.A. Chanmugam (LSU)
E.B. Churchwell (U. Wisconsin)
M. Cohen (NASA/Ames)
P.S. Conti (JILA)
J.J. Cox (New Mexico Tech)
D.J. Doiron (Clemson)
J.G. Doyle (Armagh Obs.)
S.A. Drake (JILA)
G.A. Dulk (U. Colorado)
M. Eder (ETH-Zurich)
E.D. Feigelson (Penn. State)
P.L. Fisher (NRAO)
D.R. Florkowski (USNO)
G. Garay (ESO/FRG)
D.E. Gary (Caltech)
D.M. Gibson (New Mexico Tech)
H. Heinrichs (JILA)
R.M. Hjellming (NRAO)
D.E. Hogg (NRAO)
G.D. Holman (NASA/Goddard)
V.A. Hughes (Queens U.)
G.J. Hurford (Caltech)
H. Itoh (JILA)
P.D. Jackson (U. Maryland)
J. Kuijpers (Sonnenborgh Obs.)
M.R. Kundu (U. Maryland)
S. Kwok (U. Calgary)
H. Lamers (Astron. Inst. Utrecht)
J.-F. Lestrade (Bur. Long. Paris/JPL)
J.L. Linsky (JILA)
A. Mackinnon (U. Glasgow)
D.B. Melrose (U. Sydney)
L.A. Molnar (Center for Astrophysics)
D.J. Mullan (Bartol Res. Fdn.)
R.L. Mutel (U. Iowa)
J.E. Neff (U. Colorado)
L.F. Oster (NSF)
I. Oznovich (New Mexico Tech)
R. Pallavicini (U. Florence)
M. Rodono (U. Catania)
L.F. Rodriguez (UNAM)
S.H. Saar (U. Colorado)
T. Simon (U. Hawaii)
A.R. Taylor (DAO/U. Gronigen)
K.C. Turner (Arecibo Obs.)
A.B. Underhill (NASA/Goddard)
D. Van Buren (JILA)
G.H.J. van den Oord (Lab. Sp. Res.)
C.M. Wade (NRAO)
F.M. Walter (LASP)
R.F. Webbink (JILA/U. Illinois)
R.L. White (STScI)
R.F. Willson(Tufts U.)
R.M. Winglee (U. Colorado)

PART I

THEORETICAL ASPECTS OF STELLAR RADIO SOURCES

RADIO OBSERVABLE PROCESSES IN STARS

Jan Kuijpers
Sonnenborgh Observatory
Zonnenburg 2
3512 NL Utrecht
The Netherlands

ABSTRACT. We review the most important physical processes in the atmospheres of (non-compact) stars that can be studied fruitfully with observations in the radio continuum. The emission mechanisms are free-free and gyrosynchrotron radiation, (inverse) Compton scattering and various kinds of plasma radiation. We briefly consider the free-free emission in stellar winds where observations of time variations are important for the study of instabilities and relaxation oscillations. The main part of the paper concerns magnetic activity and in particular stellar flares. Mostly from a theoretical point of view we consider the problems of the density of the flare plasma, the flare energy, dMe flare stars, detached (RS CVn stars) and semi-detached (Algols) close binaries, the role of duplicity, magnetic interactions and mass transfer, T Tauri stars, particle acceleration in shocks and in the unipolar inductor. Finally we review the various kinds of plasma radiation and point out the importance of establishing the brightness temperature of flares with VLBI and of observing the nature and degree of polarization to find out the flare plasma physics.

1. RADIO EMISSION PROCESSES

The various radio emission processes (Rybicki and Lightman 1979; Melrose 1980) allow us to trace the physical conditions in the dilute ionized regions surrounding stars. The power radiated by an accelerated particle with energy E, momentum $\underline{p}$, charge q, rest mass m and Lorentz factor γ ($E = \gamma mc^2$) is

$$P = \frac{2}{3}\frac{\gamma^2 q^2}{m^2 c^3}\left\{ \left|\frac{d\underline{p}}{dt}\right|^2 - \frac{1}{c^2}\left(\frac{dE}{dt}\right)^2 \right\} . \qquad (1)$$

It follows from (1) that for a given force value an extremely relativistic particle ($E \simeq pc$) radiates most efficiently when the force is applied in a direction perpendicular to its motion. Further for the same energy the ratio of the powers radiated by an electron and a proton equals the fourth power of the inverse mass ratio. Therefore in

R. M. Hjellming and D. M. Gibson (eds.), Radio Stars, 3–31.

general it is much more difficult to trace energetic protons from the observed radiation than electrons.

1.1. Free-Free Radiation or Bremsstrahlung

The radiation arises from electron-ion collisions and permits us to determine the ionized gas densities at different radii from the star. Together with an outflow (or infall!) velocity obtained from (UV) lines one can obtain the mass loss (or inflow) rate. In the case of **stellar winds** the method is a reliable one since in the radio (and infrared) one probes the outer regions of the wind where it supposedly has reached a constant outflow velocity. By mistake such radiation is sometimes called thermal; of course a thermal (i.e. Planckian) character of the radiation has nothing to do with the sort of emission process.

1.2. Gyro- and Synchrotron Radiation

In these processes the radiation is emitted by nonrelativistic and, respectively, relativistic particles in magnetic fields. From the observed spectrum and polarization it is possible in principle to derive magnetic field strengths and (individual and total) particle energies or at least a combination of both in the source. The power emitted per unit volume is proportional to $n_f(\gamma^2-1)$ whereas for free-free radiation (a collisional process) it is proportional to $n^2(\gamma^2-1)^{\frac{1}{2}}$ where n is the particle density, n_f is the fast particle density and γ is the Lorentz factor. Therefore the gyrosynchrotron process is more important in **dilute and energetic magnetoplasmas.**

1.3. Thomson-, Compton- and inverse Compton scattering

By these processes the angular and spectral distribution and polarization properties of radiation traversing an (electron) plasma are modified. In the case of Compton scattering hard photons give up part of their energy to the scattering electrons whereas in the case of inverse Compton scattering the reverse occurs and soft photons are upconverted in frequency by scatterings with energetic electrons (Both processes are of course related via Lorentz transformations. In the rest frame of the electron there is only one physical process, elastic Thomson scattering). The scattering processes are of great importance when the radiation energy density becomes substantial or even approaches the energy density in the magnetic field e.g. near **compact objects** (accretion, comptonization in disks, radiatively driven winds). In case of energetic electrons the inverse Compton emissivity can be obtained from the gyrosynchrotron emissivity by replacing the magnetic field energy density with the radiation energy density.

Comptonization of infrared stellar photons by flare generated electrons of 1.5 MeV has been proposed by Gurzadyan (1980) to explain the observed flares (especially in the blue) of dMe stars. A strong observational support to the mechanism appeared to be given by the observed negative flares in the near infrared shortly before the optical flare. A difficulty with this model is that a large amount of flare energy is required, 10^{36} ergs in MeV electrons only. Further the negative optical and near IR burst can be explained "conventionally" by weak impulsive heating, changing the ionization state and thereby increasing the H^- opacity (Grinin 1983; Byrne 1983; see also Andersen 1983)

1.4. Collective or Plasma Radiation

Included in this group are all processes that depend in an essential way on the presence of a plasma. They do not occur in vacuo as the above mechanisms. Further they are of interest mainly in **non-thermal plasmas** with a source of free energy driving some of the plasma modes to high energy densities. Whereas for an equilibrium plasma the brightness temperature (in energy units) can never exceed the characteristic energy of the radiating particles this limitation need not exist in non-equilibrium. Therefore plasma radiation is of special importance in the case of explosions around stars allowing us to find out the energy flows and partitioning in the course of the energy release. A particular form is the so-called transition radiation which occurs when an energetic particle encounters some inhomogeneity in the dielectric properties of the medium and thereby radiates (Ginzburg 1982).

In the following we shall review applications of these continuum emission mechanisms in the radio domain to the surroundings of stars and discuss the underlying physical processes.

2. STELLAR WINDS

Perhaps since the free-free emission from a wind is only weakly dependent on temperature (source function $\alpha\ T^{-0.35}$) and all ion species contribute (Chiuderi and Torricelli Ciamponi 1978) the estimates of mass loss rates from the observed radio and infrared (30-100 μ) spectrum seem rather accurate. Often the spectrum is characterized by a frequency dependence of the flux $S_\nu\ \alpha\ \nu^{0.6}$. The spectral index 0.6 arises for an ionized uniform spherically symmetric mass loss which is sufficiently large so that the optical thickness is much larger than unity at the observing frequencies (Wright and Barlow 1975). For a few stars the interpretation of the emission as free-free radiation from a wind has been directly confirmed by spatially resolved observations (White and Becker 1982, 1983).The main problem is the physical origin of the winds, that is the relative importance of radiation driving, Alfvén waves, shock waves and heating in stars of various types. Since the driver operates mainly at the base of the wind while the radio emission originates in the layers farther out the radio domain is perhaps not the most direct way to reveal the cause of the wind. On the other hand observations indicate winds to be clumpy or in the form of shells and recently several theoretical papers have appeared on a variety of **instabilities in winds** (Martens 1985; Owocki and Rybicki 1984). The physical nature of these instabilities and the existence of relaxation-type coronae-winds (Hearn et al. 1983) can be fruitfully studied in the radio by (quasi-)simultaneous observations at different frequencies. Characteristic growth times for instabilities are a few to ten thousand seconds and for relaxation oscillations a few thousand to a few million seconds.

Recently it has become clear that an appreciable fraction of the early type stellar radio sources is not produced by free-free emission (Abbott 1985).

A special case of interest for the study of winds are **semi-detached binaries.** It is generally found that the final wind velocity is of order of the escape velocity at the photosphere. Now suppose the rate of energy (or momentum) deposition per unit area of a star of given type is determined by the properties of the star independent of its surroundings. Then if the star is in a binary and fills its Roche lobe the mass loss rate per unit area near the inner Lagrangian point will be greatly enchanced in comparison with the loss elsewhere on the star. Let us use a value of ρc_s for the maximum rate of mass transfer per unit area; the radius of the area near the inner Lagrangian point with such a high mass loss can be estimated from $c_s^2 \simeq \Delta\Phi \simeq v_{esco}^2 (r/R_*)^2$. Here ρ is the density, c_s the sound speed, $\Delta\Phi$ the effective gravitational potential near the inner Lagrangian point and $v_{esco}^2 = 2\, GM_*/\, R_*$ is the escape velocity of the (single) star of mass M_* at radius R_*. We then find a mass loss rate of

$$\dot{M} \simeq \pi\, r^2\, \rho c_s = 1.55\ 10^{-10}\ \left(\frac{R_*}{R_\odot}\right)^3 \frac{n}{10^{14}\mathrm{cm}^{-3}} \frac{M_\odot}{M_*} \left(\frac{T}{8000\mathrm{K}}\right)^{3/2} M_\odot \mathrm{yr}^{-1}. \quad (2)$$

If the Roche lobe filling component posesses magnetic structure this may modulate the mass loss rate. Drifting in of a coronal hole above the inner Lagrangian point allows a strong mass loss while a closed magnetic loop structure in that region temporarily cuts the loss down (Fig. 1).

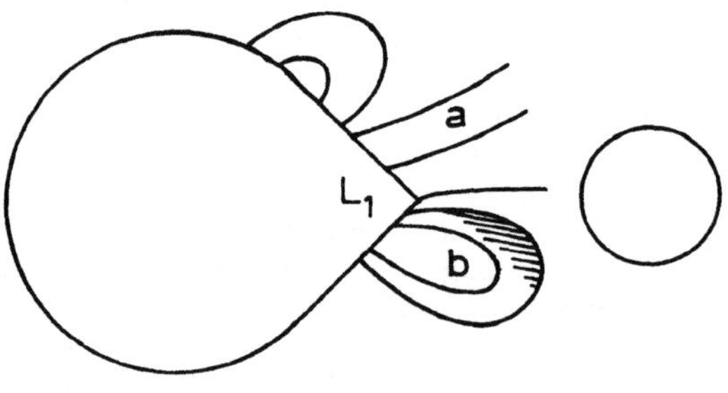

Figure 1. The mass transfer in a semi-detached binary can be controlled by magnetic structures drifting in towards the inner Lagrangian point (L_1). "Open" field structures such as coronal holes (a) permit uninhibited mass-loss. "Closed" structures such as loops and arcades (b) possibly allow for repetitive transfer of cool material assembled at the top of a loop system above L_1 via thermal and subsequent magnetic interchange instabilities.

In the case of **novae** the radio emission (interpreted as free-free radiation) shows the prolonged ejection of gas during years after the initial outburst; possibly in this later phase a radiatively driven wind is set up by a sub-Eddington luminosity powered by hydrogen burning on the surface of the white dwarf (Kwok 1983).

The radio emission from **slow novae** ("radio symbiotic stars") can be understood as free-free emission from the interacting winds of the hot and cool components in a binary (Kwok 1982).

Finally a combination of the mass loss rates of red (super)giants in the optical and the radio allows the determination of the **ionization fraction** in the wind as has been done for α Ori (ionized fraction 10^{-2}, Altenhoff et al. 1979) and for α^1 Her (ionized fraction 2.10^{-3} - 2.10^{-2}, Drake and Linsky 1984).

3. MAGNETIC ACTIVITY

In the presence of magnetic fields the free-free opacity can be increased by **gyroresonance absorption** and a (moderate) degree of circular polarization can be produced as has been shown for active regions in the solar corona (Alissandrakis et al. 1980; Shibasaki et al. 1983).

The radiation observed from active stars such as dMe dwarfs and RS CVn binaries is in general difficult to explain as free-free radiation (Hjellming 1974; Hjellming and Gibson 1980; Gibson 1985a): the required amount of gas, if at X-ray temperatures, is far less than the observed X-ray emission measure (Braes 1974); on the other hand, if the gas were in the form of a cool stellar wind mass loss rates far above conventional values are required (e.g. $10^{-7.5}$ $M_\odot$ yr^{-1} for the case of σ^2 CrB (F6V + G1; Kuijpers and van der Hulst 1984).

Gyroemission of a hot (↑ 10^8 K) thermal gas radiating at the lower harmonics of the electron gyrofrequency is a plausible candidate for some observed so-called quiescent emission at 6 cm from late-type stars requiring a field strength $B \simeq 300$ G at a distance $R/R_* \simeq 3$-4 (Gary and Linsky 1981; Topka and Marsh 1982; Linsky and Gary 1983; Gary et al. 1983). However the process cannot be the cause of the strong radiation from active stars (which usually varies on time scales of hours to days). The reason is that large volumes are required leading to unacceptably large field strengths on the surface of the star if the gas is non-relativistic and therefore radiating at the lower harmonics of the electron gyrofrequency. Further for radio emission from HR 1099 Mutel and Weisberg (1978) found a circular polarization of 70% which is difficult to explain as gyrosynchrotron emission from a thermal source with the required temperature $T = 10^9$ - 10^{10} K (Dulk and Marsh 1982; Dulk 1985). The same conclusion holds for radio flares from YZ CMi which showed $\gtrsim 75\%$ circular polarization (Gibson and Fisher 1981) and 100% circular polarization (Gibson 1984).

Rather several authors (Owen et al. 1976; Spangler et al. 1977; Spangler 1977; Feldman 1983; Borghi and Chiuderi Drago 1984) arrived at the conclusion that these outbursts are **synchrotron radiation** (modified by synchrotron and free-free absorption) from electrons with

Lorentz factors $\gamma \lesssim 10$ in magnetic fields of strengths $B \simeq 30 - 200$ G or, in case of 100% circular polarization, some form of **plasma radiation** (see section 4). So the questions arise how are these fast particles produced and can radio observations discover their origin?

As for the first question we know that in the case of the sun particle acceleration occurs also outside flares and shocks and away from large spots in some sources of microwave emission (Gaizauskas and Tapping 1980; Chiuderi Drago and Melozzi 1984; Fürst et al. 1982) and in type I and low-frequency type III sources.

As to the second question it is improbable that radio observations alone are sufficient to understand the physics of the acceleration regions since only a tiny amount of the flare energy is radiated at radio frequencies. The few existing coordinated observations of flares from stars in X-rays and at UV, optical and radio wavelengths show the importance of simultaneous observations over the entire spectrum (Worden 1983; Haisch 1983) as has been demonstrated of course before for solar flares (e.g. Hoyng et al. 1983; Simnett 1983; Tanaka 1983). Nevertheless radio observations may be essential for our understanding of these processes on stars in the following way.

VLBI observations could establish the **high brightness temperatures** which conclusively prove the operation of coherent plasma mechanisms during the outbursts. So far these high brightness temperatures are only arrived at indirectly. One estimate is obtained by assuming a source area equal to the stellar surface, e.g. Davis et al. (1978) found $T_b = 3\ 10^{12}/\beta^2$ K with $\beta \equiv$ source dimension/$2R_*$ for a flare on a dMe star at 408 MHz detected with an interferometer; Slee et al. (1981) found for a flare on a dMe star at 5 GHz $T_b = 8.5\ 10^9/\beta^2$ K; for RS CVn stars Slee et al. (1984) found at 5 GHz $T_b = 1.1 - 9.8\ 10^9/\beta^2$ K. Another estimate is obtained from the time variability using a signal travel time argument, e.g. Lang et al. (1983) found a lower limit to the brightness temperature of $T_b \gtrsim 10^{13}$ K for fast ($\lesssim 0.2$ s) fluctuations in a radio flare from a dMe star at 1.4 GHz. Evidently one needs sensitive telescopes and long observing times to catch the flares with VLBI. This may not be feasible within the standard network periods; however initially one does not need to map and one ad-hoc baseline during one week outside the network periods is easier to organize (this has been done for the sun successfully by Tapping et al. (1983)). The observed brightness temperature is given by

$$T_b(\nu) \simeq 2.4\ 10^{10}\ \frac{S_\nu}{100\text{mJy}} \left(\frac{4.8\ 10^{-4}{}''}{\alpha}\ \frac{5.10^9\ \text{Hz}}{\nu}\right)^2\ \text{K} \qquad (3)$$

for a source with a circular projection of radius α in arcsec and an unpolarized flux density S_ν at frequency ν. (The apparant radius of a star with $R_* = R_\odot$ at a distance of 10 pc is $\alpha = 4.8\ 10^{-4}$ ''). Alternatively VLBI observations can resolve incoherent gyrosynchrotron sources, as has been done recently for some RS CVn stars by Lestrade et al. (1984) and by Mutel et al. (1984).

Concerning the problem of the **origin of hard X-rays and microwaves**, also relevant for the solar case (beams, conduction fronts

(Brown et al. 1979), shocks), simultaneous observations of flares in RS CVn binaries in X-rays and in the radio domain may be important considering the relatively large ratio of magnetic scale (and flare) heights to stellar radius in these systems and the possible occurrence of eclipses of microwaves or X-rays (see below).

3.1. Density of Flare Plasma

Explosive releases of stored energy can occur in magnetically dominated dilute plasmas with $\beta \ll 1$ ($\beta = p_{gas}/p_{mag}$) which are adjacent to plasmas with $\beta \gtrsim 1$ (cf. the solar atmosphere). Now the value of photospheric magnetic fields in stars with convective mantles is probably determined by flux expulsion in the granular convection and a downdraft instability, leading to (Spruit 1984)

$$B_o \simeq 3\, B_e\, (g/g_{ms})^{\frac{1}{2}}. \qquad (4a)$$

Here g is the acceleration of gravity (ms is main sequence) and B_e is the strength of the field in equipartition with the granular flow

$$B_e = (4\pi\rho)^{\frac{1}{2}}\, v_c, \qquad (4b)$$

where v_c is the convective velocity and ρ the density in the photosphere. Therefore the value of the field strength does not vary much in the photosphere for cool stars (> F 2) on the main sequence. On the other hand the total flux concentrations may differ greatly from the solar values. In fact observations of photometric waves in RS CVn stars and dMe dwarfs (Vogt 1981) indicate that up to 40% of the stellar surface is covered by magnetic fields. Further, observations of Zeeman "broadening" of line profiles in G and K stars (Marcy 1983) show that up to 75% of the area is sometimes covered by fields above 10^3 G. As a consequence the magnetic scale height in the stellar corona can reach a value comparable to the stellar radius, or ten times the solar value. Using a value of 3000 G for the maximum coronal field and a coronal temperature of $2\; 10^6$ K the maximum flare density compatible with the condition $\beta \ll 1$ in the stellar corona is given by

$$n \ll 5\; 10^{14}\ \mathrm{cm}^{-3} \qquad (5)$$

corresponding to a plasma frequency $f_{pe} \ll 200$ GHz. Therefore it could very well be that in some stars one usually cannot observe the flare directly in microwaves since the observing frequency is below the plasmafrequency in the flare (Of course in general one expects to see radio emission later on as a result of an outward travelling transient or shock wave). Since VLBI observations up to 100 GHz are now becoming possible such observations of flaring stars would be of great interest (preceded of course by single dish observations to prove emission at these frequencies). Further note that an upper limit of 3000 G to the coronal field gives a maximum cyclotron frequency of $f_{ce} \simeq 9$ GHz. Consequently the characteristic behaviour of the ratio f_{ce}/f_{pe} in the flare region and higher as a function of height might be opposite to the solar case in the sense that in the stellar flare region $f_{ce}/f_{pe} \ll 1$ but at larger altitudes $f_{ce}/f_{pe} \gtrsim 1$, the magnetic scale height being larger or comparable to the density scale height. As a result plasma emission from runaway acceleration (Kuijpers et al. 1981) and cyclotron masers (Melrose et al. 1984) may be a much more common phenomenon at lower frequencies for these stars than in the solar case.

3.2. Flare Energy

It is well known that the largest flares observed on stars other than the sun are at least a factor one thousand more energetic (both in luminosity and total energy) than on the sun, e.g. up to 10^{31} erg s^{-1} in X-rays for the dMe stars YZ CMi (Heise et al. 1975; Heise 1984) and AT Mic (Kahn et al. 1979) and for the G-K binary HD 27130 (Stern and Zolcinski 1983, period 5.6 d). In fact most of them could not have been observed had they been much smaller. **Why are they so much more energetic?** Three groups of non-compact flaring objects stand out (cf. Feldman and Kwok (1979) and Hjellming and Gibson (1980)): 1. red-dwarf stars, in particular main-sequence dKe and dMe stars, 2. RS CVn close binaries, slightly evolved off the main sequence and 3. T Tauri stars, pre-main sequence objects. The flaring objects in these groups have in common a rather high rotational velocity ($v \geq 5$ km s^{-1}, period few days) and a convective mantle. Both properties are traditionally believed to cause the operation of a stellar dynamo and the occurrence of flares (Rosner 1983; Schüssler 1983; Weiss 1983; Ruzmaikin 1985); the more pronounced rotation and convection are, the stronger the expected activity is (this may not be true if the star is fully convective due to the absence of a stably-stratified region; see also Giampapa (1983)). For a possible connection to activity in cool supergiants we refer to Stencel (1983) and Hjellming and Gibson (1980). For flare activity from contact binaries we refer to Egge and Pettersen (1983), Hughes and McLean (1984), Xuefu and Chengzhong (1983) and Rucinsky et al. (1983).

3.2.1. dKe and dMe stars. This group must be extended to encompass all main-sequence stars with type later then F2 which possess convective mantles and the spotted pre-main sequence dwarfs (BY Dra-type), together the so-called red-dwarf stars. It is mainly for historical and observational reasons that dMe stars stand out since flares on stars were discovered in the optical and it is only in the (nearby) late-type dwarfs that the quiescent stellar optical emission is sufficiently small to allow optical flare detection. Although the majority of stars probably go through the flare star phase after the period of their formation (T Tauri stage) so far the theory of stellar evolution has not been able to explain this important phenomenon (Ambartsumian 1980).

The M stars form 80% of the galactic population; 5 % of this group are dMe stars of which at least 25% show flare activity (Lovell 1971). The characteristic flare energy as derived from optical measurements is $3.10^{31 \pm 2}$ erg (Gershberg and Chugainov 1967a,b); the largest estimate is $10^{34.8}$ erg for YZ CMi (Kunkel 1969); the rise time in the optical is $1-10^3$ s and the decay time $1-10^2$ min (Byrne 1983). A review of the radio emission from dMe stars is given by Gibson (1983).

Simultaneous observations of dMe flaring stars at radio, optical and other frequencies show a broad time correlation to exist between the occurrence of flares in the radio (5 GHz, Haisch et al. 1981, 1978; 430, 318, 196 MHz, Spangler and Moffett 1976; 240, 480 MHz, Lovell 1971) and optical domains with a tendency for the peak radio emission at low frequencies to arrive 5 min later or more. However there is **no detailed correllation.** Often flares occur only in one spectral domain (Gibson 1983). One X-ray flare from Prox Cen (Haisch et al. 1981) showed no UV, optical or radio (5 GHz) emission. The absence of UV and optical emission could be understood if the flare occurred in a **large loop** (of length $\pi\ 10^{10}$ cm $\simeq 2.5\ R_*$) at a characteristic density of $n_e \simeq 10^{11}$ cm^{-3} (note that f_{pe} = 3 GHz!) so that the energy loss was mainly by X-rays (radiative cooling time of 20 min comparable to the observed decay time at 0.2 - 4 keV). Haisch et al. (1983) arrived at a similar conclusion for another flare from the same star. Alternatively **occultation** of the feet of the flaring loop can explain the observed absence of optical and UV radiation. The absence of observed radio emission can be understood if the radio flux is strongly **anisotropic** (e.g. coming from an electron cyclotron maser, section 4.1 (Haisch et al. 1981)). Alternatively the explanation could be the relatively high value of the flare plasmafrequency.

Recent high time and spectral resolution spectroscopy of M-dwarf flares both in the optical and UV reveal that these stellar flares are remarkably similar to solar flares in spectroscopic phenomenology (Worden 1983). If stellar flares originate from the catastrophic loss

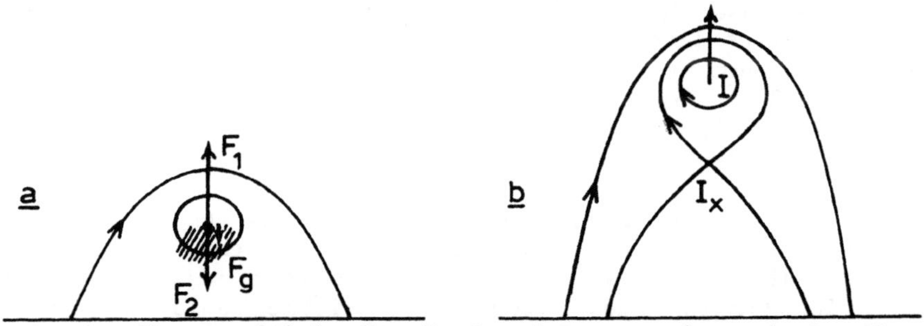

Figure 2. In the flare model by Van Tend and Kuperus (see a) the flare is initiated by the loss of mhd stability when a direct filament current I surpasses a threshold determined by the background field structure; F_1 is the upward Lorentz force on the current filament due to the conducting boundary, F_2 is the downward Lorentz force due to the background field and F_g is the surplus gravity force. Following Kaastra (1985) the upward motion of the "flaring" filament (see b) causes heating and runaway accelaration in a current layer I_x near the X-type neutral line.

of stability when a coronal current surpasses a critical value as in the model of Van Tend and Kuperus (1978) for solar flares a simple argument relates the flare energy to the stellar atmospheric parameters. In their model a direct current perpendicular to the background magnetic field is kept at an equilibrium height h above the photosphere (Fig. 2a). The repulsive Lorentz force between current filament and virtual mirror current at depth h below the photosphere is balanced by the downward Lorentz force of the filament in the background (spot) magnetic field (and the gravitational force on the filament which can be neglected). Then the equilibrium current at height h is

$$I(h) = chB(h). \qquad (6)$$

The flare energy is given by the first maximum in (6), where the equilibrium becomes unstable (note that $I(h) \downarrow 0$ for $h \to \infty$). Writing for the circuit inductance $L = \eta\, \ell/c^2$, where ℓ is the length of the circuit and η is determined by the current profile and is of order unity, the free energy is

$$W = \tfrac{1}{2} L I_{max}^2 = 4.63 \quad 10^{36}\ \eta \frac{\ell}{R_*} \left(\frac{H}{R_*}\right)^2 \left(\frac{R_*}{0.3R_\odot}\right)^3 \left(\frac{B}{1000G}\right)^2 \text{ erg}. \qquad (7)$$

Here we have substituted the magnetic scale height H for the height of the maximum current.

Further the power dissipated during the rise of the current filament (the "flare") has been calculated by Kaastra (1985) for solar two-ribbon flares (see Fig. 2b). From his results we find

$$P = I_x E \ell \simeq 4.4\ 10^{32} \frac{\ell}{R_*} \frac{H}{R_*} \left(\frac{R_*}{0.3\ R_\odot}\right)^2 \left(\frac{B}{1000G}\right)^2 \left(\frac{v}{100\ \text{kms}^{-1}}\right) \text{ergs}^{-1}. \qquad (8)$$

In (8) have used h = H, a value $I_x = 0.1$ I for the induced current near the X-type line and an electric field $E = v\, B_f/c$, where B_f is the magnetic induction at the X-type line due to the current I at a distance 2/3 h. From (7) and (8) one sees that the **flare energy and the dissipated power increase with increasing magnetic scale height (or spot size!).**

3.2.2. RS CVn stars. The second group consists of detached close binaries, both components cool main-sequence stars(> G0) or only one component of that class while the other is a slightly evolved (sub) giant. The systems show strong and variable radio emission (Feldman 1983) with e-folding variations within 1.5^h to one day (Owen and Gibson 1978; Feldman et al. 1978; Mutel and Lestrade 1984; Slee et al. 1984). Since their separation is only a few stellar radii both components rotate nearly synchronously with the orbital motion. As a result the effective rotation velocities are large and an efficient dynamo may be working in both stars. **Interaction** between the magnetic fields of both stars is expected (Bahcall et al. 1973; DeCampli and Baliunas 1979; Simon et al. 1980; Uchida and Sakurai 1983) due to slight asychronism ($\lesssim 10^{-2}$, as inferred from migration of the light curve outside-eclipse with respect to the orbital phase (Catalano 1983)) or due to differential rotation of individual spots in an "Active Longitude Belt" (Uchida and Sakurai 1983; Vogt and Penrod

1983). Note that differential rotation is not yet detected directly in stars other than the sun (Gray 1982; Wöhl 1983) and that observed changes in the migration rate of the photometric wave only lead to a differential rotation $\Delta\Omega/\Omega \simeq 1.4\ 10^{-3} - 8.2\ 10^{-3}$ (Catalano 1983); however this value can very well represent the drift of the active longitude belt while the individual spots move faster. Such an interaction between the magnetic coronae is expected to lead to dissipation of more magnetic energy, larger flares and more frequently occurring activity. However from the present observations it is not clear that duplicity really increases the flare activity (Gibson 1985a,b). In any case since the magnetic scale height can go up to $R_{\theta} \simeq R_*/3$ Eqs. (7) and (8) show that large flares are possible for the separate components.

Pulsations. The RS CVn binary HR 1099 has shown quasiperiodic radio pulsations at 2.7 GHz and 4.9 GHz with a period of 4 min and, respectively, 5 min (Brown and Crane 1978; Newell et al. 1980). If the pulsations are due to standing Alfvén waves in a flux tube (Fig. 3a) containing the radio emitting electrons and connecting both components the average Alfvén speed would be of order 10^4 $(L/2.10^{11}$ cm) km s^{-1}, which does not seem an unrealistic value. Quasiperiodic changes in the radio emission at 4.9 GHz were observed for a flare on L726-8A (dM5.5e) with a period of 56 $\pm$ 5 sec (Gary et al. 1982). In this case the pulsations cannot be connected with the presence of the second component (separation 5 AU). Since the circular polarization during the brightest peak was more than 82% the emission was ascribed to an electron cyclotron maser (see section 4.1) although this does not explain the observed time scale.

In the solar context pulsating radio sources are known to occur with typical periods of the order of seconds, often associated with the passage of a coronal shock wave (Krüger 1979). Also quasiperiodic oscillations with a period of one second have been observed without a flare connection (Gaizauskas and Tapping 1980). Quasiperiodic oscillations have been observed in hard X-ray bursts with a time constant down to 0.1 s (Brown and Loran 1984; Takakura et al. 1983).Even smaller time scales ($\lesssim$ 0.02 s) have been observed at radio frequencies (Slottje 1980; Kaufmann et al. 1980; Kaufmann et al. 1985).

The solar radio pulsations of seconds period have been suggested to arise from **radial fast mode oscillations** ($\tau = 2\pi r/v_{fast} \simeq 1.8$ s , Rosenberg 1970) possibly continuously stimulated (required because of radiative damping) by trapped energetic protons (Meerson et al. 1978) or from **trapped fast mode oscillations** in density enhancements in a low β plasma (Fig. 3b) excited by an impulsive (flaring) source ($\tau = (1 - \rho_e/\rho_o)^{\frac{1}{2}}\ 4a/v_A$, 2a slab width, v_A Alfvén speed, ρ_e external and ρ_o internal density; Roberts et al. 1984). For L726-8a the radial oscillation model leads to a tube radius of $r \simeq 16\ v_A = 0.27\ v_A\ (3000\ \mathrm{km\ s^{-1}})^{-1}\ R_*$.

The subsecond time structure may be due to the acceleration process itself (Kuijpers et al. 1981) or to a maser-like emission process (Melrose and Dulk 1982).

The minutes oscillations could result from torsional Alfvén waves

(Tapping 1983; Fig. 3a).

Finally an oscillation period of order 15 min was predicted for coronal loops by Martens and Kuin (1983); under certain conditions the combined heating and loss processes lead to a **limit cycle** whereby the loops are periodically evacuated (Fig. 3c).

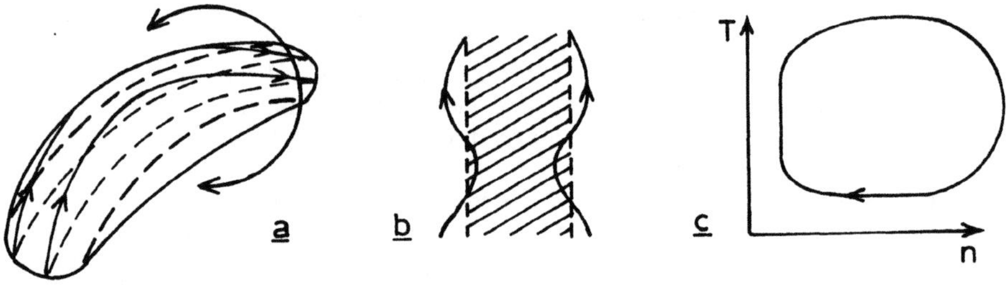

Figure 3. Different kinds of pulsation in a magnetic flux tube: a. standing torsional Alfvén waves (Tapping 1983); b. local radial pulsations (Rosenberg 1970) or fast body waves (Roberts et al. 1984) and c. cyclic behaviour of background density and temperature (Kuin and Martens 1983).

Therefore although the observed quasiperiodic changes and high degree of circular polarization point to a structural order in the source the nature of the physical cause remains at present obscure.

Semi-detached close binaries. The prototype Algol (Wade and Hjellming 1972) is known to be a strong flaring source at radio wavelengths. It is not clear whether the activity has an origin similar to that in RS CVn binaries or whether **mass transfer** due to Roche lobe overflow forms an important extra ingredient (Woodsworth and Hughes 1976). The quiescent X-ray spectrum of Algol is similar to those of the RS CVn's (Charles 1983). Recent EXOSAT observations of an X-ray flare in σ^2 CrB (a RS CVn binary) and in Algol itself (Brinkman et al. 1985; White et al. 1985) look very similar.

The mass transfer leads to shocks on the (non-compact) accreting star (Fig. 4). For a typical mass transfer of 10^{-11} $M_{\odot}$ per event

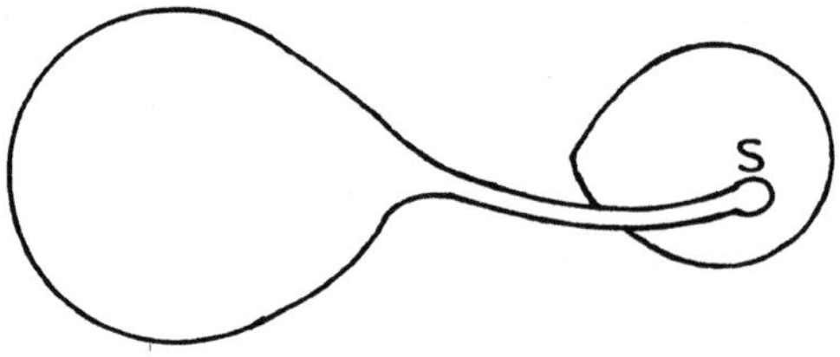

Figure 4. Roche lobe overflow in a binary with a non-compact companion leads to a shock S on the surface of the comparison.

we estimate a post shock density of
$n = 4\Delta M\ (81\ R_{\odot}^{3}\ \mu m_p)^{-1} \simeq 2.8\ 10^{12}\ cm^{-3}$ and a maximum temperature of 3.10^{6} K (corresponding to the free-fall energy). The Debije number of such a shock is of order 10^{6}. Therefore perhaps such a shock leads to (collisionless) radio emission at relatively high frequencies (15 GHz).

3.2.3. T Tauri stars are known to show the strongest X-ray flares observed in non-compact sources (up to 10^{32} erg s^{-1} or 10^{5} the X-ray luminosity of the largest solar flare for objects in the Rho Ophiuchi dark cloud region; Montmerle et al. 1983a,b; Bertout 1984; Feigelson 1984). Flares from T Tauri stars are much less reminiscent of solar flares than dMe stellar flares (Gershberg 1983). For the object DoAr 21 in the ρ Oph cloud also strong and variable ($\tau \simeq 24^{h}$) radio emission at 6 and 20 cm has been observed (Montmerle 1983); its interpretation in terms of a variable mass loss ↑ $10^{-6} M_{\odot}\ yr^{-1}$ is uncertain; a flare interpretation is favoured (Feigelson and Montmerle 1985). In the case of T Tauri South the radio emission can be explained as free-free radiation from infalling gas which is ionized by X, UV radiation from an accretion shock in the form of a compact H II region (Bertout 1983). Whether non-thermal radio emission is absent is not clear. A combination of observations at radio, CO-line, infrared and optical wavelengths (Cohen et al. 1982; Mundt and Fried 1983; Genzel and Downes 1983) suggests the appearance of bipolar flows and possible accretion occurring in a **disk** around the pre-main sequence star ($0.6-3M_{\odot}$)in agreement with a theoretical proposal by Lynden-Bell and Pringle (1974). These authors associate flares with density clumps from the disk releasing their kinetic energy in the lower atmosphere. Alternatively if the disk has a corona with magnetic fields anchored in the disk the situation is similar to a magnetic atmosphere surrounding a convective star and flare activity is expected (Fig. 5a). However many more observations are needed to substantiate such a picture. Again radio observations are important in establishing the non-thermal character of the explosion and the occurrence of particle acceleration. Further in view of a possible

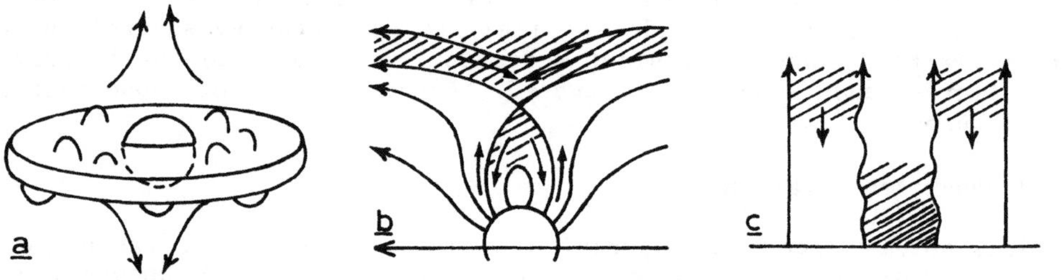

Figure 5. Activity in T Tauri stars may be caused by flares in a disk (a), by ongoing reconnection between stellar mangnetic field and the field in a surrounding accretion cloud (b, Uchida 1983) or by nonspherical clumpy accretion (c is for the case of sub-alfvénic flow).

flaring disk one would like to establish the relative location of the flares with respect to the star.

The observations of T Tauri stars are consistent with the (simultaneous) occurrence of **both outflow an inflow.** An interesting model to explain such flows has been developped by Uchida (1983) and Uchida and Shibata (1984) and is based on ongoing forced reconnection of the magnetic field in the accreting matter with the protostellar parallel (dipole) field (Fig. 5b). Accreting gas falls down freely along the field when reconnection occurs, is being shocked upon hitting the star and partially escapes again along the magnetic field which is now open to the surrounding cloud.

An entirely different model explaining in- and outflow is based on the possible existence of an **Alfvén-Eddington limit** to accretion. The idea is that asymmetric infall of blobs along the magnetic field of a protostar could lead to an appreciable conversion of the infall energy into outgoing Alfvén waves (Fig. 5c). Now the momentum in Alfvén waves is a factor of c/v_A higher than in high-frequency electromagnetic waves for the same net wave energy. Further $v_A \simeq$ few hundred km s^{-1} in T Tauri stellar atmospheres. So the wave pressure in Alfvén waves may be sufficient to stop further infall for some time for quite moderate values of the average accretion rate, smaller than the critical Eddington rate by a factor v_A/c. For a T Tauri star with mass $M = 1M_\odot$ and $v_A = 300$ km s^{-1} this Alfvén-Eddington limit is of order

$$L_{AE} = L_E v_A/c = 1.3\ 10^{35} \frac{M}{M_\odot} \frac{v_A}{300\ \text{kms}^{-1}}\ \text{erg s}^{-1}. \tag{9}$$

To arrive at such a luminosity (in Alfvén waves) the accretion must exceed

$$\dot{M} > 1.5\ \ 10^{-6} \frac{R}{10^{11}\text{cm}} \frac{v_A}{300\ \text{kms}^{-1}} M_\odot\ \text{yr}^{-1}, \tag{10}$$

which compares well with 1 $M_\odot$ in 10^5 yr, believed to be the characteristic assemblage rate for T Tauri stars. Of course the above mechanism requires a regular magnetic field, that is to say sub-alfvénic accretion. If the inflow is occurring at speeds above the local Alfvén speed the field is likely to be distorted severely by the flow. Perhaps such a turbulent accretion could lead to the generation of flares and should be studied theoretically for observational signatures.

3.3 Shock Acceleration

Collisionless shocks at appreciable distances from the parent stars can probably best be studied in the radio domain. Such shocks can be generated by explosive transients (as has been suggested for example by Lovell (1971) who derived for one flare a drift of − 2.8 MHz s^{-1} between 480 and 240 MHz, comparable to solar (shock) type II bursts). Alternatively shocks may constitute steady components in a wide binary where each of the components has a stellar wind, as is the case for many dMe stars but also in systems as α Sco which consists of a M1.5

(Iab) supergiant and a B main sequence star (Klinkhamer and Kuijpers 1981; Huang and Weigert 1982). In the solar case shocks are known to be associated with particle acceleration and radio emission at the first and the second harmonic of the electron plasmafrequency (Dulk 1985). However it is not clear if the acceleration occurs in quasi-longitudinal shocks (shock normal parallel to the magnetic field) by a **first-order Fermi mechanism** (Achterberg and Norman 1980; Webb 1983) or by **drift wave acceleration** in quasi-perpendicular shocks (Holman and Pesses 1983 ; Webb et al. 1983). In both cases only suprathermal particles are accelerated ($v_{//} \gtrsim v_A\ m_i/m_e$ resp. $v_{//} \gtrsim v_s \sec \Psi$, m_i ion mass, v_s shock velocity, Ψ angle between shock normal and field). The injection energy needed for Fermi type acceleration of electrons in shocks is so high under solar circumstances that preacceleration or injection is required. If both components in a binary have a stellar wind a **double shock system** is formed between the stars (Fig. 6); it then seems plausible that the shocks are longitudinal of the kind where Fermi acceleration takes place. Now if a "conventional" flare is occurring

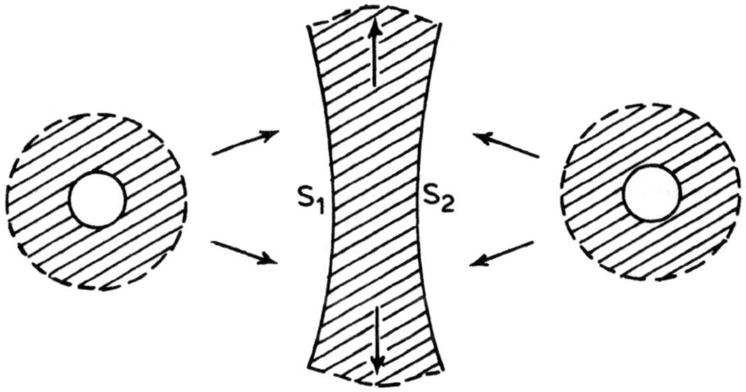

Figure 6. Standing shock system (S_1, S_2) in a wide binary where each component has a wind. The subsonic regions are shown hatched.

near one of the components part of the accelerated electrons will travel out in the wind and meet the standing shocks where they might be further accelerated to cosmic ray energies. Such an acceleration of cosmic rays may be very efficient because of the continuous presence of the shock system. Its existence should be easily detectable in the radio especially in the case of the many comparatively wide dMe binaries. Radio emission in between the components of a binary was for the first time discovered in α Sco ; however the radio emission from this system can be understood as free-free emission from the super-giant chromosphere and from an ionized cavity around its hot companion (Hjellming and Newell 1983; see Hjellming (1985) for similar systems).

3.4. Unipolar Inductor

A special kind of particle acceleration along field lines has been proposed by Piddington and Drake (1968) and Goldreich and Lynden-Bell (1969) for the Io-Jupiter system and applied by Chanmugan and Dulk (1982, 1983) and Lamb et al. (1983) to binaries such as AM Her consisting of a white and a red dwarf. The idea is that one of the components has a magnetosphere or rigidly corotating magnetic field up to the distance of the companion and further that the companion is permeated by a magnetic flux tube that corevolves with the companion (by implication a perfect conductor) at a different rate around the primary. If the flux tube keeps the same form a potential difference is set up in the frame corotating with the primary across the secondary, of magnitude

$$\Delta\Phi = -\int_1^2 \underline{v} \times \underline{B} \cdot d\underline{s} , \quad (11)$$

where $\underline{v}$ is the velocity of the companion with respect to the frame corotating with the primary. Then a current will be set up from one side of the companion along the magnetic field towards the primary where it can cross the flux tube at a sufficiently large density, leaves the primary again along the other side of the flux tube to be closed through the companion (Fig. 7). If the plasma between both components is sufficiently dilute the current can only be carried by high-energy electrons and some form of electrostatic acceleration **(double layers)** is expected to occur in the system. Now several conditions must be fulfilled for the above mechanism to work. First the magnetic field of one of the stars must be sufficiently strong to ensure **corotation** up to a distance equal to the separation of the binary; therefore only close binaries containing a magnetized white dwarf (for which the magnetic moment is large and the magnetosphere reaches far out) or of

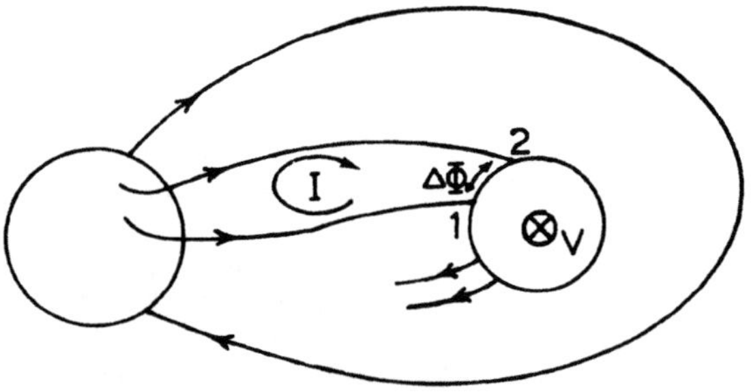

Figure 7. A motion v of the companion relative to the primary creates a potential difference ΔΦ across a connecting magnetic flux tube 1-2 and a current I in the indicated sense.

the RS CVn or Algol type (with a huge "spot") come into consideration. Further a magnetic **connection** must exist between both companions. In the stellar case this probably implies reconnection between the primary field and the local fields of the secondary. Next the components should **not corotate exactly.** Further the flux tube connected to the secondary must be able to **"slide"** over the primary: in the case of the Jupiter-Io system this sliding can occur below the ionosphere where the frozen-in condition is not satisfied; in the stellar case it is not a priori clear whether such sliding actually occurs through Ohmic diffusion in a thin boundary or via reconnection. Finally the resistance of a realistic companion is non-zero but should be sufficiently small so that the distance over which the connecting flux tube diffuses through the companion during an Alfvén crossing time of the binary separation is much less than the diameter of the companion. Probably the latter condition is easily satisfied in close binaries in contrast with the Io-Jupiter system (Hill et al. 1983). The total current can now be expressed as

$$I = \Delta\Phi\ (2R_c + R_p + R_s)^{-1} \simeq \Delta\Phi\ (2R_p)^{-1}, \qquad (12)$$

where R_c is the total resistance along one half of the flux tube, R_p the total resistance of the primary and R_s that of the secondary. The current can run perpendicular to the magnetic field in the stellar atmosphere near the transition region (Spicer 1982) where the electron-ion collision frequency becomes comparable to the ion cyclotron frequency (unmagnetized ions). The voltage available for acceleration of particles is much less than the value $\Delta\Phi/2$ (cf (11) and (12)) and is determined by the coronal density; for σ^2 CrB $\Delta\Phi$ would be of order 100 MeV while the accelerated particles have energies of only 10 keV! (Kuijpers and van der Hulst 1984). Finally the power dissipated is of order (Neubauer 1980; Lamb et al. 1983)

$$P \simeq A\,\frac{B^2}{8\pi}\,v \simeq 1.9\ 10^{29}\ \frac{A}{R_\odot^2}\left(\frac{B}{100G}\right)^2 \frac{v}{1\ \mathrm{km\ s^{-1}}}\,\mathrm{erg}, \qquad (13)$$

where A is the projected area of the satellite magnetic connection perpendicular to its motion with velocity v relative to the rotating primary.

An indication that the radio flaring of AM Her (Chanmugan and Dulk 1983) might not be related to the unipolar inductor is given by the dwarf nova SU UMa. This close binary (consisting of a white dwarf and a late type star with Roche lobe overflow) showed radio emission during optical outbursts (Benz et al. 1983). The radiation is not Bremsstrahlung and somehow connected to the process of **mass transfer,** here in a disk (cf. also Bath and van Paradijs 1983, Osaki 1983).

4. PLASMA EMISSION

The most powerful method of disentangling the physics of stellar flares would be by observing the spectrographic structures in the radio domain resulting from the various waves and instabilities arising during and after the flare (simultaneously with X-ray observations that determine the energy budget). In principle **very**

high brightness temperatures are possible of order

$$T_b{}^t \uparrow T^\sigma(\underline{k}) \simeq \frac{(2\pi)^3\, W^\sigma}{V_k}\,, \tag{14}$$

where W^σ is the energy density of waves of kind σ (t for escaping transverse waves), V_k their effective volume in $\underline{k}$ - space and the temperatures are expressed in energy units. Within an order of magnitude the right-hand side of (14) can be written as $6\pi^2\ N_D\ T\ \varepsilon_1$ with N_D the Debije number, T the plasma temperature and $\varepsilon_1 \equiv W^\sigma_1/nT$ the relative wave energy density.

Apart from a high brightness temperature a further indication for the existence of a plasma radiation process is the observation of completely (circularly) **polarized emission** (the reverse is not true).

One can distinguish between two kinds of instabilities. Those occurring in the primary acceleration (heating) region of the flare and those occurring as a result of propagation effects of the accelerated particles. Many processes of both kinds have been proposed in the context of solar radio flares and bursts. All cases involve a **deviation from thermodynamic equilibrium** either in the form of a beam (energy inversion) of non-thermal particles, some other form of

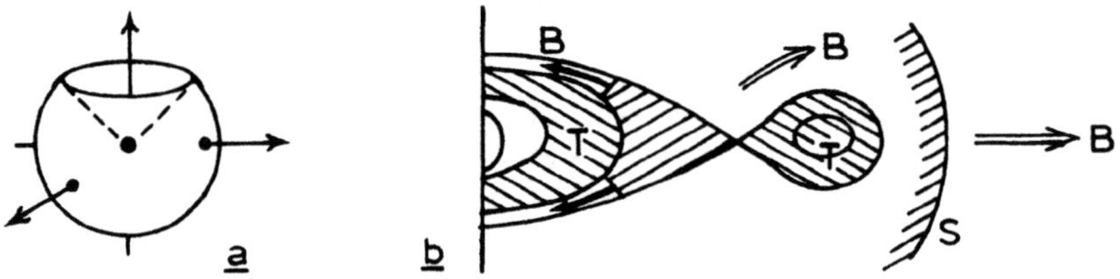

Figure 8. In a. a one-sided loss-cone distribution is sketched in velocity space; here the vertical direction coincides with the outward magnetic field direction. Fast particles are absent within a vertical loss-cone as a result of collisional degradation of energetic particles mirroring at low-lying dense atmospheric levels. In b. propagation of heated gas and accelerated particles (here caused by a filament flare moving to the right) leads to typical classes of instabilities; beam instabilities (B), loss-cone instabilities due to trapped fast particle distributions (T) with two-sided loss-cones or due to mirroring beams (B) with one-sided loss-cones and drift instabilities in shocks (S).

anisotropic velocity distribution or strong spatial gradients. Before we look in more detail into some of these instabilities we consider briefly the various ways of producing observable electromagnetic radiation from them:

1. **Plasma masers** or direct linear instabilities of high-frequency electromagnetic waves ($dW^t/dt \propto W^t$; t transverse waves).

2. **Linear conversion** of plasma waves (generated by an instability) into electromagnetic waves due to inhomogeneities.

3. **Nonlinear conversion** of plasma waves into electromagnetic

waves ($dW^t/dt \propto W^t W^\sigma$ or $dW^t/dt \propto W^\sigma W^{\sigma'}$; σ, σ' plasma waves). Since in general the energy density in plasma waves is much smaller than the kinetic energy density of the plasma ($\varepsilon_1 \ll 1$) the various nonlinear processes can be ordered according to powers of the field amplitudes. Further if the wave packets are sufficiently broad in range of phase velocities the **random phase approximation** can be used. More precisely (Davidson 1972; Porkolab and Chang 1978) the condition is $\varepsilon_2 \equiv \tau_{ac}/\tau_{int} \ll 1$. Here $\tau_{ac} \equiv (k\,\Delta(\omega/k))^{-1}$, the autocorrelation time of a wave packet, is the time during which a resonant particle feels a force exerted by the waves of that packet; τ_{int} the interaction time in which the orbit of a characteristic particle is essentially changed by the wave field, is defined as $\tau_{int} \equiv \min(\tau_{tr}, \tau_D)$ where $\tau_{tr} \equiv (e\,k\,E/m)^{-\frac{1}{2}}$ is the trapping time of a particle of mass m and charge e in a wave with wave vector $\underline{k}$ and electric field amplitude E and $\tau_D \equiv (k^2 D)^{-1/3}$ is the diffusion time of a (resonant) particle under the action of the waves. In this case of so-called **weak turbulence** the processes to second order in the field energy densities are

a **nonlinear scattering and double emission** (Melrose 1982); the waves satisfy the conservation relation

$$\omega_1 - \underline{k}_1 \cdot \underline{v} = \pm (\omega_2 - \underline{k}_2 \cdot \underline{v}) \,, \tag{15}$$

where $\underline{v}$ is the particle velocity and the plus sign is for scattering, the minus sign for double emission (for the magnetized case, $k_\perp v_\perp/\omega_c \lesssim 1$, the conservation relation is $\omega_1 - k_{1//} v_{//} = \pm (\omega_2 - k_{2//} v_{//} + N\omega_c)$ with ω_c the cyclotron frequency and N any integer (Kuijpers (1985)).

b **coalescense** of waves, satisfying

$$\omega_1 + \omega_2 = \omega_3, \tag{16a}$$

$$\underline{k}_1 + \underline{k}_2 = \underline{k}_3. \tag{16b}$$

There is no second-order process other than the above ones, that is turbulent Bremsstrahlung does not exist (Kuijpers and Melrose 1984, Melrose and Kuijpers 1984).

As soon as the wave amplitude becomes so large that $\varepsilon_2 \gtrsim 1$ the random phase approximation breaks down and one should consider the **coherent interactions** between the wave amplitudes and particles (e.g. particle trapping). Again in this case the couplings are most efficient if the waves satisfy relations (15) or (16). From published studies (Goldman et al. 1980; Freund and Papadopoulos 1980; Vlahos et al. 1983) it seems that **strong turbulence** effects only enhance the conversion efficiency into electromagnetic waves.

Mostly in analogy with solar flare physics I shall review the important plasma radiation processes.

4.1. Plasma Masers

The past few years much work has been done on the **cyclotron maser** (for a review see Melrose et al. 1984). When energetic electrons travel from the flare region downwards along a coronal loop they are partially reflected and develop a one-sided loss-cone distribution (Fig. 8a). Then if the ratio ω_{ce}/ω_{pe} is sufficiently high ($\omega_{ce}/\omega_{pe} > 1$) a direct instability can occur of electromagnetic radiation at the first harmonics of the electron cyclotron frequency. Which mode shows the strongest growth rate depends on the ratio ω_{ce}/ω_{pe}. Calculating the quantity temporal growth rate divided by group velocity and multiplied by magnetic scale height Melrose et al. (1984) find that for $\omega_{pe}/\omega_{ce} < 0.3$ the X-mode at the first harmonic grows fastest and for $0.3 < \omega_{pe}/\omega_{ce} < 1.3$ the longitudinal electron plasma wave (Z-mode, modified Langmuir wave, upper hybrid). Although the Z-mode cannot escape directly (being a longitudinal wave) the s = 1,2 electromagnetic modes are expected to suffer severe absorption at the overlying layers where the wave frequency equals two or three times the local electron cyclotron frequency, while coalescence of two upper hybrid waves can generate an escaping high-frequency electromagnetic wave (via (16)) above twice the electron cyclotron frequency so that the radiation is not seriously attenuated on its way out of the corona (absorption $\gamma_{\frac{1}{2}}(s = 2) \simeq (kT_e/mc^2)^{\frac{1}{2}}\ \omega_{pe}^2/\omega$ over a band $\Delta\omega \simeq (kT_e/mc^2)^{\frac{1}{2}}\omega$). The theory has been proposed to explain the terrestrial kilometric radiaton (TKR, X and Z mode), the Jovian decametric radiation (DAM) and the Saturnian kilometric radiation (SKR) (Hewitt et al. 1983) and for solar and stellar microwave (spike) bursts (Holman et al. 1980; Melrose and Dulk 1982 with $T_b = 10^{15} - 10^{20}$ K ; note however that the solar spikes (Slottje 1980) are not always 100% circularly polarized).

Thus we conclude that the direct generation of escaping electromagnetic waves can occur for $\omega_{ce}/\omega_{pe} \gtrsim 3$ although they can suffer severe gyro-resonance absorption on their way to a non-stellar observer.

4.2. Linear Conversion

At sharp **density contrasts** (longitudinal) plasma waves can be linearly converted into electromagnetic waves. In a small β situation under stellar coronal conditions the particle density can vary greatly on different field lines as is known from direct observations of the solar corona and also from the inferred existence of underdense ducts (Duncan 1979). The scale for density variations perpendicular to the field might be as small as a few proton Larmor radii ($r_L = 30\ (T/10^7\ K)^{\frac{1}{2}}\ (100\ G/B)$ cm). Appreciable conversion occurs for scales of $10^2 - 10^3$ km (Melrose 1980).

4.3. Nonlinear Conversion

Here we only treat the basic instabilities for plasma waves since it is often not established which of the conversion processes (15) and (16) actually operate (Kuijpers 1985).

4.3.1. Beams. For $\omega_{ce}/\omega_{pe} \gtrsim 1$ electron beams (Fig. 8b) are unstable for the generation of slow plasma waves through anomalous cyclotron resonance with electrons with velocity components parallel to the beam above a value $v_{//} = [\ 1 + k\ k_{//}^{-1}\ \omega_{ce}\ \omega_{pe}^{-1}\ ]\ v_b$, where v_b is the minimum velocity of the beam particles (Kuijpers et al. 1981; Holman et al. 1982). As a result the beam opening angle is broadened and a bump-in-tail instability for high-frequency Langmuir waves arises at velocities below $v_{//}$.

For $\omega_{ce}/\omega_{pe} \lesssim 1$ electron beams can be unstable for the generation of Langmuir waves through Landau resonance $(v_{//} = \omega/k_{//})$ as is known for type III solar radio bursts although the details are still not clear (Goldman 1983).

Apart from Langmuir waves a variety of low-frequency waves (ion-acoustic, ion- and electron cyclotron, lower hybrid waves and whistlers) can be excited from drifting (Maxwellian) distributions in large electric currents as occur possibly in regions with magnetic reconnection.

4.3.2. Trapped Particle Distributions. Fast electrons accelerated within a coronal loop develop a loss-cone distribution because of colisions at the dense footpoints (Fig. 8b). For $\omega_{ce}/\omega_{pe} \lesssim 1$ such a distribution is unstable for the generation of upper hybrid waves (especially at the locations of double resonance $\omega_{UH} = N\omega_{ce}$, where N is a natural mumber), electron Bernstein waves and low-frequency whistlers (Kuijpers 1980). They are all driven by the **anisotropy** of the electron distribution which is quickly reduced by the effect of pitch angle scattering on the growing waves (scattering time << particle bounce time), very similar to the isotropization occurring for the electron cyclotron maser. They are especially relevant after the flare and in its later phase when ω_{ce}/ω_{pe} becomes much less than unity (Zaitsev and Stepanov 1983).

For $\omega_{ce}/\omega_{pe} \gtrsim 1$ the case has been treated under 4.1. above. Again up to $\omega_{ce}/\omega_{pe} \simeq 3$ the instability of longitudinal waves is the most important.

4.3.3. Shocks. Only for the case of a weak Mach number quasi-perpendicular shock the collisionless shock structure is known with reasonable certainty (e.g. Klinkhamer and Kuijpers 1981). In this case ion-acoustic waves form an essential ingredient of the dissipation in the shock. Perhaps Langmuir or transverse waves are excited simultaneously with the ion-acoustic waves via the double emission-process (15). In any case for the sun flare-generated shocks (Fig. 8b) are known to produce radio emission at the first and second harmonic of the electron plasmafrequency (Dulk 1985). Acceleration by shocks in the form of the first order Fermi process or the drift mechanism (section 3.3) only presupposes some general properties of the shock leaving the detailed wave and particle distributons in the front unspecified. As a result no specific form of plasma radiaton is as yet predicted for strong shocks.

Summarizing we conclude that under a variety of flare related conditions such as beams, loss-cones and shocks both low- and high-frequency longitudinal waves are to be expected at a considerable level, eventually leading to emission of radiation through processes (15) or (16). In the special case $\omega_{ce}/\omega_{pe} \gtrsim 3$ also a direct instability for escaping electromagnetic waves exists. Although this electron cyclotron maser has been proposed several times in the context of emission from stellar flares in my opinion no strong observational support exists to single out this specific instability (see also Holman 1983); any plasma process which leads to a high brightness temperature, fully circularly polarized radiation , strong anisotropy on the angular scale of one radian (fine structures in solar type IV radio continua have a comparable directivity (Slottje 1982)) and strong variations on time scales below 10 s can explain the observations such as by Gibson (1984).

It is true that radiation at the fundamental of the electron plasmafrequency can be seriously attenuated in a smooth corona at higher frequencies by electron-ion collisions (e.g. at f_{pe} = 5 GHz, T = 2 10^6 K and a scale height of 7.5 10^4 km the optical depth for free-free absorption is ~ 10^3 (Dulk 1985)). However, in the case of a magnetically dominated corona abrupt density changes occur perpendicular to the loops and a corresponding scale length below 75 km (making $\tau < 1$) is well possible. Besides a well-documented case exists in the form of solar type IV dm bursts (Kuijpers 1980) where the observations conclusively show that radiation is produced at the fundamental plasmafrequency for $\omega_{ce}/\omega_{pe} \ll 1$. .

5 CONCLUSION

Observations of the time variability of the radio emission from **stellar winds** at multiple frequencies are important in sorting out the dynamic behaviour of the wind, the operating **instabilities** and a possible relaxation - like behaviour. In this case the radio emission is probably largely **Bremsstrahlung** (but see White (1985)).

Gyro-synchotron radiation is important in the presence of coronal **magnetic fields** and hot plasma or accelerated particles. This "quiescent" magnetic activity again can best be studied at multiple frequencies. The nature of the required **heating or acceleration** is not clear. We have pointed out that genuine acceleration can occur also in the absence of flares.

Finally **stellar flares** are best studied with **simultaneous observations** from the X-ray to the radio domain. Aport from gyro-synchrotron emission various kinds of **plasma processes** can be important at radio wavelengths. To study such processes one would require sensitive radio spectrographs as have been used for some decades for the sun. Because of the expected high brightness temperatures **VLBI observations**, possibly up to 100 HGz, are of the utmost importance. Observations of **circular polarization** can give constraints on the kinds of plasma emission, while the detection of **linear polarization** is important in locating the source origin sufficiently far from the star(s). Probably many aspects of stellar flares can be understood with solar flare physics. However it would be surprising if the rich variety of stellar flare conditions such as duplicity, magnetic interactions, filling of a Roche lobe, mass accretion and infall and the presence of stationary shocks were not essential in determining the various flare causes and evolutions.

REFERENCES

Abbott, D.C.: 1985, these proceedings.
Achterberg, A. and Norman, C.A.: 1980, Astron. Astrophys. **89**, 353.
Alissandrakis, C.E., Kundu, M.R. and Lantos, P.:1980, Astron. Astrophys. **82**, 30.
Altenhoff, W.J., Oster, L. and Wendker, H.J.: 1979, Astron. Astrophys. **73**, L21.
Ambartsumian, V.A.: 1980, Ann. Rev. Astron. Astrophys. **18**, 1.
Andersen, B.N.: 1983, in P.B. Byrne and M. Rodonò (eds.) Activity in Red-Dwarfs Stars, I.A.U. Coll. 71, D. Reidel Publ. Cy., Dordrecht, Holland, p. 203.
Bahcall, J.N., Rosenbluth, M.N. and Kulsrud, R.M.: 1973, Nature Phys. Sci. **243**, 27.
Bath, G.T. and van Paradijs, J.: 1983, Nature **305**, 36.
Benz, A.O. Fürst, E. and Kiplinger, A.L.: 1983, Nature **302**, 45.
Bertout, C.: 1983, Astron. Astrophys. **126**, L1.
Bertout, C.: 1984, Rep. Prog. Phys. in press.
Borghi, S. and Chiuderi Drago, F.: 1984, Astron. Astrophys. submitted.
Braes, L.L.E.: 1974, in F.J. Kerr and S.C. Simonson (eds.), Galactic Radio Astronomy, I.A.U. Symp. 60, D. Reidel Publ. Cy., Dordrecht, Holland, p. 377.
Brinkman, A.C., Gronenschild, E.H.B.M., Mewe, R., McHardy, I., and Pye, J.P.: 1985, Adv. Space Res., submitted.
Brown, J.C. and Loran, J.M.: 1984, preprint.
Brown, J.C., Melrose, D.B. and Spicer, D.S.: 1979, Astrophys. J. **228**, 592.
Brown, R.L. and Crane P.C.: 1979, Astron. J. **83**, 1504.
Byrne, P.B.:1983, in P.B. Byrne and M. Rodonó (eds.), Activity in Red-Dwarf Stars, I.A.U. Coll. 71, D. Reidel Publ. Cy., Dordrecht, Holland, p. 157.
Catalano, S.: 1983, in P.B. Byrne and M. Rodonó (eds.), Activity in Red-Dwarfs Stars, I.A.U. Coll. 71, D. Reidel Publ. Cy., Dordrecht, Holland, p. 343.
Chanmugan, G. and Dulk, G.A.: 1982, Astrophys. J. (Letters) **225**, L107.
Chanmugan, G. and Dulk, G.A.: 1983, in M. Livio and G. Shaviv (eds.), Cataclysmic Variables and Related Objects, I.A.U. Coll. 72, D. Reidel Publ. Cy., Dordrecht, Holland, p.223.
Charles, P.A.: 1983, in P.B. Byrne and M. Rodonó (eds.) Activity in Red-Dwarfs Stars, I.A.U. Coll. 71, D. Reidel Publ. Cy., Dordrecht, Holland, p 415.

Chiuderi, C. and Torricelli Ciamponi, G.: 1978, Astron. Astrophys. **69**, 333.
Chiuderi Drago, F. and Melozzi, M.: 1984, Astron. Astrophys. **131**, 103.
Cohen, M., Bieging, J.H. and Schwartz, P.R.: 1982, Astrophys. J. **253**, 707.
Davidson, R.C.: 1972, Methods in Nonlinear Plasma Theory, Academic Press, New York.
Davis, R.J., Lovell, B., Palmer, H.P., and Spencer, R.E.: 1978, Nature **273**, 644.
DeCampli, W.M. and Baliunas, S.L.: 1979, Astrophys. J. **230**, 815.
Drake, S.A. and Linsky J.L.: 1984, Astrophys. J. (Letters) **274**, L77.
Dulk, G.A.: 1985, Ann. Rev. Astron. Astrophys. **23**, in press.
Dulk, G.A., Bastian, T.S. and Chanmugan, G.: 1983, Astrophys. J. **273**, 249.
Dulk, G.A. and Marsh, K.A.: 1982, Astrophys. J. **259**, 350.
Duncan, R.A.: 1979, Solar Phys. **63**, 389.
Egge, K.E. and Pettersen, B.R.: 1983, in P.B. Byrne and M. Rodonò (eds.), Activity in Red-Dwarfs Stars, I.A.U. Coll. 71, D. Reidel Publ. Cy., Dordrecht, Holland, p. 481.
Feigelson, E.D.: 1984, in L. Hartman and S. Baliunas (eds.), Cool Stars, Stellar Systems and the Sun, Springer Verlag.
Feigelson, E.D. and Montmerle, T.: 1985, these proceedings.
Feldman, P.A.: 1983, in P.B. Byrne and M. Rodonò (eds.), Activity in Red-Dwarf Stars, I.A.U. Coll.71, D. Reidel Publ. Cy., Dordrecht, Holland, p. 429.
Feldman, P.A. and Kwok, S.: 1979, J.Roy. Astron. Soc. Canada, **73**, 271.
Feldman, P.A., Taylor, A.R., Gregory, P.C., Seaquist, E.R., Balonek, T.J., and Cohen, N.L.: 1978, Astron. J. **83**, 1471.
Freund, H.P. and Papadopoulos, K.: 1980, Phys. Fluids **23**, 732.
Fürst, E., Benz, A.O. and Hirth, W.: 1982, Astron. Astrophys. **107**, 178.
Gaizauskas, V. and Tapping, K.F.: 1980, Astrophys. J. **241**, 804.
Gary, D.E. and Linsky, J.L: 1981, Astrophys. J. **250**, 284.
Gary, D.E., Linsky, J.L. and Dulk, G.A.: 1982, Astrophys. J. (Letters) **263**, L79.
Gary, D.E., Linsky, J.L. and Dulk, G.A.: 1983, in J.O. Stenflo (ed.), Solar and Stellar Magnetic Fields, I.A.U. Symp. 102, D. Reidel Publ. Cy., Dordrecht, Holland, p. 387.
Genzel, R. and Downes, D.: 1983, in R.M. West (ed.), Highlights of Astronomy **6**, D. Reidel Publ. Cy., Dordrecht, Holland, p. 689.
Gershberg, R.E.: 1983, in P.B. Byrne and M.S. Rodonò (eds.), Activity in Red-Dwarf Stars, I.A.U. Coll. 71, D. Reidel Publ. Cy., Dordrecht, Holland, p. 487.
Gershberg, R.E. and Chugainov, P.F.: 1967a, Soviet Astron. **10**, 934.
Gershberg, R.E. and Chugainov, P.F.: 1967b, Soviet Astron. **11**, 205.
Giampapa, M.S.: 1983, in J.O. Stenflo (ed.), Solar and Stellar Magnetic Fields, I.A.U. Symp. 102, D. Reidel Publ. Cy., Dordrecht, Holland, p. 187.
Gibson, D.M.: 1983, in P.B. Byrne and M. Rodonò (eds.), Activity in Red-Dwarf Stars, I.A.U. Coll. 71, D. Reidel Publ. Cy.,

Dordrecht, Holland, p. 273.
Gibson, D.M.: 1984 in P.F. Gott and P.S. Riherd (eds.), Proc. Southwest Regional Conf. for Astron. Astrophys. **9**,35.
Gibson, D.M.: 1985a, in A. Michalitsianos (ed.), Proc. Workshop on Non-Radiative Heating/ Momentum in Hot Stars, Goddard Space Flight Center, June 1984, in press.
Gibson, D.M.: 1985b, these proceedings.
Gibson, D.M. and Fisher, P.L.: 1981, in P.F. Gott and P.S. Riherd (eds.), Proc. Southwest Regional Conf. for Astron. Astrophys.**6**, 33.
Ginzburg, V.L.: 1982, Physica Scripta **T2/1**, 182.
Goldman, M.V.: 1983, Solar Phys. **89**, 403.
Goldman, M.V., Reiter, G.F. and Nicholson, D.R.: 1980, Phys. Fluids **23**, 388.
Goldreich, P. and Lynden-Bell, D.: 1969, Astrophys. J. **156**, 59.
Gray, D.F.: 1982, Astrophys. J. **258**, 201.
Grinin, V.P.: 1983, in P.B. Byrne and M. Rodonò (eds.), Activity in Red-Dwarf Stars, I.A.U. Coll. 71, D. Reidel Publ. Cy., Dordrecht, Holland, p. 613.
Gurzadyan, G.A.: 1980, Flare Stars, Pergamon Press, Oxford.
Haisch, B.M.: 1983, in P.B. Byrne and M.Rodonò (eds.), Activity in Red-Dwarf Stars, I.A.U. Coll. 71, D. Reidel Publ. Cy, Dordrecht, Holland, p. 255.
Haisch, B.M., Linsky, J.L. Bornmann, P.L., Stencel, R.E., Antiochos, S.K., Golub, L., and Vaiana, G.S.: 1983, Astrophys. J. **267**, 280.
Haisch, B.M., Linsky, J.L., Slee, O.B., Hearn, D.R., Walker, A.R., Rydgren, A.E., and Nicolson, G.D.: 1978, Astrophys. J. (Letters) **225**, L35.
Haisch, B.M., Linsky, J.L., Slee, O.B., Siegman, B.C., Nikoloff, I., Candy, M., Harwood, D., Verveer, A., Quinn, P.J., Wilson, I., Page, A.A., Higson, P., and Seward, F.D.: 1981, Astrophys. J. **245**, 1009.
Hearn, A.G., Kuin, N.P.M. and Martens, P.C.H.: 1983, Astron. Astrophys. **125**, 69.
Heise, J.: 1984, Physica Scripta **T7**, 39.
Heise, J., Brinkman, A.C., Schrijver, J., Mewe, R., Gronenschild, E., and den Boggende, A.: 1975, Astrophys. J. (Letters) **202**, L73.
Hewitt, R.G., Melrose, D.B. and Dulk, G.A.: 1983, J. Geophys. Res. **88**, 10,065.
Hill, T.W., Dessler, A.J. and Goertz, C.K.: 1983, in A.J. Dessler (ed.), Physics of the Jovian Magnetosphere, Cambridge University Press, Cambridge, U.K., p. 368.
Hjellming, R.M.: 1974, in G.L. Verschuur and K.I. Kellermann (eds.), Galactic and Extragalactic Radio Astronomy, Springer Verlag, Berlin, p. 159.
Hjellming, R.M.: 1985, these proceedings.
Hjellming, R.M. and Gibson, D.M.: 1980, in M.R. Kundu and T.E. Gergeley (eds.), Radio Physics of the Sun, I.A.U. Symp. 86, D. Reidel Publ. Cy., Dordrecht, Holland, p. 209.
Hjellming, R.M. and Newell, R.T.: 1983, Astrophys. J. **275**, 704.
Holman G.D.: 1983, Adv. Space Res. **2**, no. 11, p. 181.

Holman, G.D., Eichler, D. and Kundu, M.R.: 1980, in M.R. Kundu and T.E. Gergeley (eds.), Radio Physics of the Sun, I.A.U. Symp. 86, D. Reidel Publ. Cy., Dordrecht, Holland, p. 457.
Holman, G.D., Kundu, M.R. and Papadopoulos, K.: 1982, Astrophys.J. **257**, 354.
Holman, G.D. and Pesses, M.E.: 1983, Astrophys. J. **267**, 837.
Hoyng, P. Marsh, K.A., Zirin, H., and Dennis, B.R.: 1983, Astrophys. J. **268**, 865.
Huang, R.Q. and Weigert, A.: 1982, Astron. Astrophys. **112**, 281.
Hughes, V.A. and McLean, B.J.: 1984, Astrophys. J. **278**, 716.
Kaastra, J.S.: 1985, Astron. Astrophys. submitted.
Kahn, S.M., Linsky, J.L., Mason, K.O., Haisch, B.M., Bowyer, C.S., White, N.E., and Pravdo, S.M.: 1979, Astrophys, J. (Letters) **234**, L107.
Kaufmann, P., Correia, E., Costa, J.E.R., Sawant, H.S., and Zodi Vaz, A.M.: 1985, Solar Phys. in press.
Kaufmann, P., Strauss, F.M. and Opher, R.: 1980, in M.R. Kundu and T.E. Gergeley (eds.), Radio Physics of the Sun, I.A.U. Symp. 86, D. Reidel Publ. Cy., Dordrecht, Holland, p. 205.
Klinkhamer, F.R. and Kuijpers J.: 1981, Astron, Astrophys. **100**, 291.
Krüger, A.: 1979, Introduction to Solar System Radio Astronomy and Radio Physics, D. Reidel Publ. Cy., Dordrecht, Holland.
Kuijpers, J: 1980, in M.R. Kundu and T.E. Gergeley (eds.) Radio Physics of the Sun, I.A.U. Symp. 86, D.Reidel Publ. Cy., Dordrecht, Holland, p. 341.
Kuijpers, J.: 1985, in Trends in Physics, Proc. Sixth General Conf. Europ. Phys. Soc., Prague, August 1984, in press.
Kuijpers, J. and Melrose, D.B.: 1985, Astrophys. J. submitted.
Kuijpers, J. and van der Hulst, J.M.: 1984, Astron. Astrophys. submitted.
Kuijpers, J., van der Post, P. and Slottje, C.: 1981, Astron. Astrophys. **103**, 331.
Kunkel, W.E.: 1969, Nature **222**, 1129.
Kwok, S.: 1982, in M. Friedjung and R. Viotti (eds.), The Nature of Symbiotic Stars, D. Reidel Publ. Cy., Dordrecht, Holland, p. 17.
Kwok, S.: 1983, Mon, Not. R. astr. Soc. **202**, 1149.
Lamb, F.K., Aly, J.-J., Cook, M.C., and Lamb, D.Q.,: 1983, Astrophys. J. (Letters) **274**, L71.
Lang, K.R., Bookbinder, J., Golub, L., and Davis, M.: 1983, Astrophys. J. (Letters) **272**, L15.
Lestrade, J.-F., Mutel, R.L., Phillips, R.B., Webber, J.C., Niell, A.E., and Preston, R.A.: 1984, Astrophys. J. in press
Linsky, J.L. and Gary, D.E.: 1983, Astrophys. J. **274**, 776.
Lovell, B.: 1971, Q. Jl. R. Astr. Soc. **12**, 98.
Lynden-Bell, D. and Pringle, J.E.: 1974, Mon. Not. R. astr. Soc. **168**, 603.
Marcy, G.W.: 1983, in J.O. Stenflo (ed.), Solar and Stellar Magnetic Fields, I.A.U. Symp. 102, D. Reidel Publ. Cy., Dordrecht, Holland, p.3.
Martens, P.C.H.: 1985, in A. Michalitsianos (ed.), Proc. Workshop on The Origin of Non-Radiative Heating/Momentum in Hot Stars,

Goddard Space Flight Center, June 1984, in press.
Martens, P.C.H. and Kuin, N.P.M.: 1983, Astron. Astrophys. **123**, 216.
Meerson, B.I., Sasorov, P.V. and Stepanov, A.V.: 1978, Solar Phys. **58**, 165.
Melrose, D.B.: 1980, **Plasma Astrophysics**, I and II, Gordon and Breach, New York.
Melrose, D.B. :1982, Australian J. Phys. **35**, 67.
Melrose, D.B. and Dulk, G.A.: 1982, Astrophys. J. **259**, 844.
Melrose, D.B., Hewitt, R.G. and Dulk, G.A.: 1984, J. Geophys. Res. **89**, 897.
Melrose, D.B. and Kuijpers, J.: 1984, J. Plasma Phys. in press.
Montmerle, T.: 1983, in J.-C. Pecker and Y. Uchida (eds.), Proc. Japan - France Seminar on Active Phenomena in the Outer Atmospheres of the Sun and Stars, Collège de France, Paris, p. 125.
Montmerle, T., Koch-Miramond, L., Falgarone, E., and Grindlay, J.E.: 1983a, Astrophys. J. **269**, 182.
Montmerle, T., Koch-Miramond, L., Falgarone, E., and Grindlay, J.E.: 1983b, Physica Scripta **T7**, 59. .
Mundt, R. and Fried, J.W.: 1983, Astrophys. J. (Letters) **274**, L83.
Mutel, R.L. and Lestrade, J.-F.: 1984, Astrophys. J. in press.
Mutel, R.L., Lestrade, J.-F., Preston, R.A., and Phillips, R.B.: 1984, Astrophys. J. submitted.
Mutel, R.L. and Weisberg, J.M.: 1978, Astron. J. **83**, 1499.
Neubauer, F.M.: 1980, J. Geophys. Res. **85**, 1171.
Newell, R.T., Gibson, D.M., Becker, R.M., and Holt, S.S.: 1980, in P.F. Gott and P.S. Riherd (eds.), Proc. Southwest Regional Conf. for Astron. Astrophys. **5**, 13.
Osaki, Y.: 1983, in J.-C. Pecker and Y. Uchida (eds.), Proc Japan-France Seminar on Active Phenomena in the Outer Atmospheres of the Sun and Stars, Collège de France, Paris, p. 156.
Owen, F.N. and Gibson, D.M.: 1978, Astron. J. **83**, 1488.
Owen, F.N., Jones, T.W. and Gibson, D.M.: 1976, Astrophys. J. (Letters) **210**, L27.
Owocki, S.P. and Rybicki, G.B.: 1984, Astrophys. J. **284**, 337.
Piddington, J.H. and Drake, J.F.: 1968, Nature **217**, 935.
Porkolab, M. and Chang R.P.H.: 1978, Rev. Mod. Phys. **50**, 745,
Roberts, B., Edwin, P.M. and Benz, A.O.: 1984, Astrophys. J. **279**, 857.
Rosenberg, H. : 1970, Astron. Astrophys. **9**, 159.
Rosner, R.: 1983, in P.B. Byrne and M. Rodonò (eds.), Activity in Red-Dwarf Stars, I.A.U. Coll. 71, D. Reidel Publ. Cy., Dordrecht, Holland, p. 5.
Rucinski, S.M., Vilhu, O. and Kaluzny, J.: 1983, in P.B. Byrne and M. Rodonò (eds.), Activity in Red-Dwarf Stars, I.A.U. Coll. 71, D. Reidel Publ. Cy., Dordrecht, Holland, p. 469.
Ruzmaikin, A.A.: 1985, in Trends in Physics, Proc. Sixth General Conf. Europ. Phys. Soc., Prague, August 1984, in press.
Rybicki, G.B. and Lightman, A.P.: 1979, Radiative Processes in Astrophysics, Wiley, New York.
Schüssler, M.: 1983, in J.O. Stenflo (ed.), Solar and Stellar Magnetic Fields, I.A.U. Symp. 102, D. Reidel Publ. Cy., Dordrecht, Holland, p. 213.

Shibasaki, K., Chiuderi Drago, F., Melozzi, M., Slottje, C., and Antonucci, E.: 1983, Solar Phys. **89**, 307.

Simnett, G.M.: 1983, in P.B. Byrne and M. Rodonò (eds.), Activity in Red-Dwarfs Stars, I.A.U. Coll. 71, D. Reidel Publ. Cy., Dordrecht, Holland, p. 289.

Simon, T., Linsky, J.L. and Schiffer F.H. III: 1980, Astrophys. J. **239**, 911.

Slee, O.B., Haynes, R.F., and Wright, A.E.: 1984, Mon. Not R. astr. Soc. **208**, 865.

Slee, O.B., Touky, I.R., Nelson, G.J., and Renie , C.J.: 1981, Nature **292**, 220.

Slottje, C.: 1980, in M.R. Kundu and T.E. Gergeley (eds.), Radio Physics of the Sun, I.A.U. Symp. 86, D. Reidel Publ. Cy., Dordrecht, Holland, p. 195.

Slottje, C.: 1982, Atlas of Fine Structures of Dynamic Spectra of Solar Type IV-dm and Some Type II Radio Bursts, Ph.D. Thesis, Utrecht University. p. 47.

Spangler, S.R.: 1977, Astron. J. **82**, 169.

Spangler, S.R. and Moffett, T.J. : 1976, Astrophys. J. **203**, 497.

Spangler, S.R., Owen, F.N. and Hulse, R.A.: 1977, Astron. J. **82**, 989.

Spicer, D.S.:1982, Space Sci. Rev. **31**, 351.

Spruit, H.C.: 1983, Mitteilungen der Astron. Gesellschaft. nr. 60, p. 83.

Stencel, R.E.: 1983, in P.B. Byrne and M. Rodonò (eds.), Activity in Red-Dwarfs Stars, I.A.U. Coll. 71, D. Reidel Publ. Cy., Dordrecht, Holland, p. 251.

Stern, R.A. and Zolcinski, M.-C.: 1983, in P.B. Byrne and M. Rodonò (eds.), Activity in Red-Dwarfs Stars, I.A.U. Coll. 71, D. Reidel Publ. Cy., Dordrecht, Holland, p. 131.

Takakura, T., Kaufmann, P, Costa, J.E.R., Degaonkar, S.S., Ohki, K., and Nitta, N.: 1983, Nature **302**, 317.

Tanaka, K.: 1983, in P.B. Byrne and M. Rodonò (eds.), Activity in Red-Dwars Stars, I.A.U. Coll. 71, D. Reidel Publ. Cy., Dordrecht, Holland, p. 307.

Tapping, K.F.: 1983, Solar Phys. **87**, 177.

Tapping, K.F., Kuijpers, J., Kaastra, J.S., van Nieuwkoop, J., Graham, D. and Slottje, C. : 1983, Astron. Astrophys. **122**, 177.

Topka, K. and Marsh, K.A.: 1982, Astrophys. J. **254**, 641.

Uchida, Y.: 1983, in P.B. Byrne and M. Rodonò (eds.), Activity in Red-Dwarf Stars, I.A.U. Coll. 71, D. Reidel Publ. Cy., Dordrecht, Holland, p. 625.

Uchida, Y. and Sakurai, T.:in P.B. Byrne and M. Rodonò (eds.), Activity in Red-Dwarf Stars, I.A.U. Coll. 71, D. Reidel Publ. Cy. Dordrecht, Holland, p. 629.

Uchida, Y. and Shibata, K.: 1984, Publ. Astron. Soc. Japan **36**, 105.

Van Tend, W. and Kuperus, M.: 1978, Solar Phys. **59**, 115.

Vlahos, L, Sharma, R.R. and Papadopoulos, K.: 1983, Astrophys. J. **275**, 374.

Vogt, S.S.: 1981, Astrophys. J. **250**, 327.

Vogt, S.S. and Penrod, G.D.: 1983, in P.B. Byrne and M. Rodonò (eds.), Activity in Red-Dwarfs Stars, I.A.U. Coll. 71, D. Reidel

Publl. Cy., Dordrecht, Holland, p. 379.
Wade, C.M. and Hjellming, R.M.: 1972, Nature **235**, 270.
Webb, G.M.: 1983, Astrophys. J. **270**, 319.
Webb, G.M., Axford, W.I. and Terasawa, T.: 1983, Astrophys. J. **270**, 537.
Weiss, N.O.: 1983, in P.B. Byrne and M. Rodonò (eds.), Activity in Red-Dwarf Stars, I.A.U. Coll. 71, D. Reidel Publ. Cy., Dordrecht, Holland, p. 639.
White, N.E., Culhane, J.L., Parmar, A.N., Kellett, B., Kahn, S., van den Oord, G.H.J., and Kuijpers, J.: 1985 preprint.
White, R.L.: 1985, these proceedings.
White, R.L. and Becker, R.H.: 1982, Astrophys. J. **262**, 657.
White, R.L. and Becker, R.H.: 1983, Astrophys. J. (Letters) **272**, L19.
Wöhl, H.: 1983, in J.O. Stenflo (ed.) Solar and Stellar Magnetic Fields, I.A.U. Symp. 102, D. Reidel Publ. Cy., Dordrecht, Holland,p. 155.
Woodsworth, A.W. and Hughes, V.A.: 1976, Mon. Not. R. astr. Soc. **175**, 177.
Worden, S.P.: 1983, in P.B. Byrne and M. Rodonò (eds.), Activity in Red-Dwarf Stars, I.A.U. Coll. 71, D. Reidel Publ. Cy., Dordrecht, Holland, p. 207.
Wright, A.E. and Barlow, M.J.: 1975, Mon. Not. R. astr. Soc. **170**, 41.
Xuefu, L. and Chengzhong, L.: 1983, in P.B. Byrne and, M. Rodonò (eds.), Activity in Red-Dwarf Stars, I.A.U. Coll. 71, D. Reidel Publ. Cy., Dordrecht, Holland, p. 485.
Zaitsev, V.V. and Stepanov, A.V.: 1983, Solar Phys. **88**, 297.

STOCHASTIC ELECTRON ACCELERATION IN STELLAR CORONAE

T. J. Bogdan
High Altitude Observatory
National Center for Atmospheric Research[1]

R. Schlickeiser
Max-Planck-Institut fur Radioastronomie

ABSTRACT. When coupled with a realistic acceleration model, the radiative signature of flare events in late-type stars is capable of giving additional, and more accurate, information concerning the nature of the accelerating regions than when the radiative signature is used alone. Here we consider the second-order Fermi, or stochastic acceleration mechanism.

1. INTRODUCTION

Typical long duration flares ($\tau \approx$ 4 min - 8 days) on RS CVn stars exhibit peak microwave emission near $\nu \approx$ 5 GHz (i.e. Gibson 1981). If the emission is due to synchrotron emission from relativistic electrons of energy E, in a magnetic field of strength B, then (Tucker 1975)

$$\nu \approx \left(\frac{B}{100 gauss}\right)\left(\frac{E}{1 MeV}\right)^2 \text{GHz} \qquad (1)$$

If one were to equate the lifetime of the flare, τ, to the synchrotron loss time, then a second independent relation is obtained and together with equation (1), E, and B can be uniquely determined.

However, an acceleration mechanism like stochastic electron acceleration (i.e. Fermi 1949; Parker 1957, 1958), gives an independent prediction of the peak electron energy, E, in terms of the physical characteristics of the accelerating region: L^3 - the volume of the region, λ - the electron scattering mean free path, n- the plasma number density, and B. One finds (Bogdan and Schlickeiser 1985)

[1] The National Center for Atmospheric Research is sponsored by the National Science Foundation.

R. M. Hjellming and D. M. Gibson (eds.), Radio Stars, 33–34.

$$\nu \approx \text{minimum of} \begin{cases} 46\left(\dfrac{L}{10^{10}\text{cm}}\right)^2\left(\dfrac{10^7\text{cm}}{\lambda}\right)^2\left(\dfrac{B}{100 gauss}\right)^3 \text{ GHz} \\ 2.2\times 10^4\left(\dfrac{10^9\text{cm}^{-3}}{n}\right)\left(\dfrac{10^7\text{cm}}{\lambda}\right)\left(\dfrac{B}{100 gauss}\right) \text{ GHz} \end{cases} \tag{2}$$

depending upon whether the upper energy cutoff in the electron spectrum is due to the escape of energetic particles from the accelerating volume (upper line) or the balance of energy gains against synchrotron losses (lower line, Schlickeiser (1984)). Similarly the decay time of a flare is the lesser of the electron escape lifetime or the synchrotron loss time.

2. CONCLUSION

The introduction of an acceleration mechanism leaves the problem under determined but brings into play additional characteristics of the accelerating volume. For particle acceleration in small volumes, electron escape considerations determine E and τ, and one would tend to overestimate B and underestimate E, if one supposed synchrotron losses were the dominant factor. (See Bogdan and Schlickeiser 1985).

As an illustrative example, consider a microwave flare of duration $\tau \approx 1$ hour and peak emission near $\nu \approx$ 5GHz. Simply equating τ to the synchrotron loss time and using equation (1), one obtains B $\approx$ 160 gauss and E $\approx$ 1.8 MeV. However, this flare signature is also consistent with the parameters $B_0 = 10_3$ gauss, E $\approx$ 6.8 MeV, L $\approx 10^{10}$ cm, $\lambda \approx 10^6$ cm, and n $\approx 10^9$ cm^{-3}. Here 1 hour is the characteristic particle escape time from the accelerating volume, and the synchrotron loss time is some 60 hours.

REFERENCES

Bogdan, T. J. and R. Schlickeiser 1985, Astr. Ap. (in press).

Fermi, E. 1949, Phys. Rev. 75, 1169.

Gibson, D. M. 1981, in Solar Phenomena in Stars and Stellar Systems, eds.: R. M. Bonnet and A. K. Dupree (Dordrecht: Reidel), p. 545.

Parker, E. N. 1957, Phys. Rev., 107, 830.

Parker, E. N. 1958, Phys. Rev., 109, 1328.

Schlickeiser, R. 1984, Astr. Ap. (in press).

Tucker, W. H. 1977, Radiation Processes in Astrophysics (Cambridge: M.I.T.).

IMPLICATIONS OF THE 1400 MHz FLARE EMISSION FROM AD LEO FOR THE EMISSION MECHANISM AND FLARE ENVIRONMENT

Gordon D. Holman[1], Jay Bookbinder[2], and Leon Golub[2]
[1]NASA Goddard Space Flight Center, Code 682, Greenbelt, Maryland 20771
[2]Harvard-Smithsonian Center for Astrophysics, 60 Garden Street, Cambridge, Mass. 02138

ABSTRACT. High brightness temperature spikes have been observed during a radio flare on the M-dwarf flare star AD Leo (Lang et al. 1983). Their high brightness temperature ($>10^{13}$ K) and circular polarization indicate that a coherent radiation mechanism must be responsible for the spike emission. The underlying flare emission, which is identified with a low polarization, gradual component, was found not to be spiky to within the 200 ms time resolution of the observations. This note is concerned primarily with this non-spiky emission.

The non-spiky emission is about 100 times more intense than the most intense solar flare emission at 1400 MHz, and about four orders of magnitude more intense than a typical solar flare. The brightness temperature of the emission is

$$T_B = 6.41\times10^{10}(S/f^2)(d/R)^2 \quad ,$$

where S is the radio flux at frequency f in Janskys, f is in GHz, d is the distance to the star in parsecs (4.9 pc for AD Leo), and R is the linear size of the source region in solar radii. Taking R to be less than the stellar radius (approximately half the radius of the sun), the brightness temperature is found to be $T_B > 3\times10^{11}$ K. Hence, the brightness temperature of the non-spiky emission is also quite high, and a nonthermal emission mechanism is required.

The emission mechanism for the non-spiky emission may be either incoherent synchrotron radiation or coherent gyrosynchrotron or plasma radiation. Independent of whichever mechanism is responsible, the 1400 MHz observing frequency puts constraints upon the density and magnetic field strength in the emission region. In order to avoid suppression of the radiation by the thermal plasma, the thermal electron density is constrained to $n < 1.24\times10^{10}f^2(\mathrm{GHz}) = 2.4\times10^{10}$ cm^{-3}. For synchrotron or gyrosynchrotron emission, the electron gyrofrequency must be smaller than the observation frequency, giving $B < 357f(\mathrm{GHz}) = 500$ G. Also, the synchrotron loss time becomes shorter than the 15 min rise time of the flare if $B > 500$ G. Even if the emission mechanism does not directly

R.M. Hjellming and D.M. Gibson (eds.), Radio Stars, 35–37.

involve the magnetic field, this condition is required to avoid thermal cyclotron absorption of the radiation. Therefore, the magnetic field strength in the emission region is comparable to or not much greater than a value that is characteristic of solar coronal loops in active regions.

If the emission is assumed to be incoherent synchrotron radiation, the source size and magnetic field are constrained by the fact that the source size cannot be smaller than the size that is required for it to become self-absorbed at the observation frequency. This constrains the linear size of the source to (cf. Pacholczyk 1970, Ramaty and Petrosian 1972)

$$R \gtrsim 3.3\ S^{1/2} B_\perp^{1/4} f^{-5/4} d \quad ,$$

where $B_\perp$ is the component of the magnetic field transverse to the direction of wave propagation and R, S, f, and d are in the same units as above. For the AD Leo flare this gives $R \gtrsim 3.4 B_\perp^{1/4}$. Hence, for $B_\perp$ = 100 G, $R \gtrsim 20$ stellar radii. Even for $B_\perp$ as low as 1 G, $R \gtrsim 7$ stellar radii is required. Taking $B_\perp$ = 1 G, the brightness temperature of the emission is $T_B \gtrsim 7\times10^9$ K and electron energies of at least 1 MeV are required. In summary, for incoherent emission, either a very large (>10 stellar radii) "magnetospheric" source region containing mildly relativistic electrons and ~100 G magnetic fields is required, or a smaller source of low magnetic field strength ($\lesssim$1 G) and fully relativistic electrons is required. In either case, the conditions are quite distinct from those that characterize the impulsive microwave emission from solar flares (cf. Kundu 1965).

An attractive coherent mechanism for the non-spiky emission, as for the spikes, is gyrosynchrotron masering from electrons reflected in a magnetic loop (Holman, Eichler, and Kundu 1980). As was noted by Lang et al., second harmonic emission requires a magnetic field strength of 250 G. An interesting possibility is that the spiky and the non-spiky (which may in fact be spiky on a time scale less than 200 ms) emissions are masering at adjacent harmonics, with the non-spiky emission arising from the higher harmonic. Alternatively, the radiation may be plasma emission at twice the electron plasma frequency or at twice the upper hybrid frequency (see Holman 1983 for a review). Emission near the plasma frequency itself is not as likely, since free-free absorption in the overlying stellar atmosphere is strong. The magnetic field and plasma parameters that are required by each of these mechanisms are comparable to those that are characteristic of flaring regions in the solar corona.

In conclusion, the AD Leo flare emission is clearly not analogous to the impulsive phase microwave emission from solar flares. On the other hand, the high brightness temperature, presence of superimposed spikes, and observing frequency are consistent with the properties of solar decimetric Type IV bursts (see Kundu 1965, Kuijpers 1980). The coherent emission mechanisms are difficult to distinguish observationally, since they all produce high brightness temperature, narrow bandwidth, highly polarized emission. The different mechanisms do depend upon different physical parameters, such as magnetic field strength for gyrosynchrotron masering and electron density for coherent emission at twice the electron plasma frequency, and could be distinguished if the values of these parameters in the emission region could be independently determined. The possibility that the emission is incoherent synchrotron radiation can be tested with polarization, timing, spectral, and high spatial resolution (VLBI) measurements, however, since incoherent emission from relativistic electrons would not be consistent with a high degree of circular polarization, rapid variability would not be consistent with the large source size that is required, and narrow-band spectral features would be indicative of a coherent mechanism.

G.D.H. is a NAS/NRC Resident Research Associate at NASA/GSFC. Studies of coronal plasma processes at the Harvard-Smithsonian Center for Astrophysics are supported by NASA grant NAGW-112.

REFERENCES

Holman, G. D., Eichler, D., and Kundu, M. R. 1980, in IAU Symposium No. 86, "Radio Physics of the Sun", M. R. Kundu and T. E. Gergeley, eds. (Dordrecht: Reidel), p. 457.

Holman, G. D. 1983, Adv. Space Res., Vol. 2, No. 11, 181.

Kuijpers, J. 1980, in IAU Symposium No. 86, "Radio Physics of the Sun", M. R. Kundu and T. E. Gergely, eds., p. 341.

Kundu, M. R. 1965, Solar Radio Astronomy (New York: Interscience).

Lang, K. R., Bookbinder, J., Golub, L., and Davis, M. M. 1983, Ap. J., **272**, L15.

Pacholczyk, A. G. 1970, Radio Astrophysics (San Francisco: W. H. Freeman).

Ramaty, R. and Petrosian, V. 1972, Ap. J., **178**, 241.

CO-ROTATING INTERACTION REGIONS IN STELLAR WINDS: PARTICLE ACCELERATION AND NON-THERMAL RADIO EMISSION IN HOT STARS

D. J. Mullan
Bartol Research Foundation of The Franklin Institute
University of Delaware
Newark, DE 19716

ABSTRACT. A co-rotating interaction region (CIR) forms in a stellar wind when a fast stream from a rotating star overtakes a slow stream. CIR's have been studied in detail in the solar wind over the past decade primarily because they are efficient sources of particle acceleration. Here, we point out the usefulness of CIR's in OB star winds to explain two properties of such winds: deposition of non-radiative energy in the wind far from the stellar surfaces and acceleration of non-thermal particles.

1. INTRODUCTION

Stars emit wind with a velocity which is not in general spherically symmetric. The origins of departure from spherical symmetry are not known, but magnetic fields probably play some role. Since all stars rotate, it is possible for faster wind to catch up with slower wind and form an interaction region which co-rotates with the star. CIR's have been the subject of extensive study for the past decade in the solar wind, mainly because they act as sources of energetic charged particles. When a CIR is fully developed, it consists of a forward shock, a stream-stream interface, and a reverse shock. The shock pair separates in time: at radial distance r, the separation of the shock pair is typically 0.1r in the solar wind.

A detector passing through a CIR in the solar wind records variations in density, temperature, and velocity as shown schematically in Figure 1. The non-linear development of the interaction creates a velocity plateau between the forward and reverse shocks (Hundhausen, 1973). The velocity plateau remains well-defined as the CIR propagates out through the wind (Holzer, 1979). The density has a marked peak while the velocity is passing through its plateau.

The temperature profile in a solar wind CIR is observed to be complicated. Particle acceleration at the shocks suggest (Smith and Wolfe, 1977) that MHD turbulence is enhanced there. The detailed temperature structure in the CIR depends on how this turbulence is dissipated. Model calculations of non-turbulent CIR evolution (Hundhausen, 1973) suggest that strong cooling is expected to occur

R.M. Hjellming and D.M. Gibson (eds.), Radio Stars, 39–42.

in the CIR as it evolves. The dashed lines in Fig. 1 are meant to indicate that this prediction may be modified in particular cases. However, the temperature jumps in the immediate vicinity of the shocks can be predicted reliably.

For the case of stellar observations, the most readily detectable features of CIR's are the velocity plateau and the shock heating. We have proposed (Mullan, 1984) that narrow absorption features observed in the spectra of various types of stars, with associated enhanced levels of ionization, are formed when light from the central star passes through the CIR on the line of sight. Here we discuss the role of CIR's in heating distant regions of OB star winds and accelerating particles.

2. VELOCITIES IN CIR'S

Narrow absorption features are observed in OB stars at velocities which are typically 0.7 times terminal speed v_t (Lamers et al, 1981). If these features are due to CIR's then the velocity jump at a CIR shock in such stars ($v_t \approx (1-3) \times 10^3$ km/sec) is $\approx$ 300–1000 km/sec. Finite lifetimes of the narrow absorption features can be ascribed to propagation time-scales of the CIR's through the wind: in the outlying regions of the wind, the CIR velocity remains essentially unchanged, but the column density along the line of sight ($\propto r^{-3}$) eventually falls below detectability.

3. DEPOSIT OF NON-RADIATIVE ENERGY IN THE WIND

At a shock where the velocity jump is Δv, the Rankine-Hugoniot relations predict the temperature jump. In the limit of a strong shock, with $\gamma = 5/3$, we find $\Delta T \approx (\mu/3R_g)(\Delta v)^2$, where μ is the mean molecular weight, and R_g is the gas constant.

In OB stars, this yields $\Delta T \approx (2-20) \times 10^6$ K, in the immediate vicinity of the CIR shocks. These temperatures are sufficient to make the CIR's act as sources of X-rays. This is an alternative to the Lucy-White mechanism of X-ray emission from "blobs" in the winds. Moreover, enhanced temperatures in CIR's explain why the ionization level is higher in the narrow absorption components (Lamers et al, 1982). However, appreciable cooling in the CIR is required if lines of OVI, NV, CIV and SiIV are to be observed.

4. DISTANCE OF CIR FORMATION

Kinematic arguments can be used to show that CIR's form at a radial distance r_i which is related to the stellar radius R_* by $r_i/R_* \approx v_t/v_r$ = wind speed/rotational speed (Mullan, 1984).

Stars have (v_o/v_r) ratios which may be much smaller than the solar value (200). Hence, CIR's are predicted to form much closer to the surfaces of other stars than in the solar wind. In OB stars, with $v_t \approx (1\text{-}3) \times 10^3$ km/sec and $v_r \approx (1\text{-}3) \times 10^2$ km/sec, v_t/v_r is typically smaller than solar by an order of magnitude. In such stars, CIR's should form at a few tens of stellar radii. At such radii, CIR's have sufficient emission measure to account for much of the Einstein X-ray emission in, say, ζ Puppis (cf. Mullan, 1984). We note that White and Becker (1983) have observed radio emission from a hot star which seems to require excess heating of the wind at many tens of stellar radii from the star. We propose CIR's as the origin of such excess heating in an otherwise (comparatively) cool wind.

5. PARTICLE ACCELERATION IN CIR'S

Fisk and Lee (1980, hereafter FL) have discussed particle acceleration in CIR shocks in terms of a diffusion coefficient K and a shock jump parameter $\beta = (V_d^2 + \Omega^2 r_s^2)^{1/2} B_u/(V_u^2+\Omega^2 r_s^2) B_d$. Subscripts u and d refer to upstream and down- stream, B is magnetic field strength, V is wind speed, r_s is thedistance of the shock from the star, and Ω is the angular velocity of the star. In OB star winds, $r_s \approx$ several tens of stellar radii (R_*). Since $v_r \equiv R_*\Omega \approx 0.1\, v_t$ in these stars, $\Omega r_s > v_t$ i.e. Ωr_s exceeds V_d or V_u. Hence $\beta \approx B_u/B_d$. If the shock is strong, $\beta \approx 1/4$ (assuming $\gamma = 5/3$).

FL consider two forms for K. First $K = K_o vr$, where v is particle speed, r is radial distance in the wind and K_o depends on the spectrum of magnetic turbulence in the wind. With $\beta = 1/4$, the velocity distribution f(v) of the particles accelerated in the shock turns out to be

$$f \propto v^{-4} \exp(-v/v_o)$$

where $v_o = 3V/8K_o$. A distribution function of exponential form fits solar wind proton data from CIR's quite well, with characteristic cut-off velocities $v_o \approx (1\text{-}2) \times 10^9$ cm/sec (corresponding to proton energies of order 1 MeV). We have no way of knowing how K_o behaves in the winds of other stars. (Its value can be evaluated in the solar wind to a fair approximation by in situ measurements of the magnetic power spectrum.) Suppose, however, that K_o is the same in the stellar wind as in the solar wind. Then in OB star winds, where wind speed V can be up to 3000 km/sec, i.e. up to ten times the normal solar wind speed, we expect v_o to be ≈ 10 times larger than in the solar wind. Hence in OB star winds, the exponential cut-off of the proton spectrum would not occur until the velocities had increased to close to the speed of light, i.e. at energies of 100-200 MeV. The possibility of accelerating near relativisitic electrons in such CIR's is of interest in the context of radio stars.

The second form of K suggested by FL is $K = Vr/2$, with particle injection at velocity $v = v_o$. Then at high energies, FL find

$$f(v) \propto v^{-16/3}$$

The differential energy spectrum $j \propto v^2 f(v)$ has the form $v^{-10/3}$ in this case, i.e., $E^{-5/3}$ (for non-relativistic particles; FL discuss only such particles). In the case of relativistic particles, $j \sim E^{-3.3}$: this is the spectrum expected for the electrons which may create non-thermal radio emission. Thus, even the electron spectrum is quite hard, and this again suggests that appreciable fluxes of highly energetic particles may be accelerated in the winds of OB stars.

Abbott (1984) has reported the detection of radio emission from several OB and Wolf-Rayet stars which is definitely non-thermal. We propose that the source of the non-thermal radio emission is near-relativistic electrons accelerated in CIR shocks in the winds, and then acting as a source of gyro-synchrotron emission.

REFERENCES

Abbott, D. 1985 (this conference).
Fisk, L. A. and Lee, M A. 1980, Ap. J. 237, 620.
Holzer, T. 1979, in Solar System Plasma Physics (eds. E. Parker et al) (North Holland), p. 101.
Hundhausen, A. 1973, J. Geophys. Res. 78, 1528.
Lamers, H. et al, 1982, Ap. J. 158, 186.
Mullan, D. J. 1984, Ap. J. 283, 303.
Smith, E. J. and Wolfe, J. H. 1977 in Study of Travelling Interplanetary Phenomena (eds. M. Shea, et al) (Reidel), p.227.
White, R. E. and Becker, R. 1983, Ap. J. (Lett.) 272, L19.

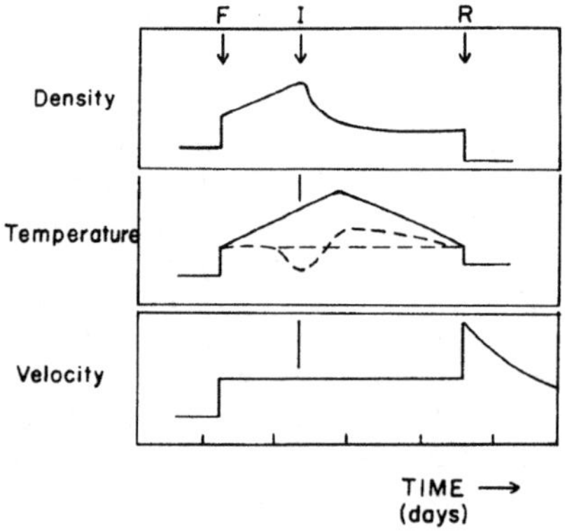

Fig. 1 Variations of solar wind parameters as CIR sweeps past a spacecraft (courtesy of E.J. Smith). The spacecraft, initially in slow wind, is overtaken by F, a forward shock. I is the interface between fast and slow wind, where density has a maximum. R is a reverse shock where the spacecraft emerges into the fast wind. Particle acceleration is observed to be most efficient at F and R.

BOMBARDMENT MODELS OF WHITE DWARF ACCRETION COLUMNS

A. M. Thompson and J. C. Brown
Department of Astronomy,
The University,
Glasgow G12 8QW, U.K.
and
J. Kuijpers,
Sterrewacht Sonnenborgh,
Zonnenburg 2,
Utrecht, The Netherlands.

The problem of the observed low ratio of hard to soft X-rays in accreting white dwarfs, with low accretion rates, (e.g. A M Her) is discussed in terms of models where the accreting matter is treated as a non-thermal stream bombarding a static atmosphere cooled by optically thin radiation (Kuijpers & Pringle, 1982). It is shown that the proton collisional mean free path used by Kuijpers and Pringle (1982) was inappropriate and that, when the correct expression is used the global mean temperature of a steady state bombardment solution is much lower ($\sim 10^5$ K) and much closer to observations.

For normal cosmic abundances, however, optically thin radiative losses cannot in fact balance locally the bombardment energy deposition unless a high gas pressure is invoked at the top of the column. Also the mean temperature is so close to the black body temperature that the optically thin assumption may be suspect. Therefore, without invoking some other kind of energy transport mechanism, bombardment solutions cannot explain the 'Soft X-ray Puzzle'.

In addition we discuss the shock solution of Frank and King (1984) and show that their hypothesis of post-shock energy transport by supra-thermal electrons does not resolve the 'Soft X-ray Puzzle' either since they over-estimate the range of such electrons due to the use of an incorrect electron collisional time scale.

REFERENCES

Frank J. and King A.R.: 1984, Astron. Astrophys. 134, 328.
Kuijpers J. and Pringle J.: 1982, Astron. Astrophys. 114, L4.

R.M. Hjellming and D.M. Gibson (eds.), Radio Stars, 43.

SYNCHROTRON EMISSION FROM CHAOTIC STELLAR WINDS

Richard L. White
Space Telescope Science Institute
Homewood Campus
Baltimore, MD 21218

ABSTRACT. A new model for the radio emission from hot stars is described. Electrons are accelerated to relativistic energies by shocks in the wind near the star and emit radio radiation through the synchrotron mechanism. The particle energy spectrum and radio spectrum for this model have been calculated; the model can account for many of the observed characteristics of some recently discovered stars which have peculiar radio emission. The magnetic fields which are required are quite small (1 to 10 Gauss at the surface of the star). High energy particles accelerated by shocks near the star may also make hot stars significant sources of γ-rays.

1. INTRODUCTION

Radio emission from hot stars usually has been interpreted as free-free emission from the massive ionized winds surrounding these stars. Recent observations suggest a nonthermal origin for the radio emission from some early-type stars (Abbott, Bieging, and Churchwell 1984, hereafter ABC). This paper presents a new model in which nonthermal radio emission is produced by the stellar wind. This paper is a brief summary of calculations which are published in full detail elsewhere (White 1985).

2. THEORY

The radiatively driven winds from O stars are subject to instabilities which Lucy and White (1980) and Lucy (1982b) suggested may lead to chaotic stellar winds with embedded shocks, large density variations, and velocities which do not monotonically increase with radius. This model for the production of X-rays in hot stars has received observational support from several directions (Long and White 1980, Cassinelli and Swank 1983, Lucy 1982a, 1983).

The shocks in winds can accelerate particles to relativistic energies (Bell 1978; Blandford and Ostriker 1978). The presence in stellar winds of many shocks changes the particle spectrum from that produced by a single shock. The particle energy spectrum is determined by the mean free path for high energy particles, the distance between shocks in the wind, and the ratio

R. M. Hjellming and D. M. Gibson (eds.), Radio Stars, 45–46.

of the particle pressure to the postshock gas pressure (White 1985).

Given the particle spectrum, the synchrotron spectrum produced by a stellar wind can be calculated:

$$L_\nu = \begin{cases} L_t \left(\frac{\nu}{\nu_t}\right)^{0.51}, & \nu < \nu_t \\ L_t \left(1 + 0.51 \ln \frac{\nu}{\nu_t}\right), & \nu > \nu_t , \end{cases}$$

where $L_t \simeq 6 \times 10^{19}\,\mathrm{ergs/s}\,(\dot{M}/2 \times 10^{-5}\,\mathrm{M}_\odot/\mathrm{yr})$ and ν_t is a complicated and strong function of many of the model parameters. The synchrotron spectrum is modified by the free-free optical depth of the wind, which is large and varies strongly with frequency.

3. COMPARISON WITH OBSERVATIONS

In general, the model is quite successful at accounting for the observed properties of Cyg OB2 No. 9 (ABC). The fluxes predicted by the model are quite similar to those which have been observed. The model spectrum is flat at high frequencies and steepens to $\nu^{0.5}$ at low frequencies, so that the infrared spectrum is dominated by thermal emission from the wind. The radio emission is compact and cannot be resolved by the VLA at 6 cm. Small changes in the model parameters can cause large changes in ν_t, which can lead to large variations in the radio flux. The stellar magnetic field required to produce the observed ν_t is about 5 Gauss, a plausible value for a hot star.

The model has more difficulty accounting for ABC's observations of 9 Sgr, which has a spectrum that is falling with frequency ($F_\nu \propto \nu^{-0.3}$), but it is likely that a similar model can produce a slowly falling spectrum.

4. CONCLUSIONS

The model makes several predictions which can be tested against future observations:

(1) Nonthermal wind sources should not vary faster than the wind flow timescale and should vary more slowly at low frequencies than at high frequencies.
(2) Most stars with rising non-thermal spectra will not be resolved by the VLA, but some flat spectrum sources might be resolvable.
(3) Hot stars should be weak, unresolved γ-ray sources.

REFERENCES

Abbott, D. C., Bieging, J. H., and Churchwell, E. 1984, *Ap. J.*, in press. (ABC)
Bell, A. R. 1978, *M.N.R.A.S.*, 182, 147.
Blandford, R. D., and Ostriker, J. P. 1978, *Ap. J. (Letters)*, 221, L29.
Cassinelli, J. P., and Swank, J. 1983, *Ap. J.*, 271, 681.
Long, K. S., and White, R. L. 1980, *Ap. J. (Letters)*, 239, L65.
Lucy, L. B. 1982a, *Ap. J.*, 255, 278.
Lucy, L. B. 1982b, *Ap. J.*, 255, 286.
Lucy, L. B. 1983, *Ap. J.*, 274, 372.
Lucy, L. B., and White, R. L. 1980, *Ap. J.*, 241, 300.
White, R. L. 1985, *Ap. J.*, 15 Feb.

DAMPING OF THE MAGNETOIONIC Z MODE

S. M. White, D. B. Melrose
Dept. of Theoretical Physics, University of Sydney, Australia, 2006

and G. A. Dulk
Dept. of Astrophysics, Planetary and Atmospheric Sciences, University of Colorado, Boulder, U.S.A.

ABSTRACT. The magnetoionic z mode has traditionally been of less interest (outside ionospheric physics) than the x and o modes because it cannot escape directly from a distant source. However recent work on the cyclotron maser instability has shown that when the electron plasma frequency, ω_p, is smaller than the electron gyrofrequency, Ω, under plausible conditions most of the free energy available to the instability may be transferred to the z mode rather than the x or o modes (Hewitt et al, 1983; Omidi et al, 1984; White, 1984). The condition $\omega_p < \Omega$ is thought to be satisfied in solar magnetic flux tubes which are the sites of flares and microwave emission. The possibility that a large fraction of the energy released by a flare may be radiated as z-mode waves leads us to consider the fate of such waves. The conclusions are important for understanding the energy balance in solar flaring regions and, by analogy, in stellar radio sources also. In particular, we wish to know if a high z-mode energy density leads to observable radiation. We consider two effects here: damping of the z-mode waves as they propagate through the ambient coronal plasma, and coalescence of z-mode waves to produce second-harmonic radiation.

Exact calculations of the damping of the z mode have been carried out for a plasma with a Maxwellian distribution corresponding to a typical coronal temperature of 2 million K, using the full relativistic gyroresonance condition. In a plasma with $\omega_p < \Omega$, the properties of the damping rate are as follows: (a)the damping rate is very large over a broad bandwidth of frequencies near $\omega = \Omega$, except for angles θ between the wavevector and the magnetic field direction which are close to 90°; (b)damping is small for $\theta \simeq 90°$ at frequencies $\omega < \Omega$; (c)there can be no resonance between the z-mode waves and particles near $\theta = 90°$ for frequencies lying between Ω and the z-mode cutoff frequency, $\omega_+(\theta) = (\Omega^2 + \omega_p^2 \sin^2\theta)^{1/2}$, and so damping is exactly zero in this range; (d)damping exactly at $\omega = \Omega$ is small at all angles θ; (e)there is a narrow band of very strong damping at $\theta \simeq 90°$ at a frequency ω just below the gyrofrequency. The z-mode waves are likely to be generated with $\omega < \Omega$ and $\theta \simeq 90°$, in a region of low damping. As

R.M. Hjellming and D.M. Gibson (eds.), Radio Stars, 47–48.

they propagate the ratio ω/Ω and the angle θ may change due to gradients in the plasma parameters. Damping near $\omega = \Omega$ will prevent such waves from reaching regions in which $\omega/\Omega > 1$. It is concluded that the z-mode waves are likely to be damped at the $\omega = \Omega$ layer, and thereby cause heating of the coronal plasma outside the flaring flux tube (cf. Melrose and Dulk, 1984).

If the z mode is the dominant emission then the brightness temperature of the z-mode waves in the source will be high. Coalescence of two z-mode waves to produce x- and o-mode waves with $\omega \simeq 2\Omega$ will then be an efficient process which could act as a sink for the energy in the z-mode radiation. For this second-harmonic radiation to be observable, it must pass through the absorption layer in the corona at $\omega = 2\Omega$. This is not likely to be possible (Melrose and Dulk, 1984). The second harmonic causes heating of the corona in essentially the same layer as does the z-mode damping. One requires some effect other than coalescence of two z-mode waves in order to produce observable radiation.

REFERENCES

Hewitt, R. G., D. B. Melrose and G. A. Dulk, J. Geophys. Res. **88**, 10065, 1983.
Melrose, D. B., and G. A. Dulk, Astrophys. J. **282**, 308, 1984.
Omidi, N., C. S. Wu and D. A. Gurnett, J. Geophys. Res. **89**, 883, 1984.
White, S. M., Ph.D. thesis, University of Sydney, 1984.

ELECTRON-CYCLOTRON MASER EMISSION DURING SOLAR AND STELLAR FLARES

R. M. Winglee
Department of Astrophysical, Planetary and Atmospheric Sciences
Campus Box 391
University of Colorado
Boulder, CO 80309
USA

ABSTRACT. Radio bursts, with high brightness temperature ($\gtrsim 10^{10}$K) and high degree of polarization, and the heating of the solar and stellar coronae during flares have been attributed to emission from the semi-relativistic maser instability. In plasmas where the electron-plasma frequency, ω_p, and the electron-cyclotron frequency, Ω_e, are such that $\omega_p^2/\Omega_e^2 << 1$, x-mode growth dominates while z-mode growth dominates if ω_p^2/Ω_e^2 is of order unity. The actual value of ω_p^2/Ω_e^2 at which x-mode growth dominates is shown to be dependent on the plasma temperature with x-mode growth dominating at higher ω_p/Ω_e as the plasma temperature increases. Observations from a set of 20 impulsive flares indicate that the derived conditions for the dominance of x-mode growth are satisfied in about 75 percent of the flares.

1. INTRODUCTION

Emission from the semirelativistic maser instability has been proposed as the source of very bright (brightness temperature greater than about 10^{10}K) and highly polarized radio bursts from the sun and flare stars (Holman et al., 1980; Melrose and Dulk, 1982a). It has also been proposed that absorption of radiation from this instability above the source region produces the heating of solar and stellar coronae during flares (Melrose and Dulk, 1982b, 1984).

The maser instability arises when electrons within the flaring flux tube are accelerated in an energy release region. Energetic electrons with small pitch angles precipitate (producing hard X rays) and are lost from the flux tube whereas electrons of higher pitch angle mirror at some point along the flux tube. The resultant distribution, f, above this mirror point has a loss-cone anisotropy. This distribution in the vicinity of the loss cone has a positive gradient with respect to $v_\perp$, the magnitude of the velocity perpendicular to the magnetic field

R. M. Hjellming and D. M. Gibson (eds.), Radio Stars, 49–53.

(i.e. $\partial f/\partial v_\perp > 0$). Waves with frequency close to the electron-cyclotron frequency, Ω_e, and possibly at harmonics of Ω_e, can tap the free energy available from a positive $\partial f/\partial v_\perp$ and grow.

The mode in which the radiation is emitted is dependent on the plasma conditions. For a plasma in which the bulk of the electrons are cold and where the energetic electrons comprise only a small fraction of the total electron density, the maser instability produces x-mode growth if $\omega_p/\Omega_e \lesssim 0.3$ where ω_p is the electron plasma frequency (Hewitt et al., 1983; Wu and Qui, 1983; Melrose et al., 1984). This radiation can escape from the source region and, depending on the plasma conditions, can be reabsorbed at a second harmonic resonance layer thereby heating the corona. For $0.3 \lesssim \omega_p/\Omega_e \lesssim 1.3$, z-mode growth dominates. This radiation, unlike x-mode radiation, cannot escape directly from the source region. Local plasma heating can occur via the absorption of the z-mode radiation but heating of the corona above the flare site must occur via some other process. One such process, suggested by Melrose and Dulk (1984), is that second harmonic x-mode radiation produced by the coalescence of z-mode waves is partially absorbed at a third harmonic resonance layer. The component of the radiation which is absorbed then produces heating of the corona while the remainder escapes to produce the observed radio bursts.

The assumption that the bulk of the electrons are cold is not valid for the source of the maser emission during solar and stellar flares. In this paper, conditions for the suppression of x-mode growth in a hot plasma are derived (Sections 2 and 3) and implications for the heating of the corona are discussed (Section 4). A summary of the results is given in Section 5.

2. CONDITIONS FOR GROWTH AND DAMPING

The semirelativistic maser instability is driven by gyroresonant electrons, i.e. by electrons with velocity, $\mathbf{v}$, such that $\omega - \Omega_e/\gamma - k_z v_z = 0$ where ω is the wave frequency, k_z is the component of the wave-number parallel to the magnetic field and $\gamma = (1 - v^2/c^2)^{-1/2}$ is the Lorentz factor. In the semirelativistic limit i.e. $0 < v^2/c^2 \ll 1$ and $\omega^2 \gg k_z^2c^2$, the resonant electrons lie in velocity space on the semi-circle with centre (Hewitt et al., 1981)

$$v_z/c = k_z c/\omega, \qquad v_\perp/c = 0$$

and radius

$$v_r^2/c^2 = (k_z c/\omega)^2 - 2(\omega - \Omega_e)/\Omega_e.$$

For there to be a finite number of resonant electrons present the radius of the resonance circle must be real i.e.

$$(k_z c/\omega)^2 > 2(\omega - \Omega_e)/\Omega_e.$$

Growth occurs if the resonance circle samples sections of the distribution, f, where $\partial f/\partial v_\perp$ is predominantly positive; damping occurs if it samples regions where $\partial f/\partial v_\perp$ is negative.

The electron distribution in the flaring magnetic flux tube is assumed to be

$$f = ((2\pi)^{1/2} v_T)^{-3} (v_\perp/\surd 2 v_T)^2 \exp(-v^2/2v_T^2)$$

where v_T is the thermal speed. This distribution has a similar form as a two sided loss-cone distribution and related types of distributions have been used in discussions of the semirelativistic maser instability (e.g. Wu and Lee, 1979; Hewitt et al., 1981).

For this distribution, (1) - (3) imply that growth occurs if

$$0 < (k_z c/\omega)^2 - 2(\omega - \Omega_e)/\Omega_e \lesssim 2v_T^2/c^2$$

and damping occurs if the reverse of the last inequality in (5) is satisfied.

3. SUPPRESSION OF X-MODE GROWTH

For certain ω_p/Ω_e, x-mode growth cannot occur because the number of electrons which lie on a resonance circle satisfying (3) and (5) is exponentially small. In particular, the x mode has a cutoff which, for $v_T^2/c^2 \lesssim \omega_p^2/\Omega_e^2 \lesssim 1$, is given by (Winglee, 1985)

$$\omega_x = \Omega_e(1 + \omega_p^2/\Omega_e^2 - 9v_T^2/2c^2).$$

This frequency is the minimum frequency obtainable for the x mode and hence, from (3) and (6), there is a finite number of resonant electrons only if

$$(k_z c/\omega)^2 \geq 2(\omega_p^2/\Omega_e^2 - 9v_T^2/2c^2).$$

The number of electrons along a resonance circle which samples regions where $\partial f/\partial v_\perp$ is positive becomes exponentially small when the centre of the resonance circle moves past the root mean square axial velocity of the electrons, i.e. if

$$|k_z c/\omega| \gtrsim \surd 2 v_T/c.$$

Hence from (7) and (8), x-mode growth occurs only if

$$\omega_p^2/\Omega_e^2 \lesssim (11/2)v_T^2/c^2.$$

In the theories of Hewitt et al. (1983), Wu and Qui (1983) and Melrose et al. (1984) the bulk of the electrons are cold and x-mode growth is restricted to values of ω_p/Ω_e less than about 0.3. Condition (9) implies that, when the plasma is hot with an average velocity, v_T, greater than about 0.13 c (i.e. $T \gtrsim 10^8$K), x-mode growth can occur at higher values of ω_p/Ω_e. Further, the maximum value of ω_p/Ω_e for which x-mode growth can occur increases with the plasma temperature.

4. OBSERVATIONS

As an indicator of whether plasma conditions during flares actually satisfy the condition (9) for x-mode growth, the inferred values of ω_p^2/Ω_e^2 and v_T^2/c^2 from the 20 impulsive solar flares reported by Batchelor (1984) are shown in the scatter plot in Fig. 1. It is seen in Fig. 1 that about 15 of the flares satisfy condition (9) for x-mode growth. In other words in about 75% of the flares, the plasma at the flare site is sufficiently hot to produce x-mode growth. In these flares, the radio frequency heating of the corona can be produced by absorption of the x-mode radiation at a second harmonic resonance layer.

In the five flares which do not satisfy condition (9), x-mode growth is suppressed and the heating of the corona must be produced by

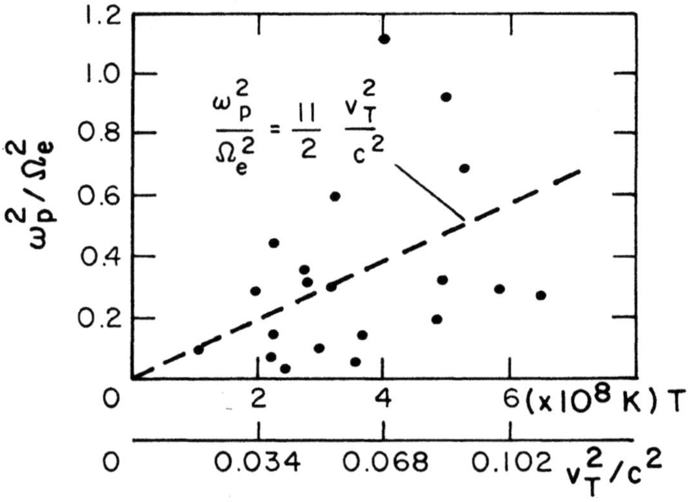

FIGURE 1. Scatter plot of ω_p^2/Ω_e^2 versus v_T^2/c^2 for the twenty impulsive flares reported by Batchelor (1984). Fifteen of the flares satisfy the condition that $\omega_p^2/\Omega_e^2 \lesssim (11/2)\, v_T^2/c^2$ required for x-mode growth while five flares do not.

some mechanism other than the damping of fundamental x-mode radiation. In particular, z-mode growth can dominate when x-mode growth is suppressed (Melrose et al., 1984; Winglee, 1985). Local plasma heating can be produced by the absorption of the z-mode radiation while heating of the corona above the flare site can be produced by the absorption of second harmonic radiation generated by the coalescence of z-mode waves (Melrose and Dulk, 1984).

5. CONCLUSION

In models for the radio frequency heating of the corona during impulsive flares, x-mode radiation emitted from the flaring flux tube is absorbed at a second harmonic resonance layer to produce the heating. It has been shown that a finite plasma temperature allows x-mode growth to occur at higher values of ω_p/Ω_e as the plasma temperature increases. Data from 20 flares indicates that in about 75% of the flares, the plasma conditions are favorable for x-mode growth.

ACKNOWLEDGEMENTS

The author wishes to thank G. A. Dulk for many valuable discussions. This work was supported by NASA's Solar Terrestrial Theory and Solar Heliospheric Physics Programs under grants NAGW-91 and NSG-7287 to the University of Colorado.

REFERENCES

Batchelor D. (1984) Ph.D. Dissertation, NASA Tech. Memorandum 86102.
Hewitt R.G., Melrose D.B and Dulk G.A. (1983) J. Geophys. Res. **88**, 10065.
Hewitt R.G., Melrose D.B. and Rönnmark K.G. (1981) Astron. Soc. Australia **4**, 221.
Holman G.D., Eichler D. and Kundu M.R. (1980) in IAU Symposium **86**, Radio Physics of The Sun, p.457, edited by Kundu M.R. and Gergely T., D. Reidel Pub. Co., Dordecht, Holland.
Melrose D.B. and Dulk G.A. (1982a) Ap. J. **259**, 844.
Melrose D.B. and Dulk G.A. (1982b) Ap. J. Lett. **259**, L141.
Melrose D.B. and Dulk G.A. (1984) Ap. J. **282**, 308.
Melrose D.B., Hewitt R.G. and Dulk G.A. (1984) J. Geophys. Res. **89**, 897.
Winglee R.M. (1985) Ap. J. (in press).
Wu C.S. and Lee L.C. (1979) Ap. J. **230**, 624.
Wu C.S. and Qui X.M. (1983) J. Geophys. Res. **88**, 10072.

DISCUSSION RELATED TO PART I

LINSKY: I would like to point out that the proposed mechanism of flaring in RS CVn systems, i.e. the interaction of magnetic loops from both stars, is very similar to that which produces flares on the Sun and, presumably, M-dwarf flare stars. In both cases flaring is the direct result of motions of the loop footpoints. In the Sun flaring often occurs when newly emerging loops interact with older loops or when sunspots move relative to each other producing loop-loop interactions. In the RS CVn systems we know from the photometry that large spots or spot groups migrate relative to the synchronous orbital-rotation system with periods of typically 10 years. As the large spots move, they must carry their large loops with them, leading to the possibility of interaction of the loops from the two stars. I therefore view flaring on single stars and RS CVn systems as the same physical process, but with a change of length scale that could also produce a change in time scale.

GIBSON: Jeff, I question whether you can get enough shear in the star-to-star loops to drive flares. The shears are effectively greater on the individual stars themselves.

UNDERHILL: Let me comment on the onset and decay of radio flares in binary systems. One may postulate that Wolf-Rayet stars have larger magnetic-loop systems than are normal for hot stars. If one observed a binary system containing such a star for a long time, it seems that one might observe a flare situation develop - i.e. trapping of a flare in the atmosphere of the WR star, followed by a disturbance of the magnetosphere of the binary system and the decay of this disturbance. Has anyone observed a WR binary system for sufficiently long times to have had a chance of detecting flares generated in this manner? The radio-active phase is at such a low density that the presence of a considerable radiation field, in the case of the WR star in contrast to small radiation fields for the G-K stars, should generate a minor perturbation of the situation.

HOGG: The data on variability of WR stars is very sparse. There are numerous observations of many stars for time scales of up to one hour. To my knowledge no flares have been seen. There are many observations separated by several months. Most WR stars appear to be unchanged, though there may be a few that have varied. There are almost no observations that cover 12 hour periods or that are daily samples over a period of several days.

MULLAN: A question for Jan Kuijpers. You drew two schematic diagrams of magnetic flux tubes - one for RS CVn stars and the other for Algol-type stars. For RS CVn systems magnetic flux tubes interconnect the two stars of the binary system, and interacting flux tubes trigger the flare mechanism. For Algol-type stars, flux tubes emerge from one star only, and they do not interconnect with the other star. What exactly is happening for flaring in the two cases? I would expect, on the basis of

R.M. Hjellming and D.M. Gibson (eds.), Radio Stars, 55–58.

Linsky's model, or Uchida and Sakurai's model, that the flaring in RS CVn stars takes place somewhere between the two stars, whereas for Algol (on the basis of your model) flaring will occur only on the star where the magnetic field emerges.

KUIJPERS
I am interested in finding out whether the observations show that a difference exits for flares in detached RS CVn binaries and semi-detached Algol systems with mass exchange. In the case of simultaneous radio and x-ray (with EXOSAT) observations of the RS CVn system σ^2CrB (Brinkman el al. 1983), the x-ray flares showed very similar time behavior.

GIBSON: I wish to bury forever the notion that Algol itself, or Algol-type binaries (here defined as an early-type plus a late-type evolved star), are different in their properties from RS CVn systems. As shown in the following histogram, which plots the number of days binaries were observed to have specific average radio luminosities at 2700 MHz, there is no significant difference between the statistics for Algol and the statistics for a sample of RS Cvn binaries.

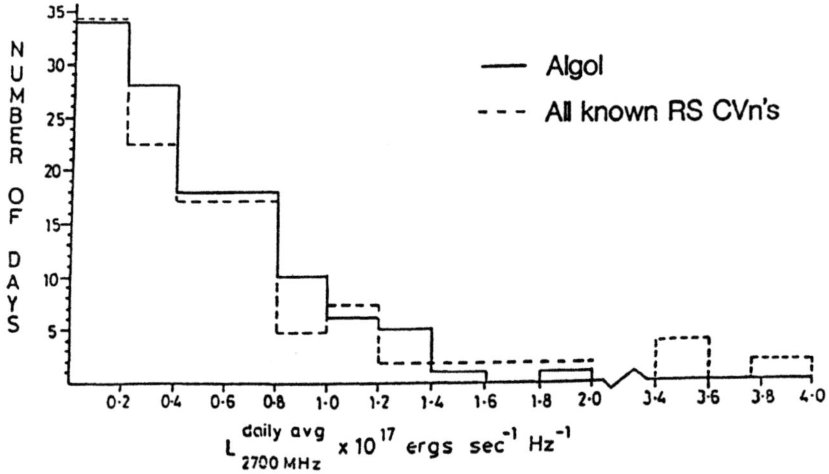

Algol's flare luminosities and the frequency of flaring are typical of those found in RS CVn's. Algol's radio spectrum is typical of eclipsing RS CVn binaries. Furthermore, we know the late-type component of Algol is a "dead-ringer" for the active components of RS CVn systems, and its x-ray luminosity and "coronal" temperatures are also typical of RS CVn binaries. In other words, an active (sub)giant doesn't care who its companion is, or for that matter, whether it has a companion.

MULLAN: Dave, are you saying that there is basically no difference between the RS CVn and Algol binaries in either flaring site or cause?

GIBSON: Yes!

UNDERHILL: With regard to Linksy and Gibson's comments on RS CVn and Algol binaries, not only does one have to have a storage of the energy seen in the radio/x-ray flare (possibly in large magnetic loops with small B) but one also has to demonstrate that the energy can be dumped as rapidly as a flare is seen to rise. This requirement should help one to differentiate between possible mechanisms and the place of origin for flares in binary G-K stars.

FLORKOWSKI: Evolutionary models of Algol binaries show that large amounts of mass ejection from the binary system are needed. Some IUE spectra of certain Algols (W Ser stars) show emission lines from a shell surrounding the binary. Such a shell could be a site for radio emission. In HR 1099 the circular polarization is phase dependent (Brown and Crane). The strongest polarization was observed when the outer Lagrangian point was directed towards the observer. This may be evidence for mass ejection from the system.

GIBSON: Jan, you discussed possible emission mechanisms quite extensively. What about absorption mechanisms?

KUIJPERS: I did not discuss the various absorption processes (such as gyroresonance absorption, collisional absorption, self-absorption, Razin cut-off) since their relative importance varies from source to source.

HOLMAN: Even for solar flares it is not easy to determine what the absorption mechanism is when a turnover is seen in the microwave spectrum of a flare. Several mechanisms are possible, but self-absorption is usually assumed to be dominant. However, absorption mechanisms can certainly be used to put constraints upon the physical properties in the source region. We did this for the AD Leo flare observed by Lang et al. (Holman, Bookbinder, and Golub - this volume). The fact that self-absorption must occur requires a large source volume if the (non-spikey) flare emission is assumed to be incoherent synchrotron radiation, for example. In general, however, much more data (such as good spectral and polarization data) are required to specify a particular absorption mechanism, or to put stronger constraints upon the emission mechanism and the properties of the source region.

HJELLMING: Is it possible that "spiky" emission may sometimes be "masked" because it occurs in dense enough environments that absorption processes wash out the short time scale effects?

KUIJPERS: What is the highest brightness temperature observed with VLBI?

MUTEL: A few times 10^{10} K.

GIBSON: A very interesting aspect of the "fact" that $T_b \sim 10^{10}$ K relates to the way flares are triggered or propagated. If we assume that the peak T_b in a small flaring region is $\sim 10^{10}$ K, then the rise time of flares is given by the rate at which the flaring volume increases (actually the surface area because the region is probably optically thick). In the case of Algol the rise times of flares are remarkably similar, typically seven hours. We also know that, near flare maximum, the size of the source is ~ 0.1 AU = 1.5×10^7 km (from VLBI observations), giving an average expansion velocity of about 600 km/s. If we assume the plasma β in the corona is ~ 0.3 (as is indicated by a number of studies, cf. Spangler A.J., **82**, 169) and $T \sim 3 \times 10^7$ K (cf. Swank et al., Ap.J., **246**, 208), then the Alfven velocity $v_A = (6kT/m)^{1/2} \approx 700$ km/s. This similarity suggests to me that successive loop arcades are involved in the flare on an Alfven time scale in a "domino effect" until the entire corona is flaring.

PART II

WINDS, INTERACTING WINDS, AND OTHER OUTFLOWS

THERMAL RADIO EMISSION FROM THE WINDS OF SINGLE STARS

David C. Abbott
Joint Institute for Laboratory Astrophysics, University of Colorado and National Bureau of Standards, Boulder, Colorado 80309

ABSTRACT. Observations of thermal emission at radio wavelengths provides a powerful diagnostic of the rate of mass loss and temperature of the winds of early-type stars. Some winds are also strong sources of nonthermal emission. Case studies of known thermal and nonthermal sources provide empirical criteria for classifying the observed radio radiation. Mass loss rates are derived for 37 OB and Wolf-Rayet stars considered "definite" or "probable" thermal wind sources by these criteria. The rate of mass loss is strongly linked to stellar luminosity in OB stars and probably linked to stellar mass in Wolf-Rayet stars, with no measurable correlation with any other stellar property. A few late-type giants and supergiants also have detectable thermal emission, which arises from extended, accelerating, partially-ionized chromospheres.

1. INTRODUCTION

The title given me by the organizing committee contains two hidden, and potentially dangerous, assumptions: (i) We know if a given star is multiple, or, in known binaries, which component emits the observed radiation. (ii) We know whether the observed radiation was generated by thermal or nonthermal processes. The brief history of stellar radio astronomy has already enforced humility on several researchers, including myself, who proceeded uncritically with one or both of these propositions.

In the present case, there is no question that luminous, early-type stars -- which are the main subject of this review -- are, in fact, strong emitters of thermal wind emission at radio wavelengths. Observations at ultraviolet, optical, and infrared wavelengths show that all OB stars more luminous than $L_* \sim 10^4\ L_\odot$ possess strong stellar winds (e.g. Lamers and Snow 1978). Further, best estimates of the mass loss rate $\dot{M}$ from UV data (e.g. Garmany and Conti 1984) imply that all OB stars with $L_* \gtrsim 10^5\ L_\odot$ have winds whose radius of optical depth unity at 6 cm, i.e., the radio photosphere, exceeds the optical radius by a factor of 10 or more. The Wolf-Rayet (WR) stars are even more

R. M. Hjellming and D. M. Gibson (eds.), Radio Stars, 61–78.

extreme. Their winds are opaque at all wavelengths, and the radio photosphere exceeds its optical counterpart by a factor on the order of 10^3. It is the large geometric extent of these radio photospheres which generates the high luminosities of the thermal wind emission.

The main question is one of contamination of these thermal wind sources by competing processes. The high radio luminosity of these sources, as well as certain stellar evolution considerations, makes contamination by unseen, low-mass companions improbable. However, the recent discovery that these winds can contain strong, nonthermal emission processes (e.g. Abbott, Bieging and Churchwell 1984a,b), creates major uncertainties about previous analysis of radio continuum emission from these stars.

2. A SIMPLE MODEL OF THERMAL WIND EMISSION

All present observational and theoretical evidence suggests that stellar winds in early-type stars reach their maximum, asymptotic velocity, V_∞, within a few stellar radii of the optical photosphere (e.g., Castor, Abbott and Klein 1975; Lamers and Morton 1976; Olson and Ebbets 1981; Barlow 1982). Thus, in the region of the radio photosphere the outflow is well-approximated by a constant-velocity, $\rho \propto r^{-2}$, density distribution. With this simplification, the transfer equation for radiation from thermal bremsstrahlung is easily solved to give the luminosity at radio frequencies. The basic model for this thermal wind emission was developed by Panagia and Felli (1975) as well as Wright and Barlow (1975). Further elaboration on the model is given in the reviews of Barlow (1979, 1982). (Note that the radio emission still originates very close to the star compared to the radius where the stellar wind impacts the interstellar medium, so it is still properly "stellar" emission rather than "H II" emission.)

The major assumptions of the basic model are isotropic and steady mass outflow with a stellar wind temperature and ionization equilibrium that is constant with radius. Given these approximations, the optical depth of free-free absorption of a ray of impact parameter p is given by (see Fig. 1)

$$\tau_\nu(p) = \int_{-\infty}^{+\infty} N_i N_e \; K(\nu,T) d\ell \propto \frac{\gamma}{p^3} K(\nu,T) \left(\frac{\dot{M}}{\mu V_\infty}\right)^2 \quad , \qquad (1)$$

where $K(\nu,T) = 0.02 \; g_{ff} \; Z^2 \; \nu^{-2} \; T^{-3/2}$ (Allen 1973), γ is the ratio of electron to ion number density, μ is the mean atomic weight, Z is the rms charge per atom, and g_{ff} is the free-free Gaunt factor. The observed flux comes from substituting equation (1) into the solution of the transfer equation along this ray, $I_\nu(p) = B_\nu(p) \; [1-\exp(-\tau_\nu(\rho))]$, and integrating over impact parameter p, which gives

$$S_\nu = D^{-2} \int_0^\infty I_\nu(p) \; 2\pi \; pdp = 23.2 \left(\frac{Z \; \dot{M}}{\mu V_\infty}\right)^{4/3} \frac{(\nu\gamma \; g_{ff})^{2/3}}{D^2} \; \mathrm{Jy} \quad , \qquad (2)$$

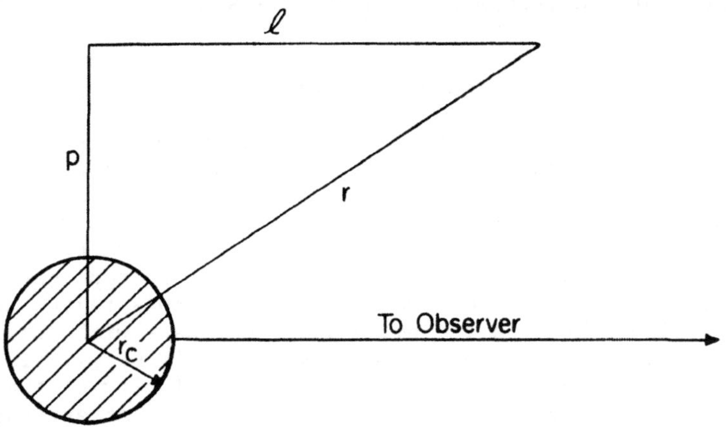

Fig. 1. Coordinate system used to calculate the radio emission.

where D is the distance to the source in kpc, $\dot{M}$ is in $M_\odot$ yr^{-1}, ν is in Hz, and V_∞ is in km s^{-1}. Note that the observed flux at a given frequency comes from a combination of optically thin and optically thick emission, which gives thermal wind emission its two distinctive characteristics: (i) a spectrum of $S_\nu \propto (\nu\ g_{ff})^{2/3} \propto \nu^{0.6}$, which is intermediate between the limits of $\nu^{0.0}$ and $\nu^{2.0}$ for optically thin and optically thick, constant-density, H II regions. (ii) The flux is temperature-independent except for a logarithmic term in the Gaunt factor, so that the concept of "brightness temperature" is not useful unless the wind is spatially resolved. These features are usually explained by picturing the observed flux as a product of the Planck function, which scales as $B_\nu \propto \nu^2 T$, and the area of the radio photosphere, which scales as $R^2_{ph} \propto \nu^{-4/3}\ T^{-1}$. While useful conceptually, this picture can be misleading if taken literally.

The stellar mass loss rate, $\dot{M}$, is determined from equation (2) by measurements of S_ν, D, and V_∞, and from the composition and ionization state of the wind. This method is independent of assumptions about the detailed velocity law of the wind, which is the major difficulty of $\dot{M}$ diagnostics at optical and infrared wavelengths, or about the ionization equilibrium of metals in the wind, which is the main drawback to the UV $\dot{M}$ diagnostics. When faced with a method that is, in principle, the ideal, model-independent diagnostic, the main question is: does it work in practice? One area of concern is the simplicity of the model. Various authors have considered modifications relaxing some of the assumptions, for example, anisotropic outflow (Schmid-Burgk 1982), time-variable outflow and simple clumping (Abbott, Bieging and Churchwell 1981), and radius-dependent temperature structures (e.g. Barlow 1979). The inferred value of $\dot{M}$ is generally insensitive to these effects, because the radio emitting region occupies a huge volume, which takes the stellar wind somewhere between 6 months and a year to replenish. The radio flux in effect measures the mass loss rate time-averaged over many dynamic time scales.

3. CASES OF DEFINITE THERMAL WIND EMISSION

A very stringent test of the thermal wind model is possible in sources whose radio emission is resolved interferometrically. Figure 2 (top panel) shows the expected contribution to the observed 6 cm luminosity as a function of radius for the example star P Cygni. The distribution peaks near 200 R_*, which is the location of the 6 cm photosphere. Interior to this radius, the wind is too opaque and photons cannot escape, so the contribution drops to zero. Beyond this radius the contribution declines because the emissivity decreases with density. The important feature to notice is the very weak tail of emission which extends to very large radii. In fact, the majority of the observed emission comes from these optically thin radii, as stressed by Panagia and Felli (1975).

White and Becker (1982) were the first to recognize that this extended tail of optically thin emission is resolvable with the VLA. Because intensity maps are not sensitive to extended regions of low surface brightness surrounding a bright core, they analyzed the structure of the wind emission by directly plotting the interferometric visibilities as a function of antenna separation. This measures the fraction of the total intensity that is within the angular diameter on the sky corresponding to the baseline of the antennas. The observed decrease in flux with radius forms a definitive test for the thermal wind model, and it also gives a first diagnostic of the wind temperature in the region of radio emission.

White and Becker's observations of P Cygni are shown in the bottom panel of Figure 2, re-plotted to conform to the scale of the top panel. The solid line shows the visibility flux predicted by the thermal wind model. The agreement is essentially exact. Interestingly, the best fit comes from a temperature distribution that is isothermal. Further, the observed emission is symmetric on the plane of the sky at an uncertainty level of roughly 10%. For this case, the simplest version of the thermal wind model seems to be the best.

Two other early-type stars have been resolved by the VLA in this manner, Cyg OB2 No. 12 (B8 Ia^+) by White and Becker (1983) and γ^2 Vel (WC8+O9I) by Hogg (1985). In both cases, the observations agree with the simple wind model with an isothermal temperature at ≈6000 K and the emission is symmetric on the plane of the sky. While these temperatures are smaller than the stellar effective temperatures, they are reasonable values for radiative equilibrium in ionizing gas surrounding a hot star. The star γ^2 Vel is a particularly important case, because the source was strong enough to be resolved with good signal-to-noise at 2 cm, 6 cm, and 20 cm and the thermal wind model provided a consistent fit at all three wavelengths. For sources such as these, where the emissivity is actually mapped with radius, there can be no doubt that the mechanism is thermal wind emission, and that the mass loss rates are reliable.

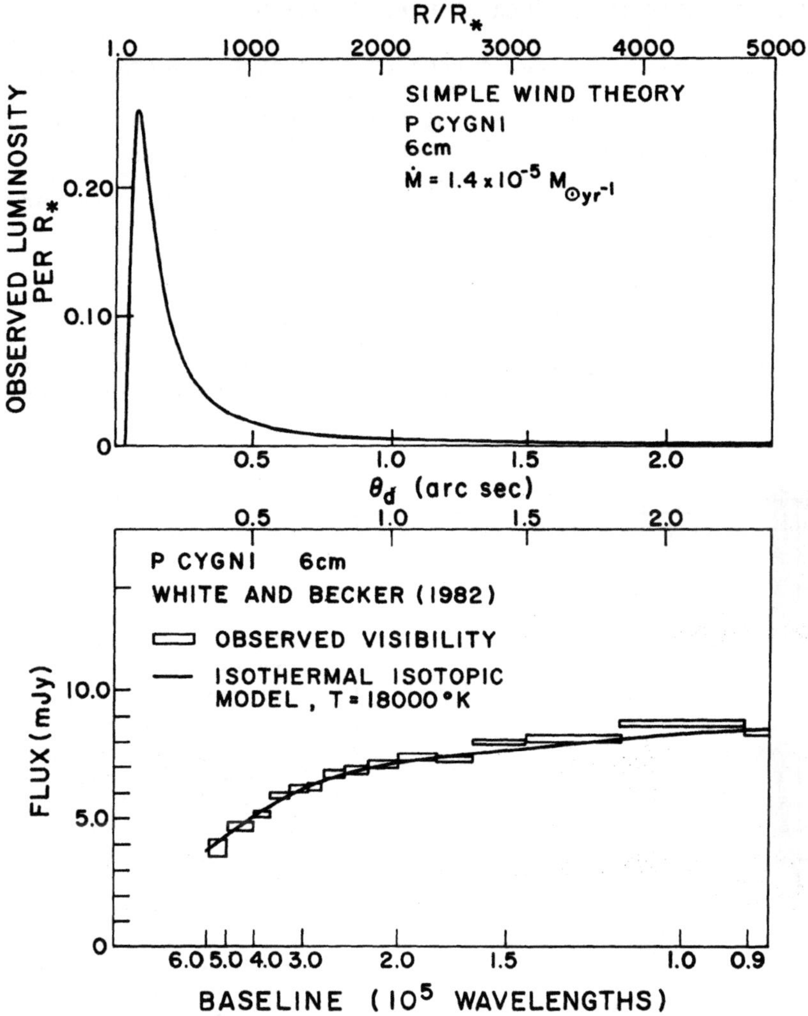

Fig. 2. (Top panel): The expected contribution to the observed 5 GHz luminosity as a function of stellar radius for the example of the star P Cygni using the simple wind model. (Bottom panel): VLA observations of the visibility function of P Cygni at 5 GHz. The boxes show the total flux versus baseline. The vertical size of the box is the 1σ error in flux and the horizontal size is the range of baselines included in the average. Solid line is the prediction of the simple wind model at the indicated temperature (from White and Becker 1982).

4. CASES OF DEFINITE NONTHERMAL WIND EMISSION

As surely as some winds are sources of thermal bremsstrahlung, others are not. The star Cyg OB2 No. 9 (O5f) provides a compelling counterpart to P Cygni as a prototype emitter of nonthermal radio radiation (e.g., Abbott, Bieging, and Churchwell 1984).

The first sign of trouble for this star was the huge excess radio flux over that expected from the $\dot{M}$ diagnostics at other wavelengths. For example, Figure 3 shows infrared photometry of Cyg OB2 No. 9 between 2 and 20 μm. The solid line shows the infrared emission expected from the hydrostatic layers of the optical photosphere. There is a detectable contribution from free-free wind emission only at 20 μm. By contrast, the dashed-dot line shows the minimum possible infrared flux if the measured radio flux was thermal wind emission. Even worse, the dotted line gives the predicted flux using standard wind models.

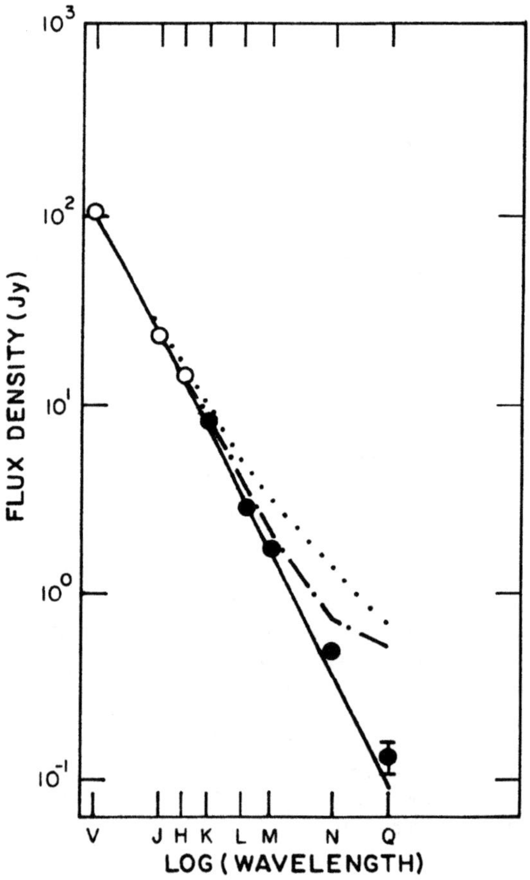

Fig. 3. Infrared photometry of Cyg OB2 No. 9. Filled circles -- measured fluxes in the K, L, M, N, and Q bands. Open circles -- catalog photometry at shorter wavelengths. Solid line -- expected emission from static photosphere. Dot-dashed line -- minimum possible infrared emission from the wind + photosphere if the observed radio flux was thermal wind emission. Dashed line -- predicted infrared emission for a wind characterized by $v(r)/V_\infty = [1-(r_*/r)]$ (from Abbott, Telesco, and Wolff 1984).

A second difficulty comes from observations of the frequency spectrum and time stability of the radio fluxes. Four years of VLA observations of Cyg OB2 No. 9 are summarized in Figure 4. The fluxes are clearly variable on time scales of months which, as discussed above, is difficult to reconcile with models of thermal wind emission. Even more critical is the frequency spectrum, which not only is time-variable, but clearly deviates from the characteristic index $\alpha = 0.6$ of thermal wind emission, on some occasions even becoming negative. To my knowledge, no configuration of gas can emit thermally with an index more negative than $\alpha \lesssim -0.1$.

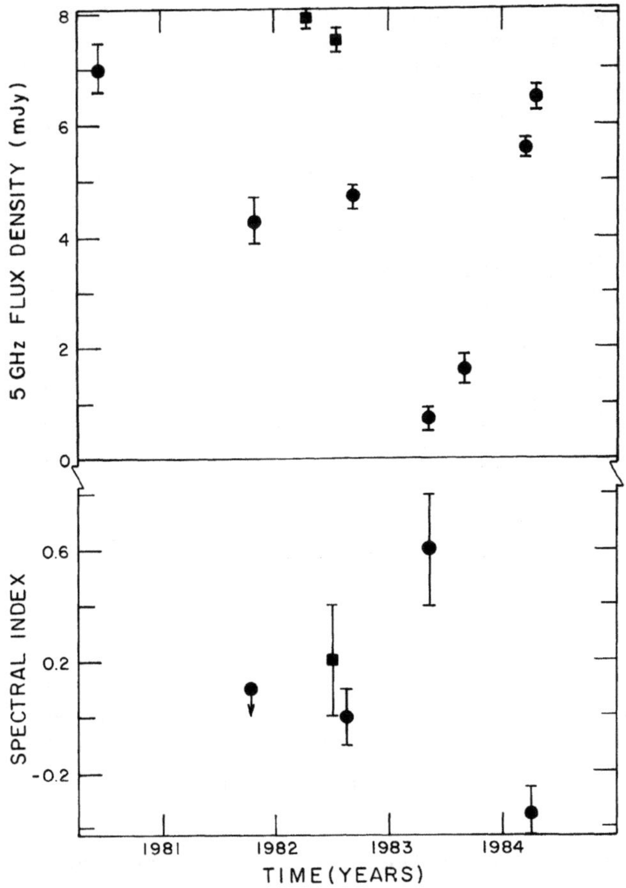

Fig. 4. Time variability of the radio flux of Cyg OB2 No. 9. Top panel -- flux density at 5 GHz. Bottom panel -- spectral index α of the frequency continuum ($S_\nu \propto \nu^\alpha$). Filled circles are from Abbott, Bieging and Churchwell (1985). Filled squares are from White and Becker (1983).

The final, and clinching, observational evidence against the thermal wind interpretation for Cyg OB2 No. 9 comes from the failure to spatially resolve the emission, as shown in Figure 5. Observations like these of other early-type stars show that Cyg OB2 No. 9 is not a unique object. Fully 25% of the OB stars with strong winds and 10% of the Wolf-Rayet stars contain strong, probably compact sources of nonthermal radio emission (Abbott, Bieging and Churchwell 1985).

5. CLASSIFYING THE OBSERVED RADIO EMISSION

These examples clearly show the importance of establishing the thermal or nonthermal character of stellar wind emission. Unfortunately, the presence of nonthermal emission is not apparent from optical/UV spectra, so the determination must come from the radio data themselves. The examples of P Cygni and Cyg OB2 No. 9 suggest several empirical tests to discriminate the thermal from the nonthermal emitters.

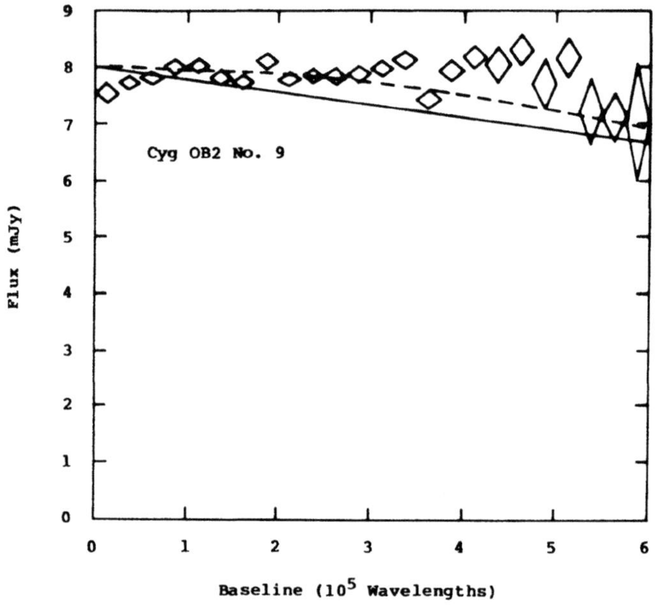

Fig. 5. VLA observations of the visibility function of Cyg OB2 No. 9 at 5 Ghz. The diamonds show the total flux versus baseline. The vertical extent of the diamond is the 1σ error in flux, while the horizontal extent is the range of baselines included in the average. Solid line -- predicted flux from thermal wind emission at $T = 3 \times 10^5$ K. Dashed line -- predicted flux from an optically thick, truncated disk. The source is unresolved and nonthermal (from White and Becker 1983).

The surest test is to spatially resolve the optically thin wind emission as was done for P Cygni. With the VLA this requires the maximum baseline provided by the "A" configuration. Optimally, the observations should include a high frequency, such as 2 cm, to avoid any possible ambiguity between compact nonthermal sources and thermal wind sources at coronal temperatures. This test requires at least a factor of 10 more telescope time than a simple detection, because good signal to noise is required for each subset of antenna pairs corresponding to a given range of baseline.

By contrast, a measurement of the frequency spectral index, α, requires only a factor of 2-3 increase in telescope time, which is feasible for essentially every source. Deviations from $\alpha = 0.6$ of the thermal model are a certain sign that the emission cannot be interpreted with the standard wind model. Negative spectral indices provide unambiguous evidence for nonthermal processes, assuming the source is unresolved at all frequencies. The detection of the expected $\alpha = 0.6$ index is positive, but not conclusive, evidence for thermal wind emission, as nonthermal processes can mimic an 0.6 index under certain conditions (White 1984).

A comparison of the observed radio flux to predictions based on independent $\dot{M}$ determinations at other wavelengths provides a valuable check for the presence of nonthermal processes without any extra expenditure of radio telescope time. Such $\dot{M}$ determinations are available, or readily obtained, for any star of reasonable brightness. Their accuracy is limited to, at best, a factor of 3, and typically an order of magnitude, depending on the diagnostic used and the quality of the data. Thus, this test can only detect gross violations of the thermal wind model. Observations of time-variability in the radio emission is another test which, like $\dot{M}$ comparisons, can provide strong evidence against the thermal interpretation, but only weak evidence for it.

To summarize, case studies of known thermal and nonthermal stellar wind emitters provides some empirical means of classifying the radio emission, which are summarized in Table 1. For completeness, I include polarization as a possible test, even though no thermal or nonthermal source detected so far exhibits measurable intrinsic polarization. As is usually the case, the reliability of the test scales with the telescope time required to obtain it, so some sacrifice of certainty for expediency is required in practice.

6. STELLAR WIND THERMAL EMISSION IN EARLY-TYPE STARS

With these working criteria, we can now discuss the topic of this review, which is thermal wind emission. The mass loss rate $\dot{M}$ is the stellar wind parameter of greatest fundamental importance. The radio method of deriving $\dot{M}$ from equation (2) works best for OB stars, because V_∞ is directly observable from the violet edge of the P Cygni profiles, the distances are well known from cluster membership or absolute magnitude calibrations, and the gas is fully ionized with normal composition. Table 2 summarizes the mass loss rates of all OB

Table 1. Working criteria to classify radio emission

1. Definite thermal wind emitters	• Resolved radio emission • $S_\nu \propto \nu^{0.6}$ • S_ν agrees with $\dot{M}$ (UV, optical, IR)
2. Probable thermal wind emitters	• No observed spectrum • S_ν agrees with $\dot{M}$ (UV, optical, IR)
3. Probably not thermal wind emitters	• No observed spectrum • S_ν exceeds $\dot{M}$ (UV, optical, IR) • Variability
4. Definite nonthermal wind emission	• Unresolved emission • $S_\nu \propto \nu^\alpha$, $\alpha \lesssim 0.0$ • Intrinsic polarization

Table 2. Mass loss rates for OB stars

Star	Spectral Type	Flux Density		V_∞	D	$\dot{M}$	Ref.[a]
		S(2 cm)	S(6 cm)	(km s^{-1})	(kpc)	($M_\odot$ yr^{-1})	
		"Definite" Thermal Emitters					
ζ Pup	O4f	3.0 mJy	1.4 mJy	2650	0.4	3.8(−6)	1,4
HD 152408	O8f	2.4	1.0	1800	1.8	1.8(−5)	4
P Cyg	B1 Ia^+	---	9.5	310	1.8	1.6(−5)	5
ζ^1 Sco	B1 Ia^+	4.3	2.0	590	1.8	1.0(−5)	4
HD 169454	B1 Ia	1.9	1.1	(1000)	1.7	9.2(−6)	4
HD 190603	B1.5 Ia	1.0	0.6	900	1.6	4.7(−6)	1,4
Cyg OB2 No. 12	B8 Ia^+	---	4.5	(1400)	2.0	5.0(−5)	2,6
		"Probable" Thermal Emitters					
Cyg OB2 No. 7	O3f	---	≲0.4 mJy	(3800)	2.0	1.9(−5)	2
HD 15770	O4f	---	0.2	2700	2.2	1.0(−5)	4
Cyg OB2 No. 9	O5f	1.2[b]	0.6[b]	(2650)	2.0	1.9(−5)	3
Cyg OB2 No. 5	O6f+O7f	---	1.8[c]	(1800)	2.0	2.8(−5)	2
HD 166734	O7.5f+O9 I	---	0.5	2600	2.3	2.1(−5)	4
HD 151804	O8f	---	0.4	2000	1.8	9.4(−6)	4
ζ Ori A	O9.5 I	---	0.7	2300	0.5	2.3(−6)	1
δ Ori A	O9.5 II	---	0.3	2300	0.4	9.3(−7)	4
ε Ori	B0 Ia	---	1.6	2000	0.4	3.1(−6)	1

[a]1. Abbott, Bieging, Churchwell, and Cassinelli (1980); 2. Abbott, Bieging, and Churchwell (1981); 3. Abbott, Bieging, and Churchwell (1984); 4. Abbott, Bieging, and Churchwell (1985); 5. White and Becker (1982); 6. White and Becker (1983).

[b]Based on one measurement when the nonthermal source appeared to be absent.

[c]Rodriguez et al. (1985) report variability for this source.

stars which are thermal wind emitters by the criteria of Table 1. I have expanded the sample to include early-type binaries, because there is as yet no evidence that the presence of companions produces measurable effect on the wind in the region of the radio photosphere. The rates of the "probable" thermal emitters are subject to revision pending further observations.

The relationship between these observed rates and the stellar luminosity L_* is shown in Figure 6. Also shown are mass loss rates derived from UV lines by Garmany et al. (1981). The methods are complementary, as the UV diagnostics are only effective for stars with low mass loss rates, whereas present radio telescopes can only detect stars with high mass loss rates. The dependence of $\dot{M}$ on L_* is quite

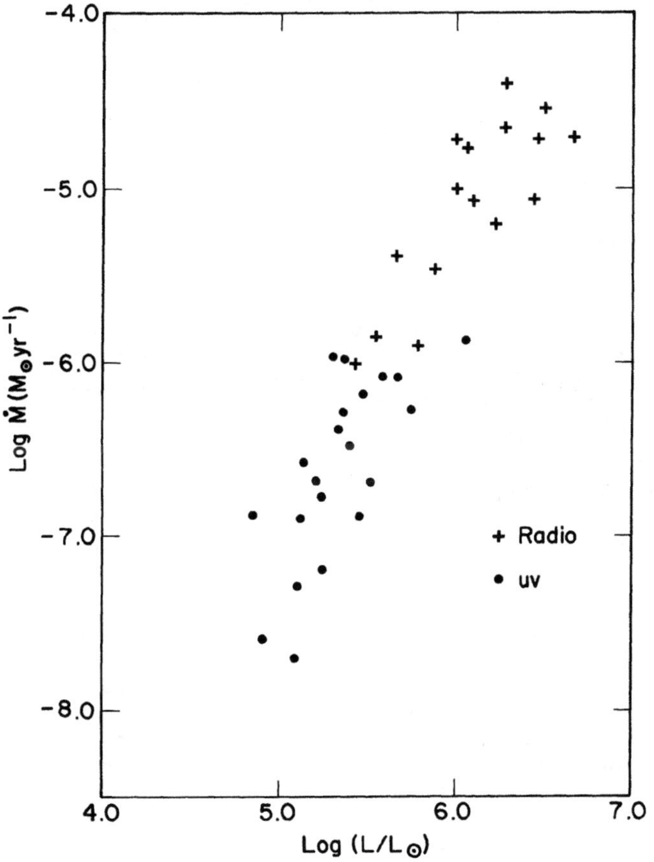

Fig. 6. The correlation between mass loss rate and stellar luminosity for the OB stars. The "radio" rates are from the VLA observations of Table 2. The "UV" rates are derived from IUE observations of N V and C IV by Garmany et al. (1981).

striking, illustrating that the winds of OB stars are a homogeneous group, probably driven by a common mechanism. Garmany and Conti (1984) have analyzed the combined set of UV/radio rates and find no measurable dependence between $\dot{M}$ and any other stellar parameter, such as R_*, M_*, or T_{eff}. They find a linear least-squares fit of

$$\dot{M} = 1.35(-7)\ (L/10^5\ L_\odot)^{1.62} \quad . \qquad (3)$$

The reliability of equation (3) for predicting $\dot{M}$ of an OB star is estimated to be ±0.5 in log $\dot{M}$.

The winds of Wolf-Rayet stars are so dense and opaque that their complex spectra yield no ready diagnostic of $\dot{M}$ from UV, optical or infrared wavelengths. There is therefore no good way to discriminate between thermal and nonthermal emission, except for the seven stars with spatially-resolved or multifrequency radio emission. [V444 Cyg (WN5+06) is the one exception. Kornilov and Cherepashchuk (1979) derived $\dot{M} = 1.2 \times 10^{-5}\ M_\odot\ yr^{-1}$ from optical measurements of a change in the orbital period.]

To provide a crude check on the expected strength of the thermal wind emission, Abbott, Bieging and Churchwell (1985) derived volume emission measures for the wind, i.e.,

$$EM \equiv \int_{r_*}^{\infty} dV\ N_e N^+ \quad ,$$

from the strengths of the stellar optical emission lines, assuming case B recombination theory. If the radio emission is thermal in origin, the measured flux is related to the emission measure by

$$S_\nu \propto (EM\ R_*\ Z\ \nu\ g_{ff})^{2/3}\ D^{-2} \quad . \qquad (4)$$

Figure 7 plots the measured fluxes at 5 GHz of 27 WR stars which have both radio and optical observations. The data generally follow the scaling of equation (4), but the scatter is large, primarily because the methods used to obtain EM and R_* have many uncertainties. The two known nonthermal emitters, MR 93 and HD 193793, are clearly discrepant, which suggests that this test may be useful in practice.

Five other labeled stars in Figure 7 have measured fluxes 1σ or more in excess of that predicted by their EM. Embarrassingly, two of them are definite thermal emitters with spatially resolved winds. The problem in γ^2 Vel is that the continuum of the O star companion masks the optically-thin emission lines, so EM is underestimated. WR 147 has a peculiar continuum spectrum in the optical and the radio, and further study is required to understand its discrepancy. The other three stars with excess fluxes -- HDE 318016, WR 105, and MR 110 -- are regarded as possible nonthermal emitters until further observations clarify their status. The remaining 15 stars in Figure 7 are classified as "probable" thermal emitters.

The mass loss rates of all WR stars deemed "probable" or "definite" thermal emitters are summarized in Table 3. The rates are

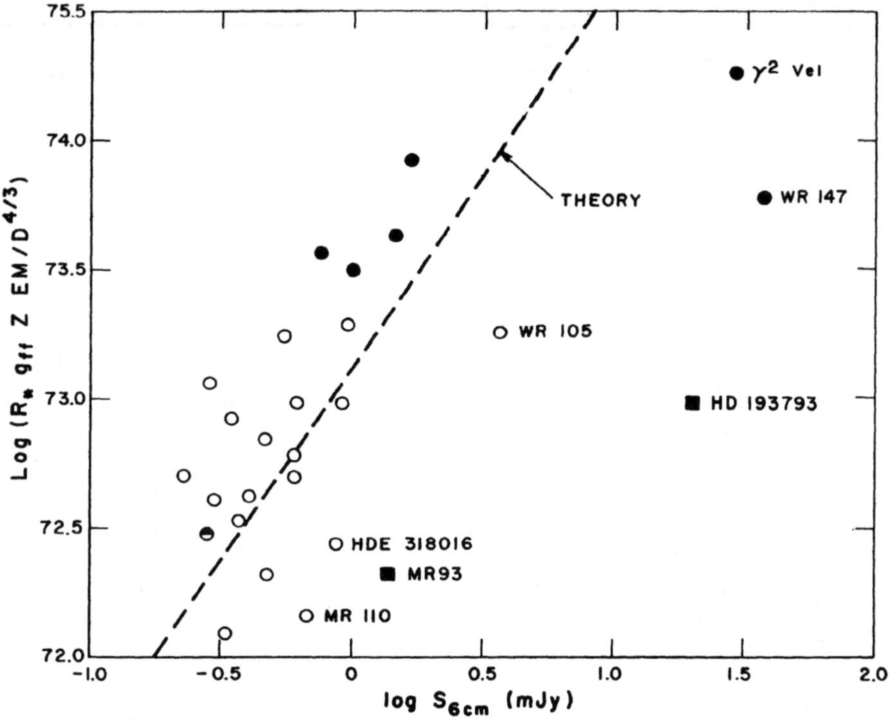

Fig. 7. A comparison of the 5 GHz flux of WR stars to the emission measure of the wind as measured from optical recombination lines. The dashed line is the expected relation for thermal wind emission as indicated by equation (4). Filled squares -- definite nonthermal sources by criteria of Table 1. Filled circles -- definite thermal sources. Half-filled circle -- the system V444 Cygni, which is very probably a thermal source. Labeled stars have radio fluxes 1σ or more above the mean relation.

plotted as a function of spectral type in Figure 8. The rates are confined to a relatively narrow range of $10^{-5} \lesssim \dot{M} \lesssim 10^{-4}$ $M_\odot$ yr^{-1}, despite large differences between the stars in properties related to spectral type, such as temperature, absolute magnitude, and chemical composition. There is also no measurable correlation of $\dot{M}$ with V_∞ or binarity. Despite this narrow range, the dispersion in $\dot{M}$ at a given spectral subtype is real (1σ is typically 0.2 dex), and it is the likely cause of a similar dispersion in line strengths demonstrated by Conti (1982).

Table 3. Mass loss rates for Wolf-Rayet stars

WR[a]	Star	Spectral Type	Flux Density S(2 cm)	Flux Density S(6 cm)	Ref.[b]	D	V_∞	C(6 cm)[c]	Log $\dot{M}$ ($M_\odot$ yr^{-1})[d]
				Definite Thermal Wind Emitters					
6	HD 50896	WN5	2.3	1.0	4	1.7	2650	8.0(-7)	-4.58 ± 0.23
11	γ^2 Vel	WC8+09 I	---	29.0	5	0.5	2000	1.3(-6)	-4.19 ± 0.12
78	HD 151932	WN7	2.8	1.5	1	1.9	2150	9.3(-7)	-4.40 ± 0.11
134	HD 191765	WN6	2.1	0.8	4	2.1	2300	8.0(-7)	-4.53 ± 0.15
136	HD 192163	WN6	3.7	1.6	1,3	1.8	2000	8.0(-7)	-4.47 ± 0.14
145	MR 111	WN-C	2.4	1.0	1,2	2.0	1500	7.5(-7)	-4.75 ± 0.28
147	NS 6	WN7	---	36.0	1	1.1	1000	9.3(-7)	-4.05 ± 0.31
				Probable Thermal Wind Emitters					
1	HD 4004	WN5	---	0.5	2	2.6	2850	8.0(-7)	-4.52 ± 0.17
79	HD 152270	WC7+06	---	1.0	1,4	2.0	3300	1.0(-6)	-4.26 ± 0.11
81	MR 66	WC9	---	0.3	1	1.9	1600	1.7(-6)	-4.79 ± 0.29
86	HD 156327	WC7+Abs	---	0.5	1	1.7	1950	1.2(-6)	-4.76 ± 0.29
89	LSS 4065	WN7	---	0.6	1	2.9	2000	9.7(-7)	-4.42 ± 0.20
93	HD 157504	WC7+Abs	---	0.9	1	1.7	2700	7.5(-7)	-4.63 ± 0.25
103	HD 164270	WC9	---	0.2	1	2.8	1400	1.7(-6)	-4.69 ± 0.26
110	HD 165688	WN6	---	1.0	2	2.1	3300	8.0(-7)	-4.36 ± 0.24
111	HD 165763	WC5	---	0.3	2	1.6	3550	7.4(-7)	-4.89 ± 0.16
113	CV Ser	WC8+08	---	≤0.4	2	2.0	2900	1.3(-6)	-4.51 ± 0.16
115	MR 87	WN6	---	0.4	1	2.2	1200	8.0(-7)	-5.10 ± 0.17
133	HD 190918	WN4.5+O	---	≤0.3	4	2.1	1750	7.5(-7)	-5.04 ± 0.16
135	HD 192103	WC8	---	0.6	4	2.1	1900	1.2(-6)	-4.59 ± 0.15
137	HD 192641	WC7+Abs	---	0.4	2	1.8	2700	1.1(-6)	-4.74 ± 0.15
138	HD 193077	WN5+Abs	---	0.6	2	1.8	1700	8.0(-7)	-4.90 ± 0.15
139	V444 Cyg	WN5+06	---	0.3	2	1.7	2500	8.0(-7)	-5.02 ± 0.16
141	HD 193928	WN6	---	0.5	1	1.8	1700	7.4(-7)	-4.96 ± 0.16

[a]WR catalog number from van der Hucht et al. (1981).

[b]1. Abbott, Bieging, and Churchwell (1985); 2. Bieging, Abbott, and Churchwell (1982); 3. Dickel, Habing, and Isaacman (1980); 4. Hogg (1982); 5. Hogg (1985).

[c]$C \equiv 0.095\ (\mu/Z)(g\gamma\nu)^{-1/2}$ from equation (2).

[d]The uncertainty is the formal error of Log $\dot{M}$ from the 1σ uncertainties in D, V_∞, and S calculated as $\sigma^2 = 1.5\ \sigma_D^2 + \sigma_V^2 + 0.75\ \sigma_S^2$.

The only positive correlation between $\dot{M}$ and an observable characteristic of a WR star is with stellar mass. Five stars in Table 3 are double-lined spectroscopic binaries with well-determined masses (e.g. Massey 1982). These are plotted in Figure 9. The statistics are limited, of course, but all five systems have reliable distances and wind velocities, so their mass loss rates are unusually good. If the observed correlation between $\dot{M}$ and M_* holds for all WR stars, two implications would follow:

1) $\dot{M}$ vs. L_*. Assuming that stellar luminosity scales with stellar mass according to a relation such as $\log(L/L_\odot) = 3.8 + 1.5 \log(M/M_\odot)$ (Maeder 1983), then the correlation of Figure 9 becomes

$$\dot{M}_{WR} \simeq 5 \times 10^{-6}\ (L/10^5\ L_\odot)^{1.55}\ M_\odot\ yr^{-1} \quad , \qquad (5)$$

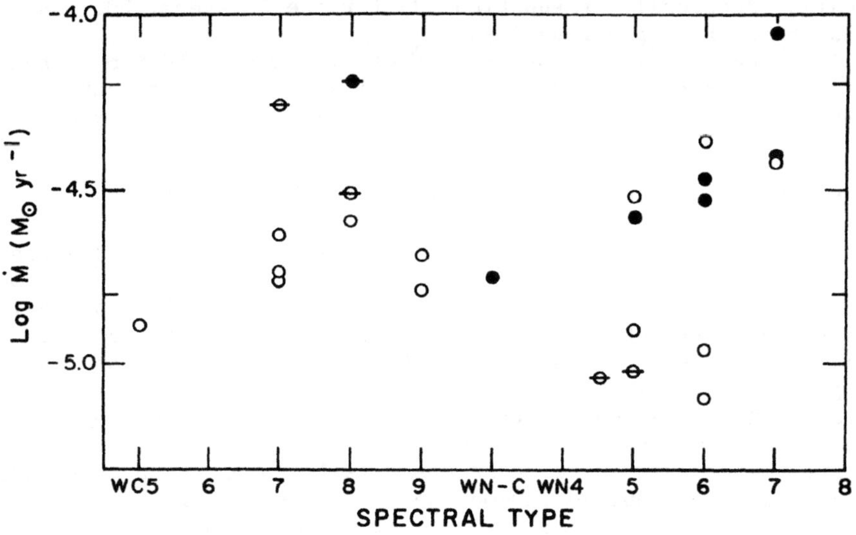

Fig. 8. Mass loss rates for the Wolf-Rayet stars in Table 3 as a function of spectral type. Filled circles -- definite thermal sources. Open circles -- probable thermal sources. Barred symbol -- double-lined spectroscopic binary.

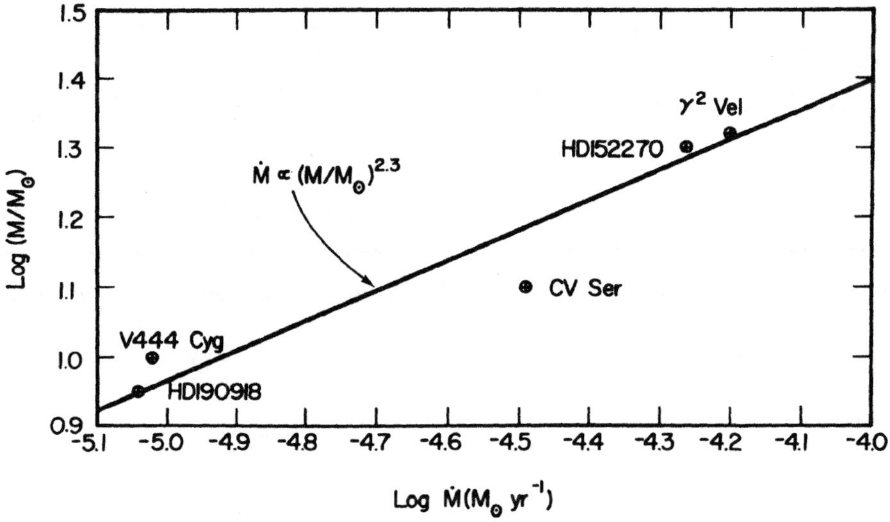

Fig. 9. Mass loss rate versus stellar mass for the five WR binaries with well-determined orbital masses.

which scales identically as equation (3) for the OB stars, but has a larger constant of proportionality.

2) The range of $\dot{M}$. The observed $\dot{M}$ of Figure 8 would imply that WR stars have a fairly narrow range of stellar masses, and that the WN and WC sequences do not differ in their average stellar masses.

7. THERMAL WIND EMISSION FROM LATE-TYPE GIANTS AND SUPERGIANTS

Three published detections of M-type giants or supergiants are strong candidates for thermal wind emission from single stars. Table 4 summarizes the observations of these systems. The thermal interpretation rests on the criteria of spectral index, agreement with fluxes predicted by models based on diagnostics at other wavelengths, and the lack of pronounced variability or polarization often seen in late-type flare stars. The observed spectral index implies that gradients in velocity and/or temperature are present in the region of the radio photosphere. The radio emission is most likely sampling an extended, expanding, partially ionized chromosphere. Drake and Linsky (1983) estimate that the radio-emitting region of α^1 Her is 0.2-2% ionized. A detailed chromosphere model by Hartmann and Avrett (1984) concludes that this is also true for Betelgeuse. Radio observation of late-type, single stars therefore gives no direct measurement of the net stellar mass loss rate, but provides a powerful constraint on model chromospheres of these stars.

8. CONCLUSIONS

Stellar winds are strong sources of thermal radio emission because their geometric extent is orders of magnitude larger than the size of the optical photospheres. Although strong compared to other stellar sources of thermal emission, winds are still difficult to detect, and

Table 4. Thermal radio emission from late-type giants and supergiants

Star	Spectral Type	Frequency Spectrum	Reference[a]
Antares	(M1.5 Iab+B2.5 V)[b]	$S_\nu = 0.47\ \nu_{GHz}^{1.05}$ mJy	2
Betelgeuse	M2 Ib	$S_\nu = 0.23\ \nu_{GHz}^{1.33}$	3
α^1 Her	M5 II	$S_\nu \simeq 0.16\ \nu_{GHz}^{0.88}$	1

[a]1. Drake and Linsky (1983, and private communication); 2. Hjellming and Newell (1983); 3. Newell and Hjellming (1982).

[b]Frequency spectrum refers to primary star only.

their systematic study has only been possible with the VLA. Observations of thermal continuum radio emission provide a means of measuring the stellar wind mass loss rate and temperature which is unobtainable by any other method. This diagnostic must be applied cautiously, however, because many winds of early-type stars are even stronger sources of nonthermal radiation.

Additional radio observations are necessary to sort the thermal from the nonthermal sources, and this work is progressing slowly. Preliminary results from 37 OB and WR stars leads to a picture of mass loss which is overall steady and symmetric, and which is strongly linked to stellar luminosity in OB stars, and probably linked to stellar mass in the WR stars. In both cases, there is no measurable correlation between the rate of mass loss and other properties of the star.

Stellar wind radio astronomy is still very much in its infancy. Further progress in the field requires instrument development in the areas of:

- Sensitivity. At most 75-100 thermal wind sources will ever be detectable with the current VLA, and most of these are too weak for any analysis of spectrum or spatial extent, which is crucial to interpreting the data. No serious systematic study of winds from late-type stars is possible at present capabilities. More sensitivity is needed. This is the major stumbling block to progress.

- Resolution. Modest gains in spatial resolution (factor of ~5-10) would be very valuable to resolve the optically thick core of hot star winds, to resolve the chromospheres of late-type stars, and possibly to resolve the nonthermal emission in hot stars.

- Spectrum. A broader range in observing frequencies would greatly enhance our ability to map temperature with radius in thermal sources, and to detect sources which are nonthermal.

This work was supported by National Science Foundation grant AST82-18375 to the University of Colorado.

REFERENCES

Abbott, D. C., Bieging, J. H., and Churchwell, E. 1981, Ap. J., **250**, 645.

Abbott, D. C., Bieging, J. H., and Churchwell, E. 1984a, Ap. J., **280**, 671.

Abbott, D. C., Bieging, J. H., and Churchwell, E. 1985, these proceedings.

Abbott, D. C., Bieging, J. H., and Churchwell, E. 1985, in preparation.

Abbott, D. C., Bieging, J. H., Churchwell, E., and Cassinelli, J. P. 1980, Ap. J., **238**, 196 (paper I).

Abbott, D. C., Telesco, C. M., and Wolff, S. C. 1984, Ap. J., **279**, 225.

Allen, C. W. 1973, Astrophysical Quantities (London: Athlone).

Barlow, M. J. 1979, in Mass Loss and Evolution of O-Type Stars, eds. P. S. Conti and C. W. H. De Loore (Dordrecht: D. Reidel), p. 119.
Barlow, M. J. 1982, in Wolf-Rayet Stars: Observations, Physics, and Evolution, eds. C. W. H. de Loore and A. J. Willis (Dordrecht: Reidel), p. 149.
Bieging, J. H., Abbott, D. C., and Churchwell, E. B. 1982, Ap. J., **263**, 207.
Castor, J. I., Abbott, D. C., and Klein, R. I. 1975, Ap. J., **195**, 157.
Conti, P. S. 1982, in Wolf-Rayet Stars: Observations, Physics, and Evolution, eds. C. de Loore and A. Willis (Dordrecht: Reidel), p. 3.
Drake, S. A. and Linsky, J. L. 1983, Ap. J. (Letters), **274**, L77.
Dickel, H. R., Habing, H. J., and Isaacman, R. 1980, Ap. J. (Letters), **238**, L39.
Garmany, C. D. and Conti, P. S. 1984, Ap. J., **284**, 705.
Garmany, C. D., Olson, G. L., Conti, P. S., and Van Steenberg, M. 1981, Ap. J., **250**, 660.
Hartmann, L. and Avrett, E. H. 1984, Ap. J., **284**, 238.
Hjellming, R. M. and Newell, R. T. 1983, Ap. J., **275**, 704.
Hogg, D. E. 1982, in Wolf-Rayet Stars: Observations, Physics, and Evolution, eds. C. de Loore and A. Willis (Dordrecht: Reidel), p. 221.
Hogg, D. E. 1985, these proceedings.
van der Hucht, K. A., Conti, P. S., Lundstrom, I., and Stenholm, B. 1981, Space Sci. Rev., **28**, 227.
Kornilov, V. G. and Cherepaschuk, A. M. 1979, Pis'ma Astr. Zu., **5**, 398.
Lamers, H.J.G.L.M. and Morton, D. C. 1976, Ap. J. Suppl., **32**, 715.
Lamers, H.J.G.L.M. and Snow, T. P. 1978, Ap. J., **219**, 504.
Maeder, A. 1983, Astr. Ap., **120**, 113.
Massey, P. 1982, in Wolf-Rayet Stars: Observations, Physics, and Evolution, eds. C. W. H. de Loore and A. J. Willis (Dordrecht: Reidel), p. 251.
Newell, R. T. and Hjellming, R. M. 1982, Ap. J. (Letters), **263**, L85.
Olson, G. L. and Ebbets, D. 1981, Ap. J., **248**, 1021.
Panagia, N. and Felli, M. 1975, Astr. Ap., **39**, 1.
Rodriguez, L. F., Roth, M., Tapia, M., Canto, J., Sarmiento, A., Persi, P., and Ferrari-Toniolo, M. 1985, these proceedings.
Schmid-Burgk, J. 1982, Astr. Ap., **108**, 169.
White, R. L. 1984, Ap. J., in press.
White, R. L. and Becker, R. H. 1982, Ap. J., **262**, 657.
White, R. L. and Becker, R. H. 1983, Ap. J. (Letters), **272**, L19.
Wright, A. E. and Barlow, M. J. 1975, M.N.R.A.S., **170**, 41.

THERMAL RADIO EMISSION FROM CIRCUMSTELLAR ENVELOPES

Sun Kwok
Department of Physics
The University of Calgary
Calgary, Alberta, Canada
T2N 1N4

ABSTRACT. The field of thermal radio star research between 1970 and 1984 is reviewed. The impact of Very Large Array telescope on high-resolution studies of stellar sources and the theoretical interpretation of the observations are discussed.

1. INTRODUCTION

It has been exactly five years since the first radio star meeting held in Ottawa in 1979 (Feldman and Kwok 1979). Looking back now, the Ottawa meeting represented a turning point in radio star research, conveniently dividing the short history of this subject into pre-VLA and post-VLA eras. The commissioning of the VLA in 1981 led to a major improvement in sensitivity and resolution over the radio telescopes of the past. Since radio stars are generally small objects, the capability of obtaining images rather than just flux densities was in part responsible for the advances in this subject in the last five years.

It was pointed out by Hjellming (1974) that for a thermal emitter to be detected, its brightness temperature (T_b) and angular radius (θ) must satisfy the following condition:

$$(T_b/10^4\mathrm{K})(\theta/'')^2 > 0.0017\ (\lambda/6\ \mathrm{cm})^2\ F_{min}\ (\mathrm{mJy}) \qquad (1)$$

where λ is the wavelength of observation and F_{min} is the minimum detectable flux density. For an ionized region with electron temperature (T_e) $\approx 10^4$ K and $F_{min} \sim 1$ mJy, Eq. (1) implies that $\theta \gtrsim 0''.04$. This corresponds to a physical radius of 6.3×10^{14} cm at a distance of 1 kpc, which is much greater than the size of stars. Therefore, for a star to be detectable it must either (i) be nearby; (ii) have a high temperature (e.g., in the corona); or (iii) possess a circumstellar envelope many times larger than the star itself. It is the third possibility that we shall be addressing today.

R. M. Hjellming and D. M. Gibson (eds.), Radio Stars, 79–91.

2. Flux Density Measurements of Thermal Radio Stars (1970-1980).

Since radio observations performed during the 70's do not have adequate angular resolution to obtain images for radio stars, structure of the sources can only be inferred by the analysis of the spectral shape and/or the time variability of the flux density. The first significant discovery in thermal emission from stars is the discovery of radio emission from nova envelopes by Hjellming and Wade (1970), who monitored the novae HR Del and FH Ser for over 2 years on the NRAO 3-element interferometer.

The observed light curves show that the radio emissions go through a maximum at $\sim$ 1-3 yr (depending on the observing frequency) after the optical outburst and the spectral index (α) evolves from $\sim +2$ through $\alpha \sim +1$ and then to $\alpha \sim 0$. These radio "lightcurves" are basically consistent with an expanding envelope evolving from optically thick to optically thin (Hjellming (1974)). A schematic diagram for the evolution of radio emission from novae is given in Figure 1.

Similar behaviour is observed for the V1500 Cygni (nova Cygni 1975, Hjellming *et al.* 1979; Seaquist *et al.* 1980). The spectral index of $\sim +1$ between the optically-thick and -thin phases is probably due to a density gradient in the envelope (Seaquist and Palimaka 1977). This can result from a sudden ejection with a velocity dispersion in

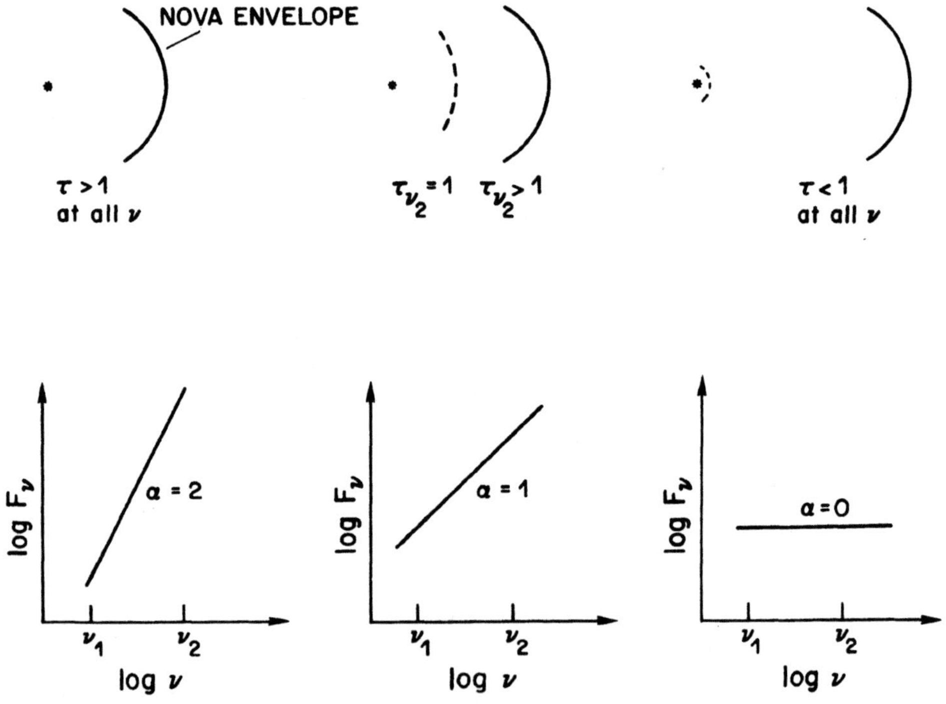

Figure 1 Schematic diagram illustrating the spectral evolution of classical novae.

the ejecta such that there are approximately equal mass at each velocity interval (Seaquist *et al.* 1980). An alternative to the velocity-dispersion hypothesis is that the novae have wind-like ejections which last over a year but the rate of ejection decreases with time (Kwok 1983) Figure 2 shows a fit to the "lightcurves" of V1500 Cygni under the wind model of novae. The nova is assumed to have an ejection velocity of 2000 km s^{-1}, a mass loss rate of 10^{-5} $(t/1\ yr)^{-2}\ M_\odot\ yr^{-1}$ and a distance of 550 pc.

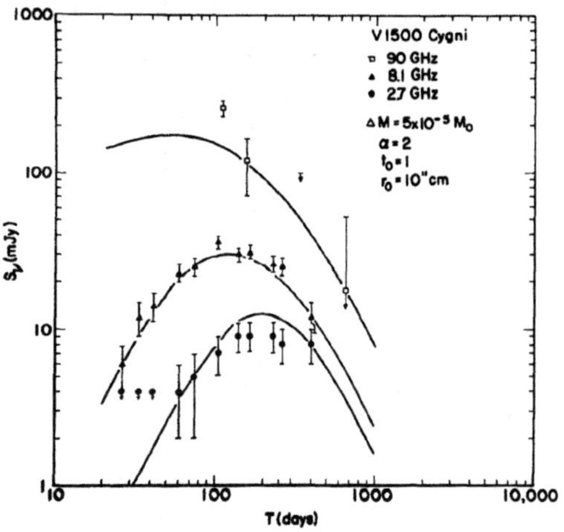

Figure 2. Radio "lightcurves" of V1500 Cygni.

Another area of active development during the decade of the 70's was the observation of emission-line stars. The first emission-line stars detected were α Sco (Hjellming and Wade 1971) and MWC 349 (Braes, Habing and Schoenmaker 1972). An extensive survey of emission line stars was performed by Purton *et al.* (1982). The objects detected in this survey of 325 stars are, however, quite diverse. It includes young stars (e.g., MWC 349 and LkHα 101), compact planetary nebulae (e.g., Hb 12 and Vy 2-2) and symbiotic stars (e.g., V1016 Cygni). Thirteen of the detected stars are considered by Purton *et al.* to be particularly interesting for they have $\alpha \sim 1$.

The best example of this class is MWC 349 which spectrum is shown in Figure 3. Its spectral shape can be approximated by (254 ± 7 mJy) x $(\nu/10\ \text{GHz})^{0.65 \pm 0.02}$ between 1 and 100 GHz and appears to be a perfect example of thermal radio emission from a stellar wind (Seaquist and Gregory 1973; Wright and Barlow 1975; Panagia and Felli 1975). For a steady stellar wind of constant temperature the observed flux density is given by:

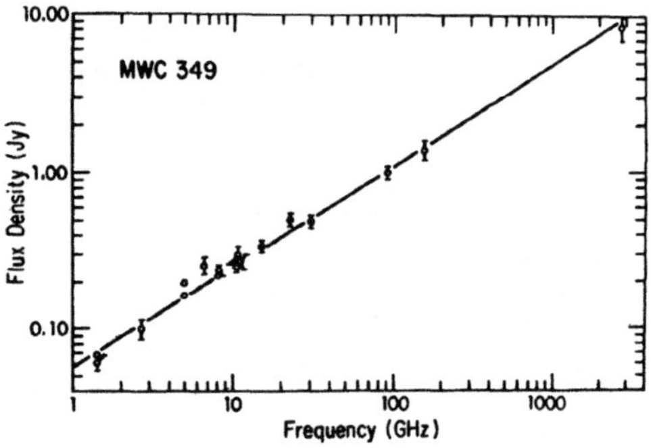

Figure 3. Radio continuum spectrum of MWC 349 (taken from Dreher and Welch 1983).

$$F_\nu = 2.32 \times 10^4 \; (\dot{M}/\mu v)^{4/3} \; \nu^{2/3} \; D^{-2} \; \gamma^{2/3} \; g^{2/3} \text{ mJy}$$

$$= 1.46 \left[\frac{(\dot{M}/10^{-5} \; M_\odot \text{ yr}^{-1})}{\mu \; (v/1000 \text{ km s}^{-1})} \right]^{4/3} (D/\text{kpc})^{-2}$$

$$\times \left[\frac{\gamma g}{(\lambda/6 \text{ cm})} \right]^{2/3} \text{ mJy} \qquad (2)$$

where $\dot{M}$ is the mass loss rate, v is the wind velocity, ν is the frequency, μ is the mean molecular weight, D is the distance, g is the Guant factor, γ is the electron to ion ratio. However, many stars of this class have the spectral indices greater than the theoretical predicted value of 0.6. It is not clear why this is the case except that the density structure of the radio emission region is probably more complicated than that given by a r^{-2} law.

Another interesting example of thermal radio emission from stars is the M2 supergiant α Ori. It has a thermal spectrum with $\alpha = 1.32$ which suggests a density distribution in the emitting region of the form $\rho \propto r^{-3.6}$ (Altenhoff, Oster and Wendker 1979; Newell and Hjellming 1982). Since the observed flux exceeds that expected from the photosphere, the radio emission probably originates from a partially ionized chromosphere.

The improved time and frequency coverage of radio observations in the 1970's have made possible the accurate determination of the "light-curves" and spectra of radio stars which in turn has improved our theoretical understanding of the sources. However, the direct imaging of the structure of radio stars had to wait until the arrival of the VLA.

3. VLA OBSERVATIONS OF RADIO STARS

In §1 we have discussed that a blackbody of 10^4 K with a 5 GHz flux density of $\sim$ 1 mJy has an angular radius of 0".04. Since the half-power synthesized beamwidth of the VLA at 5 GHz is 0".4 ("A" configuration) and the sensitivity is better than 1 mJy/beam, there are many potential compact sources which can be detected but cannot be resolved by the VLA.

Quite often one encounters sources which appear to be pointlike in the intensity maps but are in fact resolved. Figures 4 and 5 show respectively the

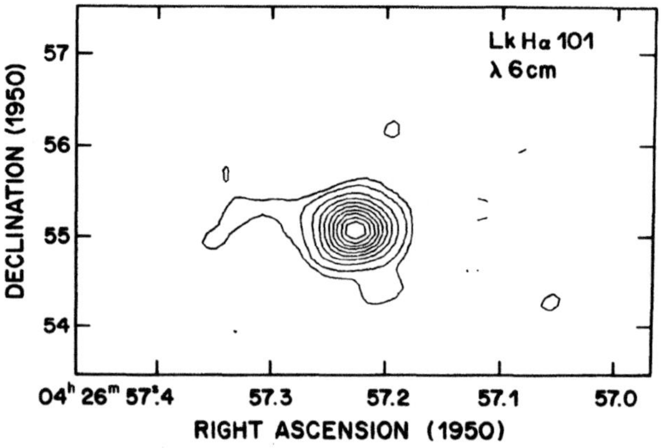

Figure 4. λ 6 cm map of LkHα 101. Contours are at intervals of 10% of peak value (12.8 mJy/beam) with an additional contour at 5%.

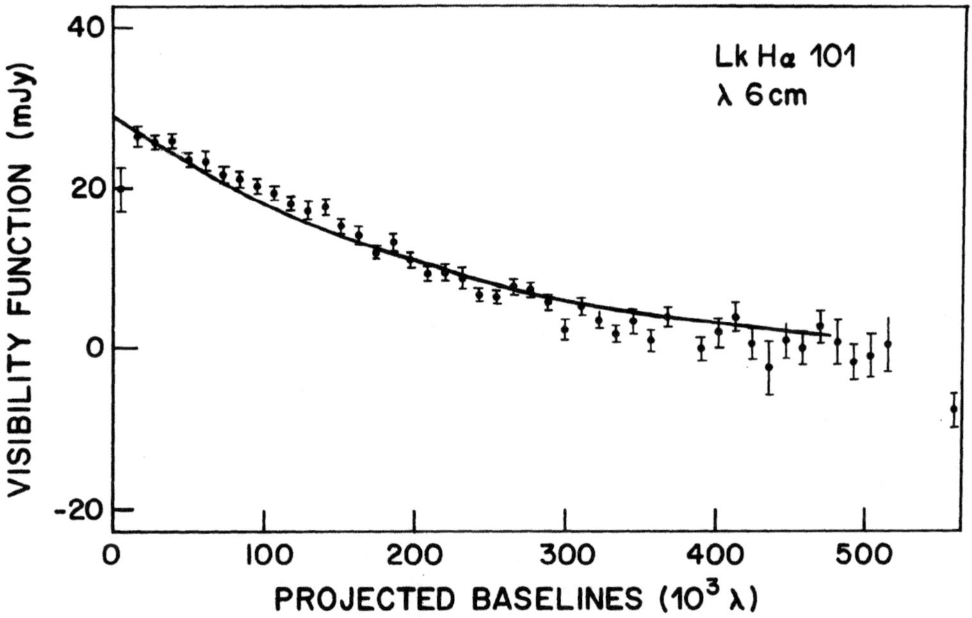

Figure 5. The real part of the λ 6 cm visibility function of LkHα 101 plotted against projected baselines.

λ 6 cm intensity map and the visibility curve of LkHα 101. We can see from Figure 5 that the source is completely resolved but this is not obvious from Figure 4. This suggests that when the source size is comparable to the beamwidth, an analysis of the visibility curve can be quite useful.

As an illustration, we have shown in Figure 6 the brightness distributions of (i) an optically-thick sphere; (ii) an optically-thin sphere; (iii) an optically thin shell and (iv) a stellar wind. Also shown in Figure 6 are the intensity maps of these sources when they are observed with a beam with FWHP equal to the diameter of the sources. For the stellar wind case, the beamwidth is equal to the full width of the 10% intensity point. In all cases, the intensity maps are similar in appearance. The λ 6 cm visibility curves of these sources are shown in Figure 7. The beam sizes as well as the source diameters are assumed to be 0."4 and the extent of the baselines corresponds to that of the VLA in the "A" configuration. In the stellar wind case, three (λ 6, 2 and 1.3 cm) visibility curves are shown. We can see that for spherically symmetric sources the visibility curves can provide valuable information on the source structure. Table 1 shows how the angular size can be determined from the half maximum point of the visibility amplitude ($b_{1/2}$) for these four geometries.

Although the wind parameter $(\dot{M}/v)^{4/3}\ D^{-2}$ can be derived from the flux density alone, the temperature of the wind can only be estimated when the wind is resolved by an interferometer. In the isothermal-wind model (White and Becker 1982), the angular radius of the $\tau = 1$ surface is given by

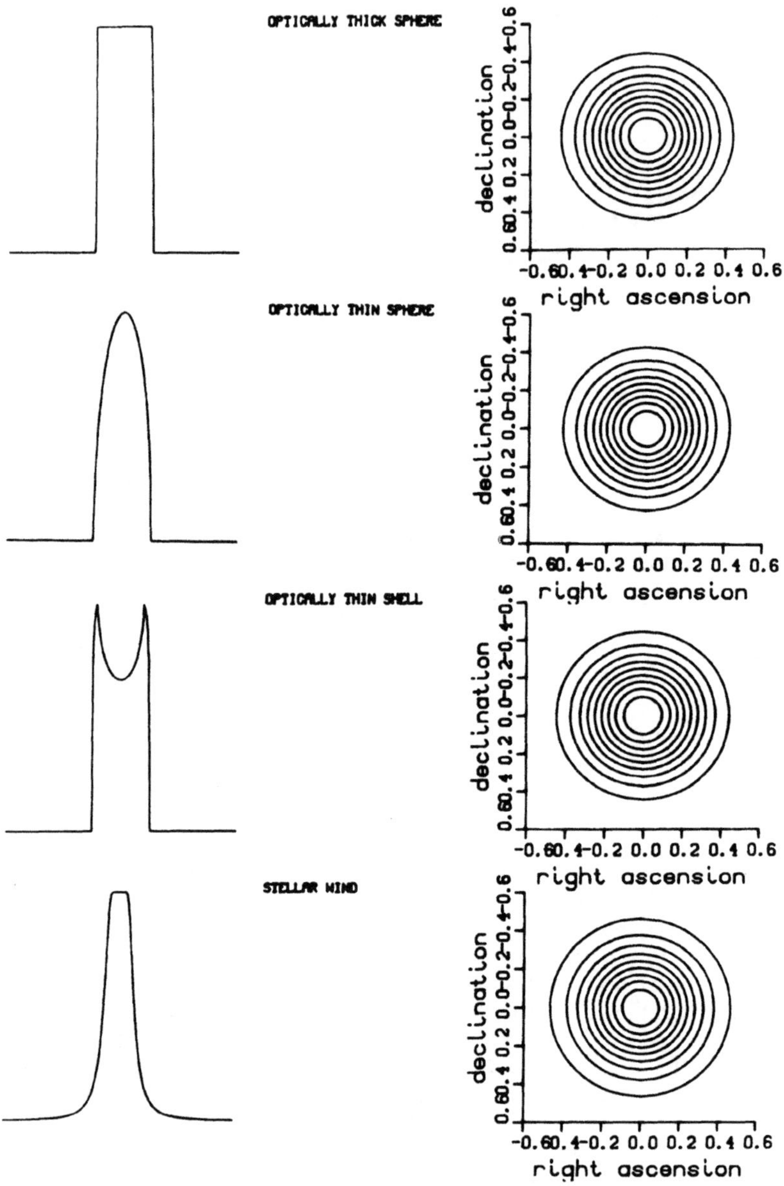

Figure 6. The brightness distributions and intensity maps of (i) an optically-thick sphere; (ii) an optically-thin sphere (τ = 0.1); (iii) an optically-thin shell (τ = 0.02); and (iv) a stellar wind. The maps are obtained by convolving the brightness distributions with telescope beams equal to the source sizes.

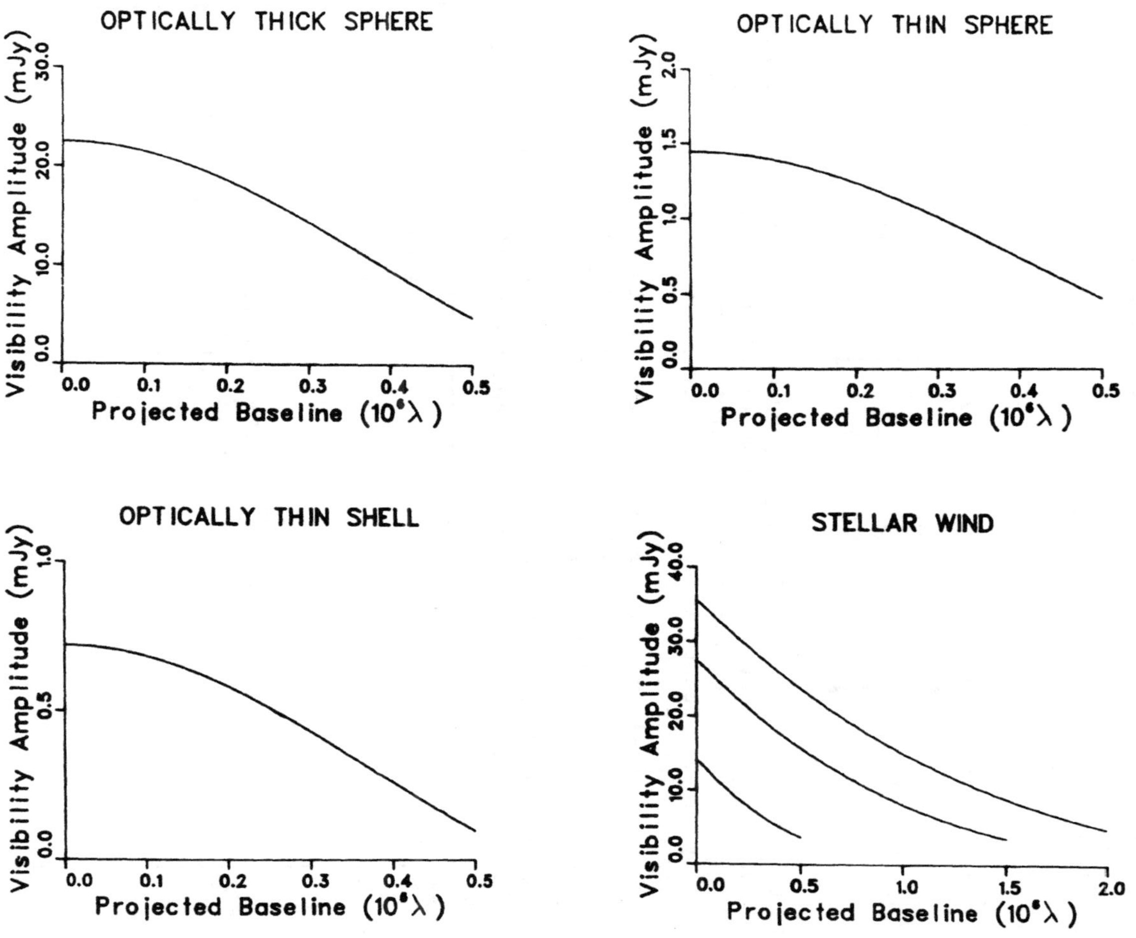

Figure 7. λ 6 cm visibility curves of the sources in Fig. 6. The stellar wind case is assumed to have $\theta_1 = 0\overset{''}{.}3\ (\nu/\mathrm{GHz})^{-0.7}$. All others have source diameters of $0\overset{''}{.}4$.

TABLE I

Geometry	Brightness Distribution	$(2\theta/\text{arc sec})(b_{1/2}/1000\ \lambda)$
-	gaussian	91
optically-thick sphere	uniform	148
optically-thin sphere	-	164
stellar wind	-	54.5*

* θ in this case is θ_1 (see Equation 3).

$$\theta_1 = 0''.031 \left[\frac{(\dot{M}/10^{-5} M_{\odot}\ yr^{-1})(\lambda/6cm)}{\mu(v/1000\ km\ s^{-1})} \right]^{2/3} (g\gamma)^{1/3} \times (T_e/10^4\ K)^{-1/2} (D/kpc)^{-1} \quad (3)$$

While Eq. (2) shows that F_ν is only dependent on T_e and the temperature cannot be obtained by flux density measurement alone, Eq. (3) implies that T_e can be derived from θ_1 and therefore from the shape of the visibility curve. Generally speaking, θ_1 decreases with increasing T_e, which results in a flatter visibility curve.

The above analysis has been applied to LkHα 101 and the result of the fit to the λ 6 cm visibility curve is shown in Figure 5. Assuming D = 800 pc and v = 1300 km s^{-1}, we find $\dot{M}$ = 5 x 10^{-5} $M_{\odot}$ yr^{-1} and T_e = 5500 K. Whether LkHα 101 can be explained by an isothermal wind can be tested by applying these parameters to observed visibility curves at other frequencies. Figure 8 shows the predicted visibility

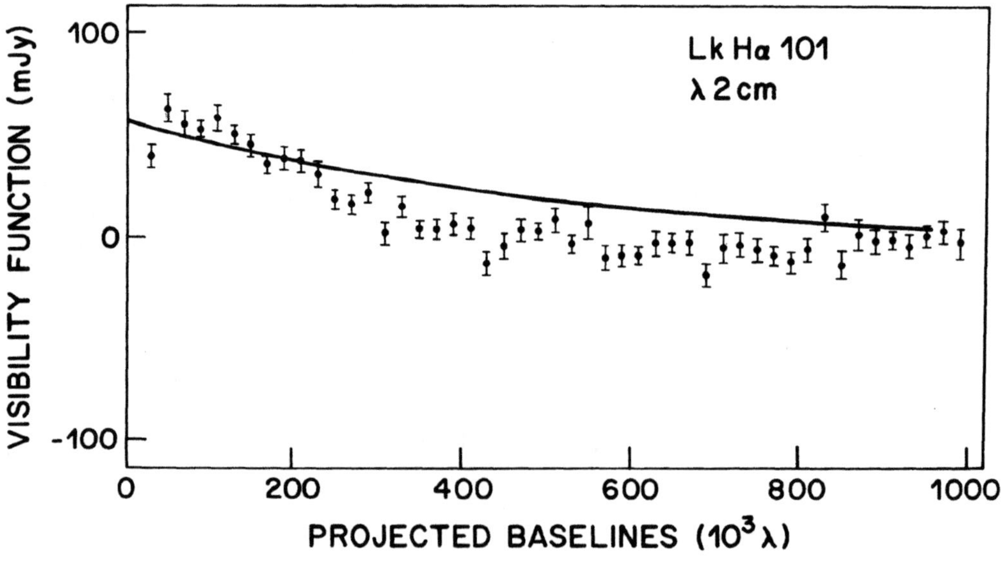

Figure 8. Model fitting to the λ 2 cm visibility curve of LkHα 101 by the isothermal wind model.

curve applied to the λ 2 cm data. The fit is not entirely satisfactory. The discrepancy between model and data is even greater at λ 1.3 cm. The higher-frequency data seem to suggest a larger source size than predicted by the model.

4. THE INTERACTION OF STELLAR WINDS

The first binary system found to have thermal and radio emission is α Sco (Hjellming and Wade, 1971). Two separate radio components are later found to be associated with the M1.5 Iab and the B2.5 V stars in the system (Gibson 1978). A complete mapping at the VLA by Hjellming and Newell (1983) find a point source at the position of the M star and an extended nebula near the B star. Analysis of the spectra of the two sources suggest that the point source arises from thermal emission in the chromosphere/corona of the M star whereas the extended emission is due to the partial ionization of the M-giant wind by the B component. A mass loss rate of $2 \times 10^{-6}\ M_\odot\ yr^{-1}$ is derived for the M-giant wind (Hjellming and Newell 1983).

Many symbiotic systems consisting of a late-type giant and an early-type star are also radio emitters (for a review see Kwok 1982). In many cases the radio emission can be understood as thermal radiation from the M-giant wind which is ionized by the hot component. A recent survey by Seaquist, Taylor and Button (1984) shows that ∿ 25% of all symbiotic stars are detectable by the VLA. They also suggest that the observed spectral indices (∿ 1) can be explained by an incomplete ionization of the M-giant wind.

A number of symbiotic stars also had episodes of optical outburst in the past. The oldest example is AG Peg which had a sudden increase of optical brightness in 1850. A detailed model of the radio behaviour of AG Peg based on recent VLA observations by Hjellming can be found elsewhere in this volume.

The most recent example of a symbiotic outburst is that of HM Sge in 1975. For this reason, HM Sge is the only symbiotic star that we have a complete radio history. Figure 9 shows the multi-frequency "lightcurve" of HM Sge. While the observed flux densities have continued increasing, HM Sge has maintained an approximate constant spectral shape (α ∿ 1) between 1977 and 1982 (Kwok, Bignell and Purton 1984). In contrast to classical novae where the envelopes become optically thin after ∿ 1 yr, the radio-emitting region of HM Sge is still optically thick 8 years after outburst.

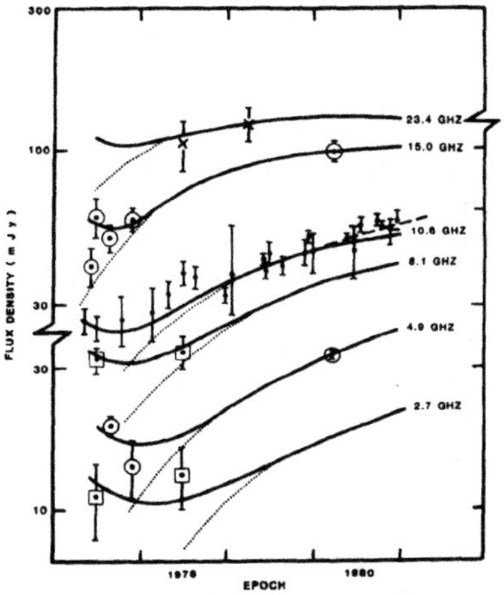

Figure 9. The multi-frequency "lightcurve" of HM Sge (taken from Purton, Kwok and Feldman 1983).

The large amount of radio data available provides a severe test for any model. Analysis of the radio light curves and the visibility curves show that HM Sge is best modelled by the interaction of a stellar wind from the hot white dwarf with a wind from the M giant (Kwok and Purton 1979; Purton, Kwok and Feldman 1983; Kwok, Bignell and Purton 1984).

The high surface brightness of HM Sge makes it possible to be observed with very high resolution and still has adequate signal to noise per beam area. Using the VLA in the "A" configuration (maximum baseline 36 km) and observing at λ 1.3 cm, angular resolution of ∿ 0".07 can be achieved. Figure 10 shows the λ 1.3 cm map of HM Sge. The dynamic range is ∿ 20 to 1 after self calibration. The superior resolving power of the VLA is clearly demonstrated.

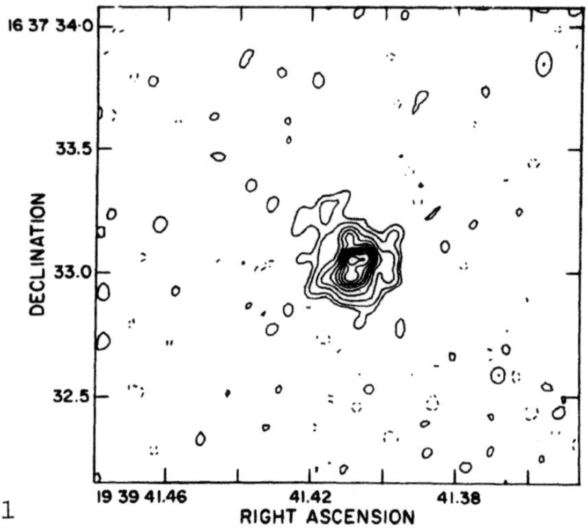

Figure 10. λ 1.3 cm map of HM Sge. Contours are at intervals of 10% of peak value (25.9 mJy/beam) with an additional contour at 5% (taken from Kwok, Bignell and Purton 1984).

5. WHAT IS A RADIO STAR?

Many of the radio sources detected in the 1970's as the result of searches through star catalogues turn out not to be stars but small nebulae. This distinction is often arbitrary for if one considers radio emission from stellar winds as radio stars then why not nebulae ejected from stars; afterall the nebular ejection process may be related to stellar winds (Kwok, Purton and FitzGerald 1978). Approximately half of all known planetary nebulae are unresolved optically and are only classified as such because of their emission-line spectra. This has led to some confusion as whether a detected radio source is a Be star, symbiotic star, or a stellar planetary nebula. Interest in radio star research has in fact results in the clarification of the nature of many of these objects.

Figure 11 shows the λ 2 cm map of Vy 2-2 obtained by Seaquist and Davis (1983). It was first detected as one of the mass-outflow objects with α = 1 (Purton *et al.* 1981) but Figure 11 shows a definite shell structure and it is likely to be a young planetary nebula. An increasing number of optically-unresolved planetary nebulae has

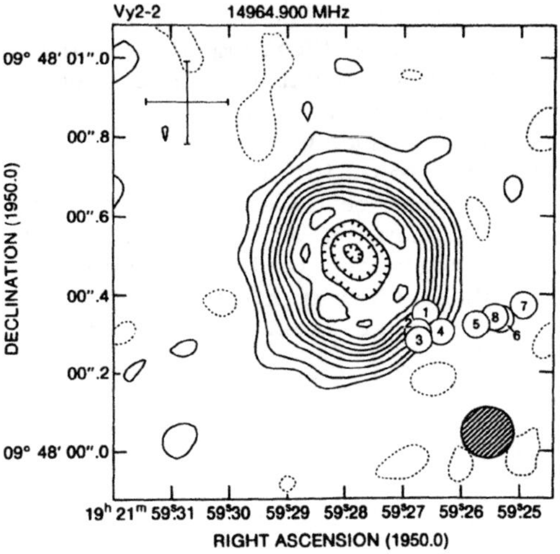

Figure 11. λ 2 cm map of Vy 2-2. Contour intervals are 2 mJy/beam with an additional contour at the 1 mJy/beam level (taken from Seaquist and Davis 1983).

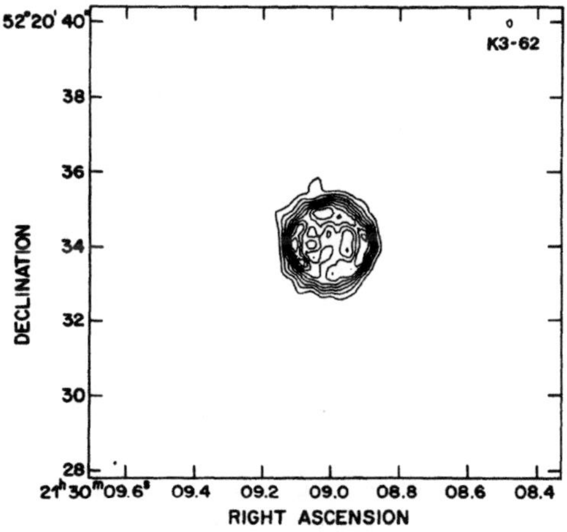

Figure 12. λ 6 cm map of K3-62. Contour are at intervals of 10% of the peak value (4.9 mJy/beam).

now been resolved by the VLA (Kwok 1984). Figure 12 shows the λ 6 cm map of K3-62 (θ ∿ 1"), the shell structure is clearly evident.

GL 618 is believed to be an object in transition between red giant and planetary nebula. Figure 13 shows the map of its radio core which is elongated along the same axis as the optical nebulosities. This suggests that the bipolar morphology of GL 618 extends over a scale of 0".3 to 7".

6. CONCLUSION

In the last 15 years we have witnessed the development of radio star research into a mature subject. Radio observations have been demonstrated to be useful probes of circumstellar regions, providing critical data to test our theoretical understanding of wind and nebular ejection processes. Thermal radio emission also have the additional advantage of being relatively easy to interpret, not being sensitive to local inhomogenities and fluctuations. The old barrier of inadequate angular resolution is now beginning to be overcome and we can expect more fruitful years ahead when improving sensitivity and resolving power lead to the imaging of more heretofore unresolved stars.

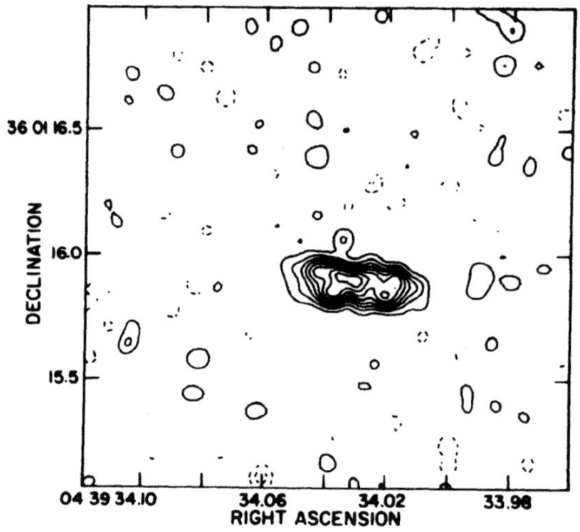

Figure 13. λ 1.3 cm map of GL 618. Contour intervals are 2.5 mJy/beam (taken from Kwok and Bignell 1984).

Acknowledgement

I wish to thank Mr. Orla Aaquist for producing Figures 6 and 7 and Dr. C.R. Purton for helpful discussions. Research in radio astronomy at The University of Calgary is supported by the Natural Sciences and Engineering Research Council of Canada.

REFERENCES

Altenhoff, W.J., Oster, L., and Wendker, H.J. 1979, *Astron. Astrophys.*, **73**, L21.

Braes, L.L.E., Habing, H.J., and Schoenmaker, A.A. 1972, *Nature*, **240**, 230.

Dreher, J.W. and Welch, W.J. 1983, *Astron. J.*, **88**, 1014.

Feldman, P.A. and Kwok, S. 1979, *J. Roy. Astron. Soc. Canada*, **73**, 271.

Gibson, D.M. 1978. *Bull. Amer. Astron. Soc.*, **10**, 631.
Hjellming, R.M. 1974, in *Galactic and Extragalactic Radio Astronomy*, ed. G.L. Verschuur and K.I. Kellerman (Springer-Verlag), p. 159.
Hjellming, R.M. and Wade, C.M. 1970, *Astrophys. J. (Lett).*, **162**, L1.
Hjellming, R.M. and Wade, C.M. 1971, *Astrophys. J. (Lett).*, **163**, L65.
Hjellming, R.M., Wade, C.M., Van den Berg, N.R. and Newell, R.T. 1979, *Astron. J.*, **84**, 1619.
Kwok, S. 1982, in *IAU Colloq. 70: The Nature of Symbiotic Stars*, ed. M. Friedjung and R. Viotti (Dordrecht: Reidel), p. 17.
Kwok, S. 1983, *Mon. Not. Roy. Astron. Soc.*, **202**, 1149.
Kwok, S. 1984, *Astron. J.*, in press.
Kwok, S. and Purton, C.R. 1979, *Astrophys. J.*, **229**, 187.
Kwok, S. and Bignell, R.C. 1984, *Astrophys. J.*, **276**, 544.
Kwok, S., Purton, C.R., and FitzGerald, M.P. 1978, *Astrophys. J. (Lett).*, **219**, L125.
Kwok, S., Bignell, R.C., and Purton, C.P. 1984, *Astrophys. J.*, **279**, 188.
Newell, R.T. and Hjellming, R.M. 1982, *Astrophys. J. (Lett).*, **263**, L85.
Panagia, N. and Felli, M. 1975, *Astron. Astrophys.*, **39**, 1.
Purton, C.R., Kwok, S. and Feldman, P.A. 1983, *Astron. J.*, **88**, 1825.
Purton, C.R., Feldman, P.A., Marsh, K.A., Allen, D.A., and Wright, A.E. 1982, *Mon. Not. Roy. Astron. Soc.*, **198**, 321.
Seaquist, E.R. and Gregory, P.C. 1973, *Nature*, **245**, 85.
Seaquist, E.R. and Palimaka, J. 1977, *Astrophys. J.*, **217**, 781.
Seaquist, E.R. and Davis, L.E. 1983, *Astrophys. J.*, **274**, 659.
Seaquist, E.R., Taylor, A.R., and Button, S. 1984, *Astrophys. J.*, **284**, 202.
Seaquist, E.R., Duric, N., Israel, E.P., Spoelstra, T.A.T., Ulich, B.L., and Gregory, P.C. 1980, *Astron. J.*, **85**, 283.
White, R.L. and Becker, R.H. 1982, *Astrophys. J.*, **262**, 657.
Wright, A.E. and Barlow, M.J. 1975, *Mon. Not. Roy. Astron. Soc.*, **170**, 41.

PROBLEMS WITH INTERPRETING THE RADIO EMISSION FROM HOT STARS

Anne B. Underhill
Laboratory for Astronomy and Solar Physics
NASA Goddard Space Flight Center
Greenbelt, MD 20771, USA

ABSTRACT. The hypothesis that the radio emission from a hot star is due solely to bremsstrahlung in a spherically symmetric wind flowing at a constant velocity and the constraint that the wind be transparent enough to allow the stationary photosphere to be seen place a limit on $\dot{M}/v_\infty$. The constraint that the momentum in the wind be provided by the radiation field places a limit on $\dot{M}v_\infty$. It is noted that both constraints are satisfied by the usually deduced values of $\dot{M}$ and v_∞ for OB supergiants. The case for early O stars is marginal, while for Wolf-Rayet stars $\dot{M}/v_\infty$ and $\dot{M}v_\infty$ are too large to satisfy the several hypotheses usually made. The trouble is due to $\dot{M}$ being too large by at least a factor 10. It is noted that postulating that part of the radio flux from Wolf-Rayet stars is caused by processes in a low-density magnetized plasma provides a solution to the dilemma. This solution offers advantages when accounting for the emission lines of Wolf-Rayet stars.

The radio flux from a hot star is usually interpreted as being due to free-free (bremsstrahlung) emission in a spherically symmetric wind moving at constant velocity v_∞ (Wright and Barlow 1975; Panagia and Felli 1975). The observed radio flux together with an outflow velocity estimated from UV spectra and a distance enable one to estimate the rate of mass loss ($\dot{M}$) in solar masses per year and the radius of the emitting region. Typically for OB supergiants $\dot{M}$ is of the order of 10^{-6} $M_\odot$ yr^{-1} and $R(6\ cm)/R_*$ is of the order of 50. For Wolf-Rayet stars $\dot{M}$ is of the order of 3×10^{-5} $M_\odot$ yr^{-1} and $R(6cm)/R_*$ is of the order of 800.

Because the winds of hot stars are fully ionized, the chief source of continuous opacity in the wind at 4500 Å is electron scattering. If the spectrum of the stationary photosphere at optical wavelengths is observed undisturbed by passing through the wind, then one may conclude that

$$\tau_e(\text{wind}) < \sim 0.3.$$

(The number 0.3 results from the quadrature procedures usually used for computing the emergent spectrum from a spherical atmosphere.)

R. M. Hjellming and D. M. Gibson (eds.), Radio Stars, 93–96.

By adopting a velocity law and assuming that the outflow occurs in spherical shells, one may deduce an upper limit for $\dot{M}/v_\infty$ which must not be exceeded if the stationary photosphere is to be visible at 4500 Å. In the case of OB supergiants and O stars, the deduced values of $\dot{M}/v_\infty$ are consistent with the observed fact that stationary photospheres are seen. In the case of Wolf-Rayet stars, one deduces that optical thickness due to electron scattering develops sufficiently far out in the wind that the deepest layers which may be seen are moving with a velocity of several hundred km s^{-1}. This deduction leads to problems with the source of momentum (Underhill 1983); it suggests that, perhaps, the underlying hypotheses are not entirely correct.

A further constraint on the inferred $\dot{M}$ from a hot star is given by the need for $\dot{M}v_\infty$ to be of the order of or less than L/c, where L is the luminosity of the star. This statement reflects the hypothesis that the radiation field is the sole source of momentum in the wind.

Some typical values of $(\dot{M}/v_\infty)_{-9}$ and $(\dot{M}v_\infty)/(L/c)$ for OB supergiants, early O-type stars, and Wolf-Rayet stars are presented in Table I.

Table I

Typical Observed Values[a] of $R_*/R_\odot$, $(\dot{M}/v_\infty)$, and $(\dot{M}v_\infty)/(L/c)$

Type of Star	Typical $R_*/R_\odot$	No. Stars	$\frac{\dot{M}}{v_\infty}$	$\frac{(\dot{M}v_\infty)}{(L/c)}$	Problems
Early B s.g.	30	5	1	0.6	No
O3 - O5	20	6	3	1.3	May be
WR with absn. lines[b]	10	3	5	(9.)	Yes
WR no absn. lines[b]	10	6	11	61.	Yes
WN7	36	3	15	7.2	Yes

[a] $\dot{M}$ is in units of 10^{-6} $M_\odot$ yr^{-1} and v_∞ in 10^3 km s^{-1}.
[b] Not types WN7 or WN8

In Table I $\dot{M}$ is the average value of $\dot{M}$ estimated (1) by means of the formula of Wright and Barlow (1975), (2) from UV resonance lines, (3) from Hα profiles, and (4) from IR fluxes; v_∞ has been found from UV resonance lines; $L/L_\odot$ has been estimated from integrated fluxes, monochromatic fluxes in the visible range, Kurucz model-atmosphere fluxes, and estimated distances.

In the case of the OB supergiants, there are no problems. The derived values for $\dot{M}$ and v_∞ are consistent with the hypotheses. The $\dot{M}$

values for the early O stars are large enough to push the constraining conditions to their limits, but with a little sharpening of the arguments and consideration of the possible errors in the observed quantities, one can say that there is no serious problem with the hypotheses used to describe the winds of these stars. In the case of the Wolf-Rayet stars, there are severe problems. The chief trouble spot seems to be the large values deduced for $\dot{M}$.

One possible solution of these problems is to postulate that some of the observed radio flux from Wolf-Rayet stars (and O stars?) is due to processes which occur in a magnetised plasma where electrons are accelerated and decelerated. Underhill (1983) has suggested that one is observing material suspended in large magnetic loops above the photospheres of Wolf-Rayet stars.

If a star possesses some patches of high-temperature plasma ($T \sim 10^7$K) confined in magnetic loops with $N_e = 10^{10}$ and the confining magnetic fields are a few hundred to 1000 gauss, the emissivity in gyroresonance radiation is of the order of 10^5 times that in bremsstrahlung (Underhill 1984). If magnetically confined plasma is present, a sizable fraction of the observed radio flux may not be due to bremsstrahlung in a spherically flowing wind.

The hot plasma would exist near the magnetic field lines; it need not be everywhere outside the photosphere. The higher the electron temperature in the loops, the smaller the loops need be to account for much of the radio flux from Wolf-Rayet stars.

The presence of local, small magnetic fields is attractive as a source for the superheated plasma observed by spectroscopy and for creating a means for placing material in a suitable position for producing shortward displaced discrete components in resonance lines (Underhill and Fahey 1984). The envisioned magnetic fields are of the order of or less than 600 gauss. Bhatia and Underhill (1984) have shown that a high-temperature, low density plasma such as might be created by the presence of closed and open field lines would account for the observed spectra of Wolf-Rayet stars. The envisioned magnetic fields are not expected to be large enough to generate a measurable longitudinal field when the star is observed from the earth.

One problem requiring solution is the following: In the case of a hot star, how does one determine how much radio flux is due to bremsstrahlung and how much is due to processes in a magnetized plasma? It seems clear that the presently used hypothesis for interpreting the radio fluxes from Wolf-Rayet stars (the hypothesis that the radio flux is solely due to bremsstrahlung in a spherically symmetric wind) leads to unsatisfactory conclusions and that one must make other hypotheses. Taking account of what may occur if a low-density, lightly magnetized plasma is present provides an attractive next step.

REFERENCES

Bhatia, A. K. and Underhill, A. B. 1984, submitted to the Ap. J. Suppl.

Panagia, N. and Felli, M. 1975, Astr. Ap., **39**, 1.

Underhill, A. B. 1983, Ap. J., **265**, 933.

Underhill, A. B. 1984, Ap. J., **276**, 583.

Underhill, A. B. and Fahey, R. P. 1984, Ap. J., **280**, 712.

Wright, A. E. and Barlow, M. J. 1975, M.N.R.A.S., **170**, 41.

THE RADIO-EMITTING WIND, JET, AND NEBULAR SHELL OF AG PEGASI

R. M. Hjellming
National Radio Astronomy Observatory*
Socorro, NM 87801 U.S.A.

ABSTRACT. VLA observations of the radio-emitting symbiotic star AG Peg show that the radio source, which has been interpreted previously as an expanding stellar wind, is primarily (80%) a resolved nebular sphere 1.5" (=1.3 X 10^{16} cm) in diameter that has had an average expansion velocity of 15 km/sec since about 1850 when ejection began. The rest of the radio emission comes from the combination of an unresolved point source with a spectral index of 1.2 and a jet-like feature extending from the point source to the nebular shell.

1. HISTORY

The AG Pegasi binary system (Boyarchuk 1967) consists of a normal red giant (M3 III) and a hot low-luminosity Wolf-Rayet (WN6) star that are imbedded in a circumstellar nebula at a distance of 600 pc. It has been called a symbiotic nova because of the combination of its spectral characteristics and a slow nova outburst. Beginning between 1850 and 1870 AG Peg increased three magnitudes in optical brightness, followed by a slow decline to nearly pre-flare magnitude in the 1960's. Boyarchuk (1966, 1967) has interpreted the optical data in terms of a slowly developing flare from the WN6 star that expanded, was optically thick to optical continuum radiation until about 1900, and by 1964 was a nebulosity with a temperature of 17000 K, a mass of 10^{-3} $M_\odot$, a size of 10^{16} cm, and a mean electron concentration of 7 X 10^{6} cm^{-3}. Forbidden emission lines in this nebulosity had a bell-shaped profile that indicated a spherical nebula rather than a thin shell, and the half-width of 114 km/sec was interpreted as consistent with the inferred size scale by assuming the nebular ejection began in 1920.

AG Peg was found to be a radio source by Gregory et al. (1977), who interpreted detections at 2.7 and 8.1 GHz and an upper limit at 10.5 GHz in terms of a hot, spherical stellar wind with an expansion velocity

*The National Radio Astronomy Observatory is operated by Associated Universities, Inc., under contract with the National Science Foundation.

R. M. Hjellming and D. M. Gibson (eds.), Radio Stars, 97–100.

of 114 km/sec and a mass-loss rate of 10^{-6} $M_\odot$/year, using the formula of Wright and Barlow (1975). Subsequent radio measurements were also made by Wright and Allen (1978), Bowers and Kundu (1979), Johnson (1980), and Ghigo and Cohen (1981). Ghigo and Cohen (1981) noted descrepancies, particularly a factor of 2 to 3 between the diameter (0.4") predicted by the stellar wind model and the diameter (1.0" ± 0.3") inferred from the 4.9 GHz visibility amplitudes.

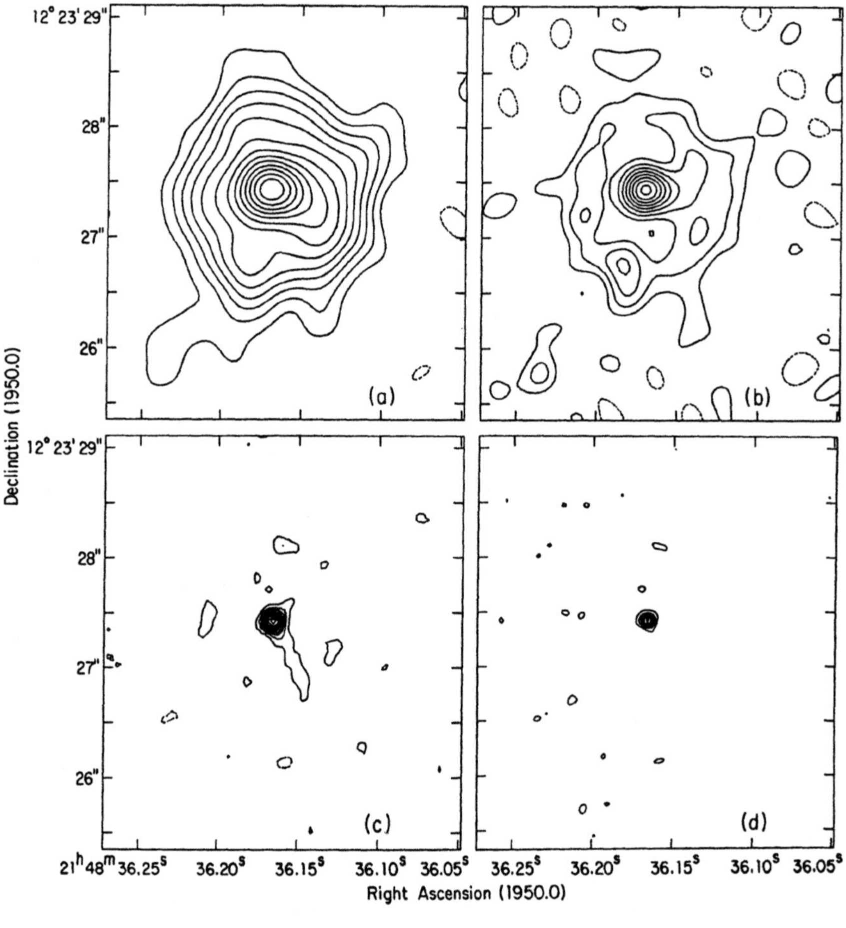

Figure 1. VLA maps of the AG Peg radio source: (a) 4.9 GHz with natural weighting, beam width (at half-power) of 0.5", and contour intervals of 0.1 mJy/beam; (b) 4.9 GHz with uniform weighting, beam width of 0.33", and contour intervals of 0.11 mJy/beam; (c) 15 GHz with natural weighting, beam width of 0.15", and contour intervals of 0.24 mJy/beam; and (d) 15 GHz with uniform weighting, beam width of 0.1", and contour intervals of 0.33 mJy/beam.

2. MAPPING OF AG PEG's NEBULAR SHELL, WIND, AND A "JET-LIKE" CONNECTION

AG Peg observations were made in Oct.-Nov. 1983 at 1.4, 4.9, and 15 GHz with the 35 km VLA configuration. Maps at 4.9 and 15 GHz are shown in Figure 1. About 80% of the 4.9 GHz flux of AG Peg comes from a mainly spherical nebulosity that is 1.5" in diameter. There is a point-source component that is unresolved at the 0.3" and 0.1" levels at 4.9 and 15 GHz respectively. The 4.9 GHz maps also show a "jet-like" extension from the central point source that extends into, and merges with, the nebulosity. This is very reminiscent of the "jet-like" feature in the symbiotic star R Aqr seen by Sopka et al. (1982) at both radio and optical wavelengths. The jet appeared between 1970 and 1977 as a central feature in a nebulosity ejected about 600 years ago (Hubble 1943, Baade 1944). Both AG Peg and R Aqr are symbiotic stars that have undergone a "slow nova" event that produced a nebular shell.

The 1983.9 VLA observations gave total flux densities of 3.8 ± 0.4, 6.3 ± 0.4, and 3.0 ± 0.4 mJy, for frequencies of 1.4, 4.9, and 15 GHz, respectively, for the sum of the nebulosity, un-resolved point source, and "jet-like" structure extending from the point source to the west in the 5 GHz maps. In Figure 2a we plot the radio spectrum of AG Peg using these data and the previously referenced radio observations. In Figure 2b the data are re-plotted in terms of a compact unresolved source with a spectral index of 1.2 (from the 1983 data at 4.9 and 15 GHz) and a resolved source, a mixture of the 1.5" spherical nebulosity and the "jet-like" feature, with a spectral index of about 0.5.

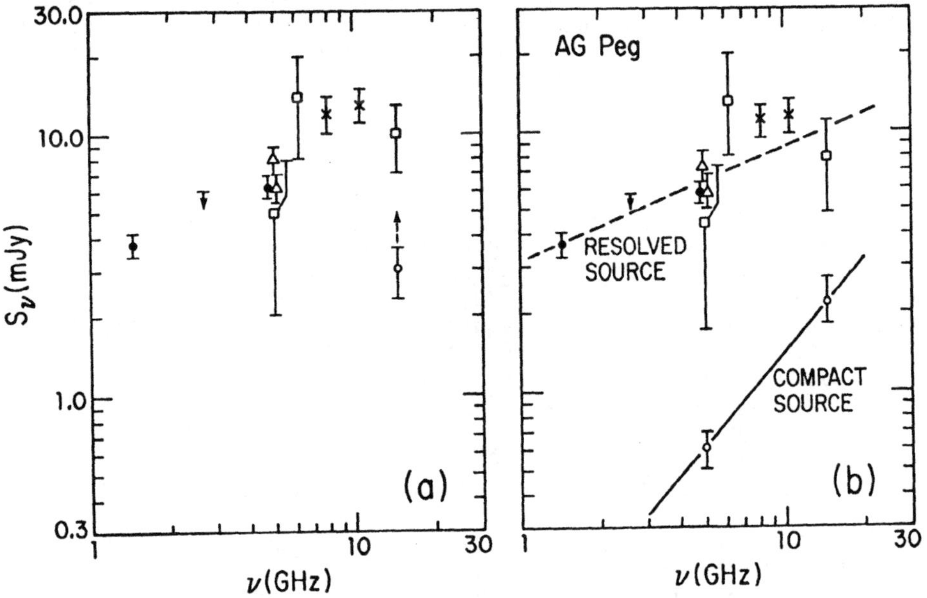

Figure 2. The radio spectrum of AG Peg: (a) raw data; and (b) the spectra of the unresolved point source and the resolved nebulosity.

3. CONCLUSIONS

The high-resolution VLA maps of the AG Peg radio source indicate that most of the radio flux exists in a nebulosity 1.5" in diameter. There is a still-unresolved component with a spectral index of about 1.2 that has a "jet-like" extension connected to the nebulosity. These facts explain the discrepancies found by Ghigo and Cohen (1981), and their reservations about an interpretation in terms of a stellar wind with a mass loss rate of 10^{-6} M☉/year were more than justified. The 1983 VLA observations shown in Figure 1 indicate that the unresolved source flux in the center is 0.6 ± 0.1 and 2.2 ± 0.5 mJy at 4.9 and 15 GHz, respectively. Although one might argue for a continuing stellar wind with a mass-loss rate of the order of 10^{-7} M☉/year, the steep spectral index of 1.2 is an indication of an optically thick source that is more complicated than a spherically symmetric stellar wind. This spectral index is, in fact, typical of the complex nebulosities found for VV Cephei binaries by Hjellming (1985) and for symbiotic stars by Taylor and Seaquist (1985). This indicates that the WN6 star is ionizing most of the wind from the red giant companion, but that there is a subsection of this wind that remains neutral. This occurs because the ionizing radiation cannot penetrate the innermost parts of the red giant wind, and so causes a neutral zone on the "other" side. The spectral index is steeper than the 0.6 index of a spherically symmetric wind because of the ionized boundary between the neutral and ionized parts of the wind.

The AG Peg nebulosity is an example of a slow nova "wind" whose expansion has been slowed by interaction with the pre-existing red giant wind. An initial ejection in 1850 and a diameter of 1.5" (=1.3×10^{16} cm) in 1983.9 corresponds to an average velocity of 15 km/sec. This is much less than the "modern" wind velocity of 114 km/sec but close to the typical expansion velocities of red giant and supergiant winds.

REFERENCES

Baade, W.A. 1944, Rep. Dir. Mt. Wilson Obs. 1943-44, **12.**

Boyarchuk, A.A. 1966, Soviet Astronomy, **10**, 783.

Boyarchuk, A.A. 1967, Soviet Astronomy, **11**, 8.

Bowers, P.F., and Kundu, M.R. 1979, A.J., **84**, 791.

Gregory, P.C., Kwok, S., and Seaquist, E.R. 1977, Ap.J., **211**, 429.

Ghigo, F.D., and Cohen, N.L. 1981, Ap.J., **245**, 988.

Hjellming, R.M. 1985, Radio Stars (Reidel: Dordrecht), ed. R.M. Hjellming and D.M. Gibson

Hubble, E.P. 1943, Rep. Dir. Mt. Wilson Obs. 1943-44, **12.**

Johnson, H.M. 1980, Ap.J., **237**, 840.

Sopka, R.J., Herbig, G., Kafatos, M., and Michalitsianos, A.G. 1982, Ap.J.(Letters), **258**, L32.

Taylor, A.R., and Seaquist, E.R. 1985, Radio Stars (Reidel: Dordrecht), ed. R.M. Hjellming and D.M. Gibson

Wright, A.E., and Allen, D.A. 1978, M.N.R.A.S., **184**, 893.

Wright, A.E., and Barlow, M.J. 1975, M.N.R.A.S., **170**, 41.

MULTI-FREQUENCY RADIO IMAGES OF L1551 IRS5

J. H. Bieging
Radio Astronomy Laboratory
University of California
Berkeley, CA 94720, USA

and

Martin Cohen
NASA-Ames Research Center
Moffett Field, CA 94035, USA

ABSTRACT. We discuss new multi-frequency VLA maps of the bipolar outflow source L1551 IRS5.

L1551 IRS5 is the prototype of bipolar outflow sources. The star IRS5 drives both a large-scale bipolar molecular outflow (Snell, Loren, and Plambeck 1980) and high-velocity Herbig-Haro objects (Cudworth and Herbig 1979). Weak radio emission (3 mJy at 5 GHz) is centered on the star and extends in both directions toward the much larger molecular lobes (Cohen, Bieging, and Schwartz 1982; Bieging, Cohen, and Schwartz 1984).

We have mapped the radio emission from L1551 IRS5 at 1.5, 4.9, and 15 GHz with the VLA in its A-configuration. The source is detected and resolved at all three frequencies, which enables us to derive spectral index information for both the core source and the extended component. At 1.5 GHz, the extended jet is only slightly resolved, and at 4.9 GHz the map is the same as those published previously, except for better signal-to-noise ratio. At 15 GHz, the central core is well-resolved into two point-like sources surrounded by weaker extended emission. The two sources lie on a line at p.a. 191°, and the northern (brighter) source is coincident with the peak at 4.9 GHz. The jet component seen at 4.9 GHz is barely detected at 15 GHz.

By smoothing the 15 GHz map to the same resolution as that at 4.9 GHz, we can derive the spectral index, α, for the source components. We find that the jet has $\alpha \sim 0.3$, while the two sources seen at 15 GHz have much steeper spectra. For the northern (brighter) source, $\alpha = 0.9$, and for the southern component, $\alpha \sim 1.1$. Since the jet is unresolved in the

R.M. Hjellming and D.M. Gibson (eds.), Radio Stars, 101–102.

4.9 GHz map, we can set an upper limit on its transverse thickness of <0.3", or <7 x 10^{14} cm at a distance of 160 pc. The total observed length of the jet is 3.2", or 7.7 x 10^{15} cm, so the length-to-width ratio, L/W, is >10. If we assume that the jet emission is by thermal processes and has a brightness temperature of ~10^4 K, the observed length and total flux density (after subtracting the contribution of the point sources) implies an even smaller thickness of ~2 x 10^{13} cm. In this case, L/W ~ 400, which requires a very highly collimated jet.

By analogy with the known binaries which are also double radio sources, namely T Tau and Z CMa, we suggest that L1551 IRS5 may also be a binary system whose orbital plane coincides with the CS molecular toroid found by Kaifu et al. (1984). L1551 IRS5 is clearly a good candidate for high-resolution IR speckle imaging.

A full discussion of these results is presented by Bieging and Cohen (1985).

REFERENCES

Bieging, J.H., and Cohen, M. 1985, Ap. J. (Lett.), in press.
Bieging, J.H., Cohen, M., and Schwartz, P.R. 1984, Ap. J. 282, 699.
Cohen, M., Bieging, J.H., and Schwartz, P.R. 1982, Ap. J. 253, 707.
Cudworth, K.M., and Herbig, G.H. 1979, A.J. 84, 548.
Kaifu, N. et al. 1984, Astron. & Ap., submitted.
Snell, R.L., Loren, R.B., and Plambeck, R.L. 1980, Ap. J. (Lett.) 239, L17.

A LUMINOSITY-LIMITED VLA SURVEY OF T TAURI STARS IN TAURUS-AURIGA

J.H. Bieging
Radio Astronomy Laboratory
University of California
Berkeley, CA 94720, USA

Martin Cohen
NASA-Ames Research Center
Moffett Field, CA 94035, USA

and

P.R. Schwartz
Naval Research Laboratory
Washington, DC 20375, USA

ABSTRACT. We summarize the results of a radio continuum survey of T Tauri stars in the Taurus-Auriga dark clouds to a limiting bolometric luminosity of $\log(L_{bol}/L_{\odot}) > 0.2$.

We have surveyed 46 pre-main-sequence stars in the Taurus-Auriga dark clouds for 6 cm continuum emission, using the VLA in its A-configuration. These stars comprise a complete sample with $\log(L_{bol}/L_{\odot}) > 0.2$. Of the six stars which were detected, five excite Herbig-Haro (H-H) nebulae, while only one star, V410 Tau, is from the typical T Tauri stellar population. Despite its very weak optical emission line character, V410 Tau has undergone dramatic radio variation over the past two years (see Table I). Its fluxes have recently decreased by factors of 50 and 20 at wavelengths of 2 and 6 cm. In February 1982, V410 Tau showed a radio spectral index of 1.3 from 20 cm to 2 cm, with a possible turnover between 2 cm and 1.3 cm. A non-thermal, flarelike mechanism may be responsible for the radio emission.

We can make a statistical statement about the connection between radio continuum emission and associations with H-H objects. Of the 46 stars in the Taurus-Auriga dark clouds, only 1 is detected of the 41 with no known association with H-H objects (~2% detection rate), while 5 stars are detected out of 5 that are associated with H-H objects (100% detection rate). These statistics suggest that by the time a pre-main-sequence star is visible as a T Tauri variable, it is very unlikely that 6 cm continuum emission will be detectable. Conversely, whatever

R. M. Hjellming and D. M. Gibson (eds.), Radio Stars, 103–104.

phenomenon that causes directed mass loss and is responsible for the production and excitation of H-H objects, also produces detectable 6 cm continuum emission.

Radio emission from DG Tau at 6 cm has been spatially resolved and shows a bipolar structure, elongated along the flow axis from the star (as indicated by the nearby H-H object--Mundt and Fried 1983), but pinched in the orthogonal direction, where the circumstellar dust disk lies. The radio emission from this star is similar to that toward L1551 IRS5, which is also associated with H-H objects and with a strong collimated bipolar molecular outflow.

Finally, we have evidence for another stellar double radio source. Z CMa shows two clearly resolved radio peaks at 6 cm. Since independent optical observations show that Z CMa is a double star, with a separation close to that observed for the radio peaks, we suggest that this star is similar to another double radio system, namely T Tauri and its binary companion.

A complete description of this survey is given by Bieging, Cohen, and Schwartz (1984).

TABLE I: Variations of Flux Density and Spectral Index for V410 Tau

Epoch	S(4.9 GHz)	S(15 GHz)	Spectral Index
22/23 Mar 1981	18.6 ± 1.0	...	...
8-13 Feb 1982	14.4 ± 1.4	64 ± 6	1.33 ± 0.17
11 Aug 1983	0.4[a]	...	...
30 Oct 1983	...	<0.9[b]	...
21 Nov 1983	0.87 ± 0.13	0.95 ± 0.16	0.08 ± 0.28

[a]M. Kutner, priv. comm. (1984)

[b]R. White, priv. comm. (1984)

REFERENCES

Bieging, J.H., Cohen, M., and Schwartz, P.R. 1984, Ap. J. 282, 699.
Mundt, R., and Fried, J.W. 1983, Ap.J. (Lett.) 274, L83.

RADIO CONTINUUM EMISSION FROM PRE-MAIN SEQUENCE STARS AND ASSOCIATED STRUCTURES

A. Brown,[1] R. Mundt[2] and S. A. Drake[1]
[1]Joint Institute for Laboratory Astrophysics, University of Colorado and National Bureau of Standards, Boulder, Colorado 80309
[2]Max-Planck-Institut für Astronomie, Königstuhl, D-6900 Heidelberg 1, Federal Republic of Germany

ABSTRACT. VLA 6 cm (4860 MHz with 100 MHz total bandwidth) C array observations of regions of recent star formation containing pre-main sequence (PMS) stars, Herbig-Haro objects, and stellar jets, are presented. Three PMS stars, HL Tau, XZ Tau and FS Tau, were definitely (6σ) detected and FS Tau B is possibly (5σ) detected. The mass loss rates implied for spherically-symmetric stellar winds with outflow velocities of 200 km s^{-1} are in the range $1\text{-}2 \times 10^{-8}$ $M_\odot$ yr^{-1}. HL Tau shows extended structure including a 3σ detection of emission possibly associated with the jet from this star. No Herbig-Haro objects were detected at a typical 3σ rms level of 0.12 mJy/beam.

1. INTRODUCTION

Remarkably few detections of radio continuum emission from T Tauri stars have yet been made with the VLA. These stars are young, pre-main sequence (PMS), cool (G-M spectral type) stars still contracting toward the main sequence. Such stars are found in regions of recent star formation in close association with Herbig-Haro objects (Herbig 1974) and other types of emission nebulosity. Previous radio observations of PMS stars were made by Bertout and Thum (1982) and Felli et al. (1982) and VLA observations have been presented by Cohen, Bieging and Schwartz (1982), Rodriguez and Canto (1983), Schwartz et al. (1984) and Bieging, Cohen and Schwartz (1984). The VLA observations have detected only three classical T Tauri stars at a high statistical significance, namely T Tau N (the optical star), DG Tau and V410 Tau, and of these only T Tau N shows a spectral index suggesting free-free emission from a stellar wind to be the emission mechanism. Most sources detected have been earlier type stars or stars driving bipolar outflows.

Mundt and Fried (1983) and Mundt et al. (1984) have described observations of optical jets from young stars. These jets seem to represent well-collimated, high-velocity flows originating from PMS stars. In this paper we report on radio observations of some of these stellar jet sources.

R.M. Hjellming and D.M. Gibson (eds.), Radio Stars, 105–110.

2. OBSERVATIONAL DATA

On 1984 April 30, May 3, and May 5 NRAO[3] VLA 6 cm C array observations were obtained of five fields containing pre-main sequence objects associated with optical jets, namely HL/XZ Tau, FS Tau (Haro 6-5), HH 33/40, HH 19 and 1548C27. In all cases, the region of interest was offset from phase center to avoid instrumental artifacts. These data were analyzed using the standard AIPS routines on the VLA VAX computers. Maps were produced using natural weighting and by CLEANing all serendipitous sources. The resulting 3σ rms noise levels were 0.09 mJy for FS Tau, 0.12 mJy for HL/XZ Tau, HH 33/40 and HH 19, and 0.13 mJy for 1548C27. In the fields of HH 33/40, HH 19 and 1548C27 no sources were detected which were related to PMS stars and nebulosity associated with such stars. Among the Herbig-Haro objects not detected were HH 19-22, 24, 26, 33, 34 and 40. Maps of the HL/XZ Tau and FS Tau regions are shown in Figs. 1 and 2. The PMS stars, XZ and HL Tau, are clearly detected with HL Tau appearing as an extended source, somewhat similar in shape to the nebulosity seen in deep optical images. Additional structure is seen in the vicinity of these stars including a 3σ source coincident with the brightest part of the optical jet from HL Tau seen by Mundt and Fried (1983). In the case of FS Tau, sources coincident with both FS Tau A and FS Tau B are seen. FS Tau B possibly shows weak extended structure. The properties of the stellar sources detected are presented in Table 1.

Table 1. Positions, Fluxes and Other Properties of Detected T Tauri Stars

Source	(1950.0) α h m s			δ ° ' "			Peak Flux (mJy)	$\dot{M}$ ($M_\odot$/yr)	Comment
FS Tau B	4	18	56.10	26	50	30.3	0.15	1.2(-8)	Jet Source
FS Tau A	4	18	57.65	26	50	29.0	0.19	1.5(-8)	
HL Tau	4	28	44.21	18	07	35.4	0.25	1.8(-8)	Jet Source
XZ Tau	4	28	46.04	18	07	34.1	0.24	1.7(-8)	

[3]The National Radio Astronomy Observatory is operated by Associated Universities Inc., under contract with the National Science Foundation.

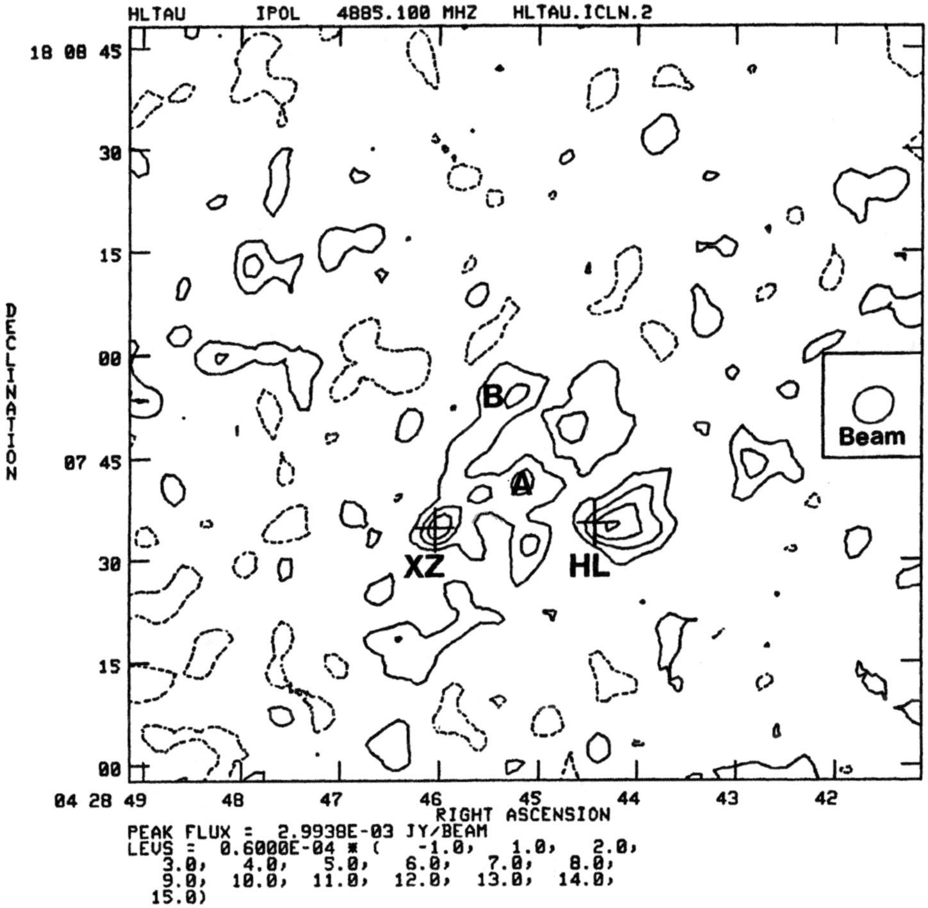

Fig. 1. VLA 6 cm (4860 MHz) map of the region around HL Tau and XZ Tau. Optical positions are from Jones and Herbig (1979) including proper motion correction to 1984.33. Source A is strongly nonthermal and is most probably a dMe or PMS M star given that it is only just visible in the images of Mundt and Fried (1983). Source B coincides with the brightest optical region in the jet from HL Tau. The contours are at 1.5σ intervals.

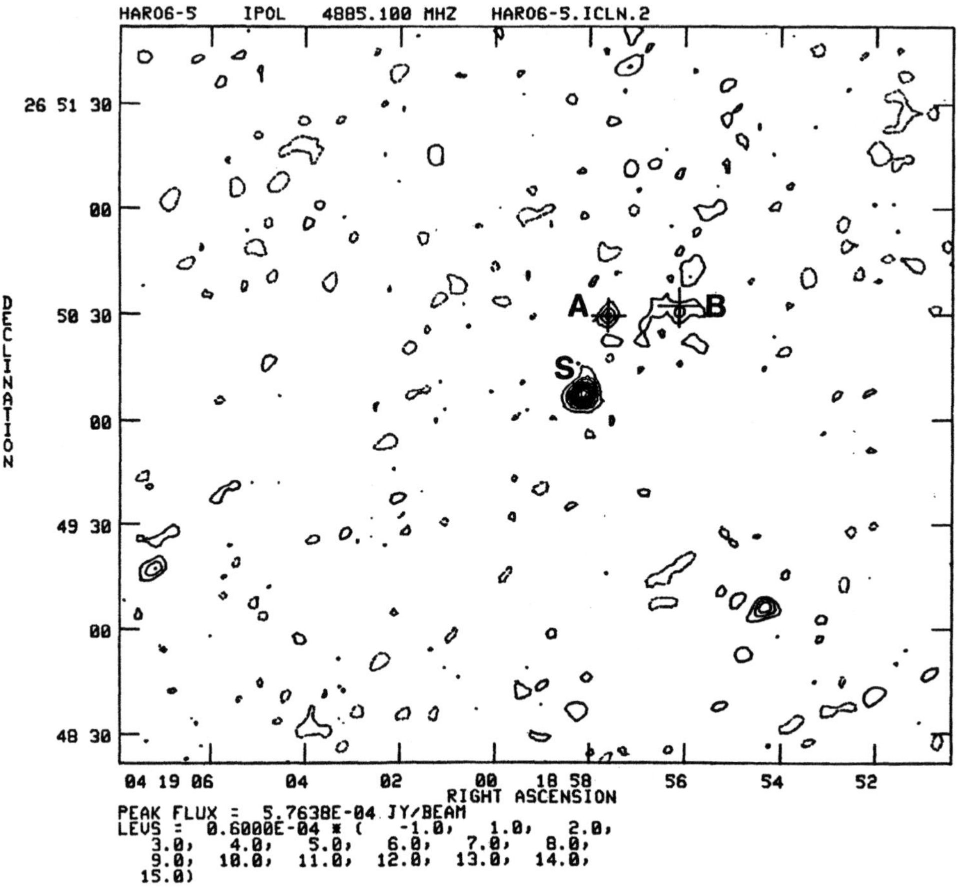

Fig. 2. VLA 6 cm (4860 MHz) map of the region around FS Tau (Haro 6-5): optical position of FS Tau A is from Jones and Herbig (1979) corrected for proper motion to 1984.33, while that for FS Tau B is based only on a rough offset from FS Tau A. Source S has no optical counterpart and may well be extragalactic. The contours are at 2σ intervals.

3. DISCUSSION

Previously published radio data for the regions that we have studied are only available for HL/XZ Tau. Rodriguez and Canto (1983) and Bieging, Cohen and Schwartz (1984) give 6 cm VLA upper limits for HL Tau of 1 mJy and 0.4 mJy respectively. (An earlier upper limit of 0.16 mJy given by Cohen, Bieging and Schwartz is presumably incorrect since this estimate was revised in their later paper.) Bertout and Thum (1982) reported a single dish 23 GHz detection of HL Tau at 4.9 ± 0.9 mJy with a half power beam width of 44 arcseconds. This detection therefore refers to an integrated 1.3 cm flux for all the central sources in our 6 cm map (Fig. 1). If the sources in this region contribute equally at 6 cm and 1.3 cm this would imply spectral indices of ~1.

In Table 1 mass loss rates are presented based on the assumption that the observed 6 cm radio emission is due to thermal free-free emission from spherically-symmetric stellar winds with outflow velocities of 200 km s^{-1} (a typical value for T Tauri stars) using the Wright and Barlow (1975) formalism. The mass loss rates fall in the range 1-2 × 10^{-8} $M_\odot$ yr^{-1}. The distance to the stars was taken to be 150 pc. Such values are consistent with the ideas of DeCampli (1981) concerning the wind driving mechanisms in T Tauri stars. The mass loss rate derived for HL Tau (1.8 × 10^{-8} $M_\odot$ yr^{-1}) is consistent with the value of 6 × 10^{-9} $M_\odot$ yr^{-1} derived by Edwards and Snell (1984) from CO observations, assuming conservation of momentum between the stellar wind and molecular gas. These values would imply that approximately a third of the stellar wind momentum is transferred to the more distant molecular outflow, provided that all the 6 cm emission arises from the stellar wind. The only T Tauri star for which the radio emission has been convincingly shown to be due to its stellar wind is T Tau N which has a mass loss rate of ~5 × 10^{-8} $M_\odot$ yr^{-1}.

4. CONCLUSIONS

From observations of five regions of recent star formation containing optical jets emanating from PMS stars, we have detected four PMS stars (HL Tau, XZ Tau, FS Tau A and B) as 6 cm radio emitters. These are the deepest VLA observations of such stars yet obtained, having typical 3σ rms limits of 0.12 mJy/beam, and perhaps indicate that many more T Tauri stars will be detectable by similar observations. (No other classic T Tauri stars were present in our fields.) In addition, these observations have shown a complex structure of extended 6 cm emission in the region surrounding HL and XZ Tau, including a possible detection of the brightest optical knot in the jet from HL Tau.

ACKNOWLEDGMENTS

This work has been supported by NASA grant NGL-06-003-057 through the University of Colorado.

REFERENCES

Bertout, C. and Thum, C. 1982, Astr. Ap., **107**, 368.
Bieging, J. H., Cohen, M., and Schwartz, P. R. 1984, Ap. J., **282**, 699.
Cohen, M., Bieging, J. H., and Schwartz, P. R. 1982, Ap. J., **253**, 707.
De Campli, W. 1981, Ap. J., **244**, 124.
Edwards, S. and Snell, R. L. 1984, Ap. J., **281**, 237.
Felli, M., Gahm, G. F., Harten, R. H., Liseau, R., and Panagia, N. 1982, Astr. Ap., **107**, 354.
Herbig, G. H. 1974, Lick Obs. Bull., No. 658.
Jones, B. F. and Herbig, G. H. 1979, A. J., **84**, 1872.
Mundt, R., Bührke, T., Fried, J. W., and Stocke, J. 1984, Astr. Ap., in press.
Mundt, R. and Fried, J. W. 1983, Ap. J., **274**, L83.
Rodriguez, L. F. and Canto, J. 1983, Rev. Mex. Astr. Astrof., **8**, 163.
Schwartz, P. R., Simon, T., Zuckerman, B., and Howell, R. R. 1984, Ap. J., **280**, L23.
Wright, A. E. and Barlow, M. J. 1975, Mon. Not. R. astr. Soc., **176**, 41.

VARIATIONS IN THE RADIO FLUX OF THE HYPERGIANT P CYGNI (B1 Ia^+)

G. H. J. van den Oord,[1,2] L.B.F.M. Waters,[1] H.J.G.L.M. Lamers,[1,2] D. C. Abbott,[3] J. H. Bieging,[4] and E. Churchwell[5]
[1]Laboratory for Space Research, Utrecht.
[2]Astronomical Institute, Utrecht.
[3]Joint Institute for Laboratory Astrophysics, University of Colorado and National Bureau of Standards, Boulder.
[4]Radio Astronomy Laboratory, University of California, Berkeley.
[5]Washburn Observatory, University of Wisconsin, Madison.

ABSTRACT. The radio flux of P Cygni is variable on a time scale of months or shorter. The variations are interpreted as variations in the degree of ionization of the wind due to the ejection of shells which shield the wind from the ionizing radiation of the star. The observed time scale agrees with the recombination time scale of the wind.

1. INTRODUCTION

The hypergiant star P Cygni (HD 193237, B1 Ia^+) is surrounded by a dense wind with a mass loss rate of 1.5×10^{-5} $M_\odot$/yr and a wind velocity of 300 km/s. The radius of the star is $R_* = 76$ $R_\odot$ (Lamers et al. 1983). It was the first single early-type star from which radio emission was detected (Wendker et al. 1973) with a flux of about 6 mJy at 6 cm. Observations of the frequency spectrum and angular distribution of the emission by Newell (1981) and White and Becker (1982) showed conclusively that the P Cygni radio source is thermal bremsstrahlung from a stellar wind whose temperature is roughly 18,000 K between 100 and 500 R_*.

Abbott et al. (1981) suggested that the radio flux of P Cygni is variable on a time scale of months. This time scale is surprisingly short, considering the fact that the emitting region is several hundred stellar radii thick and the time needed to change the density in such a large volume by means of variations in the mass loss rate is of the order of a year or more. This would indicate that some other mechanism is responsible for the change of the radio flux.

R. M. Hjellming and D. M. Gibson (eds.), Radio Stars, 111–116.

In this paper we report the study of the variations in the radio flux of P Cygni at 6 cm. We will demonstrate that the variations are real and that they likely result from changes in the degree of ionization in the wind.

2. THE OBSERVATIONS

The flux of P Cygni at 5 GHz (6 cm) has previously been measured with the WSRT and VLA by Wendker et al. (1973; WSRT), Abbott et al. (1980, 1981; VLA), and Newell and Hjellming (1980; VLA). We report here new observations by van den Oord and Waters (WSRT) and Abbott, Bieging, and Churchwell (VLA). The fluxes observed since 1978 are plotted in Figure 1. The new VLA data (since JD 4800) were obtained with compact array configurations (maximum baseline of 3.4 km) which do not resolve the source. The WSRT observations between JD 5440 and 5490 consist of a series of 15 measurements with a mean flux of 6.4 ± 0.6 mJy. This agrees very well with the flux of 6.7 ± 0.2 mJy measured*with the VLA during the same period. We confirm that the spectral index was 0.6 during this period. Figure 1 clearly shows that the flux varies between 1.6 and 10.8 mJy. The fastest variations observed occur on a time scale of one to three months. The actual time scale of the variations might even be shorter.

3. INTERPRETATION OF THE VARIATIONS

3.1. Density Variations

The UV spectrum of P Cygni has shown that the mass loss rate is variable, and that a shell is ejected about every year with a typical mass $M_{shell} \gtrsim 10^{-8} M_\odot$. The velocity of the shells at large distance from the star is 200 km/s (Lamers et al. 1984). Calculations of the variations of the radio flux due to the density variations in optically thick and thin shells show that the radio flux can vary on a time scale of 3.5 years. This is the time needed for a shell to travel a distance of 400 R_*. The observed variations occur on a time scale which is much shorter.

3.2. Variable Ionization of the Wind

For a stationary wind the radio flux S_ν depends on the ion-density as $S_\nu \propto (n_i \cdot n_e)^{2/3}$, where n_i and n_e are the ion and electron density. For a wind consisting mainly of hydrogen, $n_i = n_e = qn_H$, where q is the ionization fraction of hydrogen, so $S_\nu \propto q^{4/3}$. Variations in the degree of ionization will cause variations in the radio flux.

If the source of ionizing radiation (Lyman and Balmer continuum photons) is switched off, the wind will recombine on a time scale $\tau_{rec} = 1/\alpha n_e$, where α is the recombination coefficient. The recombination time scale was calculated for a model of the stellar wind of

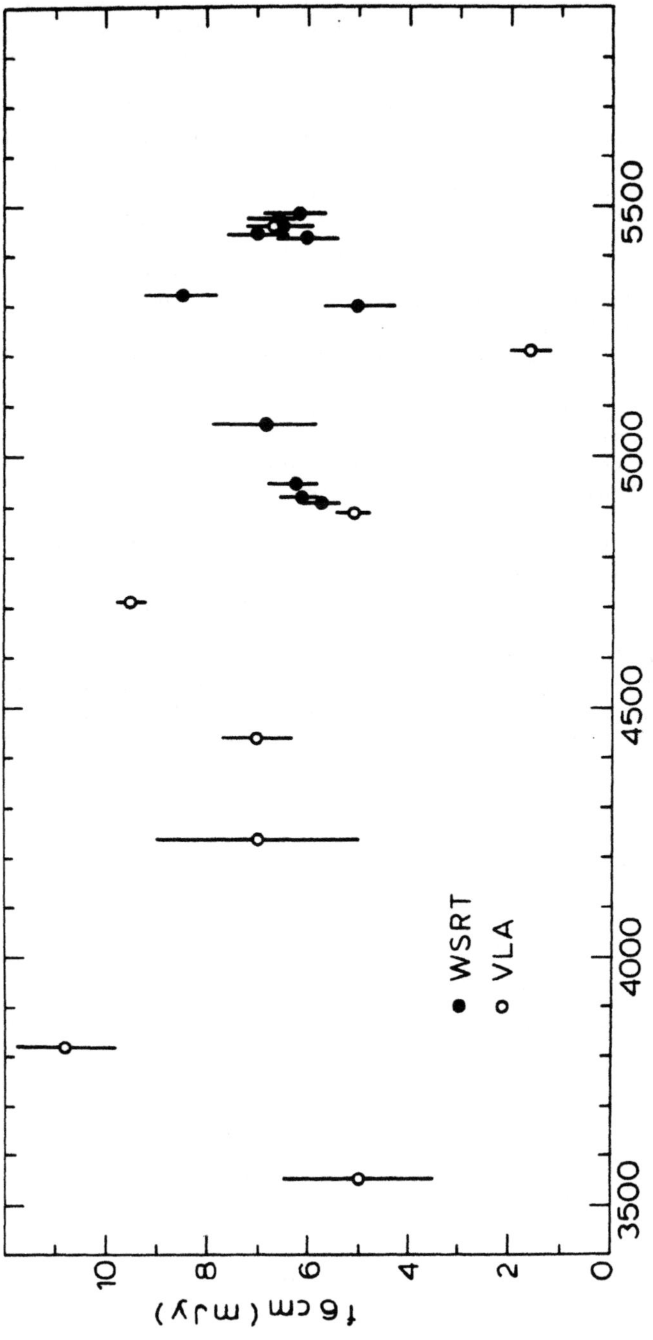

Fig. 1. The radio flux of P Cygni at 6 cm (5 GHz) as a function of time, since 1978.

P Cygni (Waters and Wesselius 1984) with a mass loss rate of 1.5 × 10^{-5} $M_\odot$/yr. The values of τ_{rec} are given in Table 1. At distances between 100 and 500 R_*, where most of the 6 cm flux is generated, the recombination time scale is between 9 and 220 days. This is approximately the time scale of the observed variations.

Lamers et al. (1984) showed on the basis of the IUE spectra that P Cygni ejects shells at intervals of approximately one year, with masses of the order of 10^{-8} $M_\odot$ or more. The hydrogen column density of such a shell is $N_H \gtrsim 10^{22}$ cm^{-2} when the shell has reached a distance of 2 R_*, shortly after its ejection. The optical depth at the Lyman and Balmer continuum edges are $\tau_L \gtrsim 8 \times 10^4$ (1-q) and $\tau_B \gtrsim 4 \times 10^2$ (1-q), where (1-q) is the fraction of neutral hydrogen in the shell. We assumed that the n = 2 and n = 1 level are in equilibrium with one another with an excitation temperature of 18,000 K, which should hold until the Lyman-alpha line becomes transparent. If the fraction (1-q) in the shell is larger than 10^{-4}, the shell will be optically thick to Lyman continuum radiation and if (1-q) > 10^{-2}, it will also be optically thick for Balmer photons.

An optically thick shell will temporarily shield the outer part of the wind from ionizing photons so that the wind will recombine, and the radio flux will decrease on a time scale τ_{rec}. When the shell moves outward and its column density decreases, it will become transparent to the Lyman and Balmer continuum, and the wind at large distances will be photoionized again. The resulting increase in the radio flux will occur on a time scale which is the sum of the time it takes for the shell to become transparent plus the time it takes to re-ionize the wind, τ_{rec}. The optical depth of an expanding shell will decrease with an e-folding time scale of roughly $\tau_{ex} \sim 0.5\, r_o/v(r_o)$, where r_o is the radius of formation of the ionizing continuum.

Table 1. Recombination time scales in the wind of P Cygni.

Distance (R_*)	n_e (cm^{-3})	τ_{rec} (days)
10	8.5×10^8	0.05
100	4.7×10^6	9.1
300	5.2×10^5	76
500	1.8×10^5	220

The appropriate value of r_o is unknown. For the typical values $r_o = 2\ r_*$ and $V(r_o) = 30$ km s^{-1} the time scale is $\tau_{ex} \sim 20$ days. Thus, the radio flux will regain its maximum value on a time scale, $\tau_{ex} + \tau_{rec}$, which is roughly a factor of 2 larger than the time scale, τ_{rec}, for the flux to drop to its minimum. Since the interval between major shell ejections is roughly 400 days (Lamers et al. 1984), it is not surprising that P Cygni is most often observed at its high level of radio emission.

This model could be rigorously tested by combined radio and UV obervations which monitor P Cygni continuously over a time period which includes a shell ejection. Present data are too incomplete and non-simultaneous to provide this direct test.

4. CONCLUSION

The radio flux of P Cygni is variable on a time scale of the order of months or shorter. These variations cannot be explained by density variations in the wind due to a variable mass loss rate, but they can be explained by a variable degree of ionization in the wind. We propose that the ionization state of the wind changes when the star episodically ejects shells of material which block the ionizing radiation from the stellar surface. These shells have been observed in the UV as they travel away from the star. Since the expected radiative flux of P Cygni is barely sufficient to maintain the ionization of the wind (e.g. Felli and Panagia 1981), small changes in the amount of flux penetrating into the wind can cause major decreases in the ionization fraction of the outer wind, where the radio flux originates. The radio flux will then drop on a recombination time scale, which at the distance of the radio photosphere ($\approx 200\ R_*$) is roughly one month. The radio flux will increase again after the shell becomes transparent, which also will occur on a time scale of roughly one month.

We acknowledge the support of the U. S. National Science Foundation Grants AST82-18375 to the Univ. of Colorado and AST81-14717 to the Univ. of California. The VLA of the National Radio Astronomy Observatory is operated by Associated Universities, Inc., under contract with the U. S. National Science Foundation.

REFERENCES

Abbott, D. C., Bieging, J. H., and Churchwell, E. 1981, Astrophys. J., 250, 645.
Abbott, D. C., Bieging, J. H., Churchwell, E., and Cassinelli, J. P. 1980, Astrophys. J. 238, 196.
Felli, M. and Panagia, N. 1981, Astron. Astrophys. 102, 424.
Lamers, H.J.G.L.M., de Groot, M., and Cassatella, A. 1983, Astron. Astrophys., 128, 299.
Lamers, H.J.G.L.M., Korevaar, P., and Cassatella, A. 1984, Proc. of 4th European IUE Conference, ESA SP-218, p. 315.
Newell, R. 1981, Ph.D. thesis, New Mexico Institute of Mining and Technology.
Newell, R. and Hjellming, R. M. 1980, Bull. Am. Astron. Soc., 12, 458.
Waters, L.B.F.M. and Wesselius, P. R. 1984, Astron. Astrophys., in press.
Wendker, H. J., Baars, J. W. H., and Altenhoff, W. J. 1973, Nature Phys. Sci., 245, 118.
White, R. L. and Becker, R. H. 1982, Astrophys. J., 262, 657.

RESOLUTION OF THE RADIO SOURCE γ^2 VEL

David E. Hogg
National Radio Astronomy Observatory*
Edgemont Road
Charlottesville, Virginia 22903 USA

ABSTRACT. The WC star γ^2 Vel has been resolved at radio wavelengths with the VLA. At a frequency of 4860 MHz, the flux density observed at a given antenna separation is independent of the position angle of the baseline, implying that the wind is spherical. The data at three frequencies suggest that the electron density decreases with distance more rapidly than r^{-2} for distances greater than 3×10^{15} cm. The wind is isothermal with temperature 5600 ± 500 K, and the mass loss rate is $(8.6 \pm 1.0) \times 10^{-5}$ $M_{\odot}$ yr^{-1}.

1. INTRODUCTION

A large number of Wolf-Rayet stars have now been detected as radio sources (Abbott and Bieging 1984) and mass loss rates have been deduced for those that are thermal emitters. The calculation of the mass loss rate assumes that the emission arises in an isothermal envelope expanding spherically at constant velocity. These assumptions can be verified for stars that are resolved at radio wavelengths, by examination of the visibility function (White and Becker 1982). Most of the detected WR stars are so weak that an accurate visibility function cannot be measured. The best candidate is the bright binary star γ^2 Vel (WC8 + 09I) which has a flux density at 6 cm of nearly 30 mJy.

2. OBSERVATIONS

The observations were made with the VLA in its largest configuration. The synthesized beamwidth at 4860 MHz is 2.9 x 0.4 arcsec, and the maximum antenna separation is 6×10^5 wavelengths. Observations in March 1982 at 4860 MHz showed that the source is heavily resolved, and

*The National Radio Astronomy Observatory is operated by Associated Universities, Inc., under contract with the National Science Foundation.

R. M. Hjellming and D. M. Gibson (eds.), Radio Stars, 117–119.

observations in November 1983 were used to derive visibility functions at 1490, 4860, and 14940 MHz.

After editing and preliminary calibration, all data were self-calibrated to remove the effects of the short-term atmospheric phase fluctuations which are a serious problem at the low elevation angles used in the observations. In addition, a strong nonthermal double source was found approximately 2 arcmin to the northeast. The contribution of this source was removed from the data at 1490 and 4860 MHz, since it distorted the visibility function at short spacings.

3. RESULTS AND DISCUSSION

The visibility curves at 4860 MHz from the two epochs agreed to within the observational errors, implying that the structure of the atmosphere remained unchanged during the 20 month interval. The data from the two epochs were combined to improve the signal-to-noise ratio, and the resultant mean visibility amplitudes are plotted in Figure 1 as a function of the antenna separation.

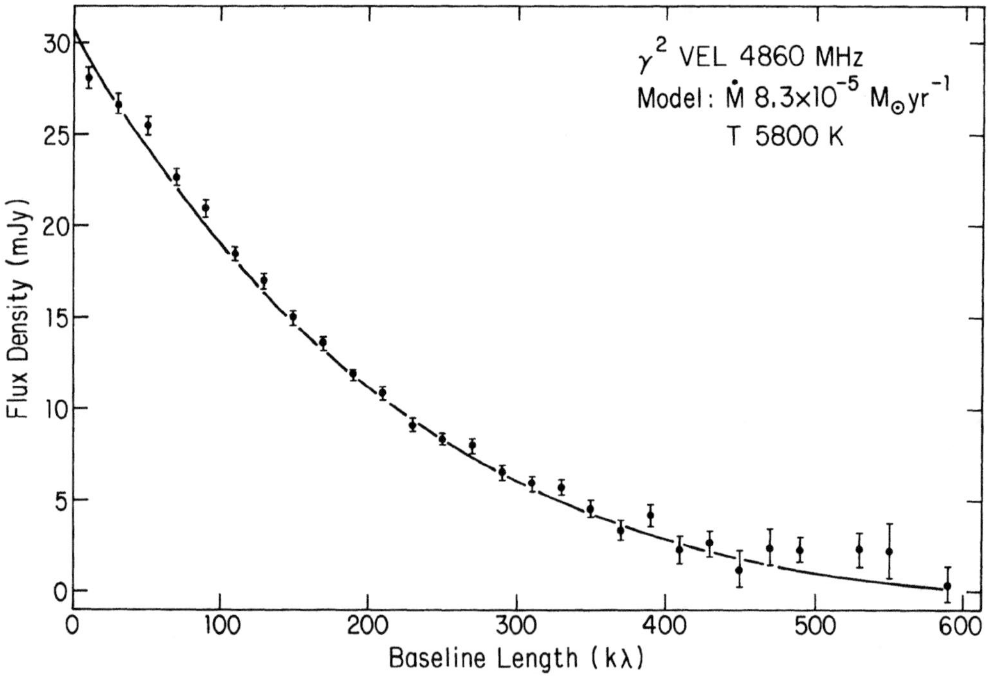

Figure 1. The observations are compared with a model wind (solid line) having parameters as shown.

Because of the foreshortening of the array, the maximum baseline which samples the visibility function at all position angles is 6×10^4 wavelengths. At this spacing, there is no significant variation of visibility amplitude with position angle. If the source is elliptical in the plane of the sky, the axial ratio must be less than 1.2. The data are consistent with the source being circular, as is expected if the outflow is spherical.

Figure 1 compares the data with the visibility function expected for an isothermal envelope expanding spherically at constant velocity. The model is characterized by an electron temperature of 5800 K, and a mass loss rate of $8.3 \times 10^{-5}\ M_{\odot}\ yr^{-1}$.

The same model is used to predict the visibility functions at the other frequencies. The agreement with the data is less satisfactory. Analysis of the variation of flux density and width of the visibility curve with frequency suggests that the wind is isothermal, but the electron density decreases more rapidly than r^{-2} for distances greater than 3×10^{15} cm. Such an effect would arise if the helium, assumed to provide most of the electrons, is beginning to recombine. This model requires a slightly lower temperature (5600 K) and a slightly greater mass loss rate ($8.6 \times 10^{-5}\ M_{\odot}\ yr^{-1}$).

REFERENCES

Abbott, D. C. and Bieging, J. H. 1984, B.A.A.S., **16**, 508.
White, R. L. and Becker, R. H. 1982, Ap. J., **262**, 657.

THE TEMPERATURE OF OUTER ENVELOPES OF WR STARS

N. G. Bochkarev
Sternberg State Astronomical Institute
Moscow, USSR

ABSTRACT. One of the important parameters for the interpretation of radio-observations of WR stars is the temperature T of their envelopes. Correct radio measurement of T can be made only if the effective diameter of the star can be measured. This method gives T ≈ 18000 for P Cyg (White and Becker, 1982). Two models of outer WR stellar envelopes are discussed: a high-temperature envelope and a low-temperature envelope with a hot core. The dependences of the width of spectral lines on ionization potential shows that in several cases, temperature is probably increasing toward the outer parts of envelopes.

1. INTRODUCTION

Analysis of the eclipses in binary systems containing WR and normal components (usually of OB spectral type) offers a good opportunity for the investigation of the structure of WR star atmospheres (Cherepashchuk, 1982). Comparison of IR, optical and UV-light curves of such binaries (Cherepashchuk et al., 1981, 1984) shows that while outflowing envelopes are optically thick they become more transparent from the IR to the UV with increasing radius. The total observed spectrum of a WR star, which can usually be approximately described by a Planck spectrum with effective temperature $T_{eff} \approx (25-35) \times 10^3$ K (Nussfaumer et al., 1982), is probably formed by radiation from a very hot core with $T \approx (80-100) \times 10^3$ K (Khaliullin, 1980; Cherepashchuk, 1982) and a cooler semi-transparent envelope with an electron temperature of $(20-30) \times 10^3$ K (and even lower in the outer parts where the effective temperature $T \approx 7000$ K; Cherepashchuk, 1982). However, examples of opposite temperature distributions may be found (Willis, 1981; Barlow, 1982).

In this work we analyze the spectrophotometric variability of binary WR+OB stars and show that in the $\lambda \lesssim 1500-2000$ Å range, the light curve will be qualitatively different from that at longer wavelengths. This difference is a result of numerous resonance lines in the far UV which are an additional opacity source in WR envelopes when compared to the $\lambda \gtrsim 1500-2000$ Å, region where spectral lines are not

R. M. Hjellming and D. M. Gibson (eds.), Radio Stars, 121–126.

important. IUE observations of V444 Cyg = HD193576 (Eaton et al., 1982, 1984) confirm this behaviour of the light curve.

2. OBSERVATIONS AND INTERPRETATION

Detailed analyses of the light curves in different filters (see, Cherepashchuk, 1982; Cherepashchuk et al., 1984) support the Beals (1944) model of WR stars and show that coronal phenomena do not appear until $R \gtrsim 20\ R_\odot$. In the IR range the envelopes of the outflowing gas are optically thick. Therefore, WR-stars are observed as objects with an effective radius $R_{eff} \approx 5-10\ R_\odot$ and with a low surface temperature $T \approx (20-30) \times 10^3$ K. Low temperatures in the outer parts of the stellar envelopes were found from radio data (Abbott et al., 1982; Hogg, 1982; White and Becker, 1982).

With decreasing λ, free-free absorption also declines and it becomes possible to see the radiation from the hotter inner parts of the star. For the WR-component of V444 Cyg R_{eff} varies from 5.6 $R_\odot$ near $\lambda = 2.8\ \mu$ to 2.9 $R_\odot$ for $\lambda = 2920$ Å (Cherepashchuk, 1982). The temperature of the core was found to be $(80-90) \times 10^3$ K. A number density $n \approx 10^{13}$ cm^{-3} was found at a radius $R = 2.9\ R_\odot$. Evidence for high temperatures of nuclei of other WR stars discussed by Khaliullin (1980).

Detailed interpretation of V444 Cyg's light curves was made by Cherepashchuk et al. (1975, 1981, 1984). The inclination of the V444 Cyg system is $i = 78° \pm 1°$. The secondary component of the binary is an O6-star, and is separated by 40 $R_\odot$ from the WR-star. The O6-star's radius is about 10 $R_\odot$. Therefore, a decreasing R_{eff} results in a reduction of the width and depth of the minima (Fig. 1). The velocity of the gas outflow increases approximately as R^{-1}.

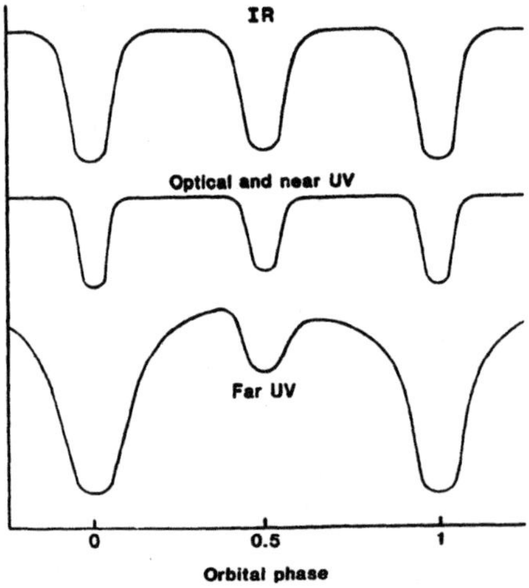

Fig. 1. Qualitative picture of variations of light curves of eclipsing WR+O binaries for IR, optical, near UV ($\lambda \gtrsim 2000$ Å), and far UV ($\lambda \lesssim 2000$ Å) ranges. For the far UV range one predicts the presence of deep eclipses and reflection effects as a result of the increasing importance of opacity in resonance lines. The reflection effect may be asymmetrical in consequence of the asymmetry of the outflowing wind in the rotating system.

Assuming that the main near-UV opacity mechanism is electron scattering, it can be easily found that the Thompson optical depth $\tau_T = 1$ at $R \gtrsim 3\ R_\odot$ corresponds to $n_e(3\ R_\odot) \approx 1.4 \times 10^{13}$ cm^{-3}, which is close to the estimate of 0.9×10^{13} cm^{-3} by Cherepashchuk et al. (1981, 1984). This means that the main source of the opacity in range 2460-3320 Å is electron scattering.

Far UV observations of V444 Cyg (Eaton et al., 1982, 1984) show that for $\lambda \lesssim$ 1400-1500 A eclipses become deeper and broader. This means that a new source of opacity appears in the far UV. It could be scattering by resonance spectrum lines (Bochkarev, 1984) or absorption from metastable levels (Eaton et al., 1984). In the first case the reflection effect must be present in the FUV and in the second case this effect is absent. Observations by Eaton et al. (1982, 1984) show a distortion of the light curve which is evidence of a reflection effect. To distinguish these two cases one needs a measurement of the variations of linear polarization in the FUV; the variations would be strong in the first case and absent in the second case.

3. THE INFLUENCE OF UV LINE OPACITY ON LIGHT CURVES OF WR+O BINARIES

It was shown in a number of fundamental papers by Kurucz (e.g. Kurucz, 1979) that spectral lines can blanket a large part of the UV flux of hot stars and must be taken into account in the calculation of adequate model atmospheres. The reason for strong UV blanketing is the numerous resonance lines of different elements in the UV range. According to Kurucz's (1979) calculations for solar abundances and $T_{eff} \lesssim (20-25)\times10^3$ K, spectral lines constitute at least 20-40% of the opacity sources. Typically the opacity coefficient ζ has a cut-off like behaviour with small ζ for long wavelengths and with $\zeta \simeq 0.1-04$ if λ is short of the cut-off.

Hydrogen is apparently almost absent in WR star atmospheres (Willis, 1982). Helium, carbon, oxygen and other heavy elements must be dominant (Willis, 1982; Nugis, 1982a,b). Therefore, the relative abundance Z of the heavy elements is higher in WR atmospheres than in normal stars and, instead of the usual value of abundance $Z_i \sim 10^{-4}$ (in number densities) for individual elements, Z_i can reach 10^{-2} and may even be as large as 0.1-1 (Willis, 1982).

A large quantity of ions of heavy elements in the atmosphere dominates the scattering in the far wings of spectral lines over electron scattering. In fact, the coefficient of scattering in the wing of an individual spectral line at frequency ν is (Lang, 1974)

$$\sigma_R = \sigma_T \, f_{ik}^2 \left[\frac{\nu^2}{\nu_{ik}^2-\nu^2}\right]^2 , \qquad (1)$$

where f_{ik} is the oscillator strength and ν_{ik} is the central spectral line frequency. According to Eqn. (1) for abundance $Z_i = n_i/n_e$ of the conforming spectral line, $\sigma_R > \sigma_T$ in the frequency range $\Delta\nu = |\nu-\nu_{ik}|$

$$\frac{\Delta\nu}{\nu} = \frac{1}{2} f_{ik} \sqrt{A} . \qquad (2)$$

For the strong lines ($f_{ik} \sim 1$) of the most abundant ions ($Z_i \sim 10^{-2}$) an individual resonance line has $\Delta\nu/\nu \sim 0.05$, i.e., the range of scattering in the line can exceed that by electron scattering for the range $2\Delta\nu/\nu \sim$ 10%.

Using the detailed calculations by Kurucz (1979) for the solar abundances and taking into account Eqn. (2) we find that an increase of the heavy element abundance by 1-2 orders of magnitudes results in the dominations of spectral line scattering over electron scattering for $\lambda \lesssim 1500$-2000 Å, if the temperature of the atmosphere does not exceed $(25$-$30)\times10^3$ K and the degree of the domination increases with decreasing atmospheric temperature.

In the outer parts of the WR star envelope of V444 Cyg ($R \gtrsim (5$-$8)R_\odot$) the necessary conditions for predomination of spectral line scattering in the range $\lambda \lesssim 1500$ Å are satisfied. When $\tau_T \sim 0.2$, $\tau_{total} \sim 1$ and slowly decreases (slower than R^{-1}) in outer parts of the envelope.

To model WR envelopes accurately, it is necessary to produce detailed calculations of the ionizational structure of the envelope or to obtain corresponding observational data. These calculations require an estimate of the nonradiative heating (coronal effects). If absent, then the larger the radius of the envelope considered, the colder its temperature.

This problem is discussed now. There is at least one example of a WR-star, where one finds that the higher the ionization potential of an element, the higher the velocity shown by its lines (HD 192103; c.f. Willis, 1981; Barlow, 1982).

X-ray observations (Sanders *et al.*, 1982) confirm that only a small number of WR stars have hot atmospheres. Moreover, X-ray flux from WR+0 binaries can be formed by a shock wave between stellar winds of two components. In most cases observations demonstrate a decreasing temperature with distance from the star (Kuhi, 1973; Willis, 1981; Barlow, 1982). Observations of 444 Cyg show the absence of a corona at least until 20 $R_\odot$ (Cherepashchuk, 1982).

Ionization equilibrium of WR envelopes is determined by radiation processes. As follows from the observations of Cherepashchuk *et al.* (1981, 1984), the outer parts of the V444 Cyg envelope, $R > 8\ R_\odot$, are relatively cool ($T \lesssim 20$-30×10^3 K). The column density of the envelope $\sim 10^{22}$ cm^{-2} can be enough to absorb ionization radiation of the core. Thus, scattering of radiation by the UV lines must result in a strong increase of the WR-star's effective radius R_{eff}.

The increase of the effective radius of the WR star in a WR+0 binary system must result in an increase of width of the eclipse minima of the light curve and depth of the minima (Fig. 1) if eclipses are not total. A large effective radius of WR stars in the far UV, a slowly decreasing optical depth with increasing R ($\sim R^{-\alpha}$, $\alpha < 1$) and a high temperature of the secondary components must result in the appearance of the reflection effect in the light curves of WR+0 binaries (Fig. 1). Therefore, the the light curve between eclipses cannot be constant; maximum occurs near orbital phase 0.5, *i.e.* when the 0-component is in front of the WR star. To distinguish between Eaton's *et al.* (1984) suggestion about appearance in the V444 Cyg envelope of FeV ion absorption in the range $\lambda < 1500$ Å and resonance scattering (Bochkarev, 1984) it is necessary to study in detail the reflection effect for $\lambda \lesssim 1500$ Å and the behaviour of the linear polarization in the range.

Other WR+O systems, which were studied in detail in the UV, optical, and IR ranges by Stickland et al. (1984), e.g. CQ Cep, show for $\lambda < 1500$ Å light curves qualitatively similar to V444 Cyg, but the components of CQ Cep are very close and tidally distorted, which complicates the interpretation.

4. CONCLUSIONS

1. For the most part, WR-stars show a decrease of electron temperature towards outer parts of their atmosphere and, therefore, the influence of the scattering in numerous resonance lines in the UV(λ 1000-2000 Å). Outflow in envelopes ($R > 5$-$10R_\odot$) can result in a strong increase of effective radius R_{eff} of the WR star between the $\lambda > 1500$-2000 Å range and the $\lambda < 1500$-2000 Å.

2. In eclipsing WR+O binaries (stars similar to V444 Cyg) as R_{eff} increases for $\lambda < 1500$ Å the depth and width of mutual eclipses of WR and O components must also increase. This effect explains systems where eclipses are absent in the optical and near UV ranges, but are present in strong resonance lines (e.g., CV Ser).

3. Growth of R_{eff} as a result of light scattering in sufficiently cold ($T < 20\text{x}10^3$ K) outer parts of WR-star atmospheres must result in the appearance of the reflection effect of O-component radiation by WR-star envelope.

4. Observations of the effects described in conclusions 1-3 of this section (see Eaton et al., 1982, 1984) support the model of WR stars (hot core and cold semi-transparent envelope) developed in a series of articles by Cherepashchuk et al. (e.g., Cherepashchuk, 1982).

5. Analysis of the effects described in 1-2 above can be used to determine the physical conditions in the outflowing atmospheres of WR-stars including the temperature distribution and ionization of outer part of WR-star atmospheres.

5. REFERENCES

Abbott, D.C., 1982, Ap.J., **259**, 282.

Abbott, D.C., Bieging, J.H., Churchwell, E., 1982 in IAU Symp. N99 Wolf-Rayet stars: observations, physics, evolution. Eds. C.W.H. De Loore and A.J. Willis (Dordrecht:D.Reidel), p. 215.

Barlow, M.J., 1982 in IAU Symp. 99. Wolf-Rayet stars: observations, physics, evolution. Eds. C.W. H. De Loore and A.J. Willis (Dordrecht:D. Reidel), p. 149.

Beals, C.S., 1944, M.N.R.A.S., **104**, 205.

Bochkarev, N.G., submitted to Ap.J.

Cherepashchuk, A.M., 1975, Astron. Zh., **52**, 81 (Sov. Astron., **19**, 47).

Cherepashchuk, A.M., Eaton, J.A., Khaliullin Kh.F., 1981 in Close Binary Stars: Observations and Interpretations, IAU Symp N88.

Eds. M.J. Plavec, D.M. Popper, R.K. Ulrich: (Dordrecht:D. Reidel).
Cherepashchuk, A.M., Eaton, J.A., Khaliullin, Kh.F., 1984, Ap.J., **281**, 774.
Eaton, J.A., Cherepashchuk, A.M., Khalliulin, Kh.F., 1984, Preprint, Indiana Astronomy Publo No. 8.
Hartmann, L., 1978, Ap.J., **221**, 193.
Hogg, D.E., 1982 in IAU Symp N99 Wolf-Rayet stars: observations, physics, evolution. Eds. C.W.H. De Loore and A.J. Willis, (Dordrecht:D. Reidel), p. 221.
Khaliullin, Kh.F., 1980, in Proc. of the meeting "Ejection and Accretion of Matter in Binary Systems," Tatranska Lomnica, p. 99.
Kuhi, L.V., 1973 in IAU Symp N49, Eds. M.K.V. Buppu and J. Sahade, (Dordrecht:D. Reidel), p. 205.
Kurucz, R.L., 1979, Ap.J. Suppl. Ser., **40**, 1.
Lang, K.R., 1974, Astrophysical formulae (Berlin-New York, Springer-Verlag).
Nugis, T., 1982a, in IAU Symp N99 Wolf-Rayet stars: observations, physics, evolution. Eds. C.W.H. De Loore and A.J. Willis, (Dordrecht:D.Reidel), p. 127.
Nugis, T., 1982b, Ibid, p. 131.
Nussfaumer, H., Schmutz, W., Smith, L.J., Willis, A.J., 1982, Astron. Astrophys. Suppl. Ser., **47**, 257.
Robberecht, W., De Loore, C., Olson, G., in IAU Symp N99, Wolf-Rayet stars: observations, physics, evolution, Eds. C.W.H. De Loore and A.J. Willis,(Dordrecht:D. Reidel), 209.
Sanders, W.T., Cassinelli, J.P., van der Hucht, K.A., 1982 in IAU Symp N99 Wolf-Rayet stars: observations, physics, evolution. Eds. C.W. H. De Loore and A.J. Willis (Dordrecht:D. Reidel) p.
Stickland, D.J., Bromage, G.E., Budding, E., Burton, W.M., Howarth, I.D., Jameson, R., Sherrington, M.R., Willis, A.J., 1984, Astron. Astrophys., **134**, 45.
Underhill, A.B., 1982, in IAU Symp. N99, Wolf-Rayet stars: observations, physics, evolution. Eds. C.W.H. De Loore and A.J. Willis (Dordrecht:D. Reidel), p. 47.
White, R.L. and Becker, R.H., 1982, Ap.J., **262**, 657.
Willis, A.J., 1982, in IAU Symp N99 Wolf-Rayet stars: observations, physics, evolution. Eds. C.W.H. De Loore and A.J. Willis (Dordrecht:D. Reidel) p. 84.

RADIO AND INFRARED OBSERVATIONS OF CYG OB2 No. 5

L.F. Rodríguez, J. Cantó, A. Sarmiento
Instituto de Astronomía, UNAM
Apdo. Postal 70-264
04510 México, D.F., MEXICO
M. Roth, M. Tapia
Instituto de Astronomía , UNAM
Apdo. Postal 877
22800 Ensenada, B.C., MEXICO
P. Persi, M. Ferrari-Toniolo
Istituto Astrofísica Spaziale, CNR
C.P. 67
00044 Frascati, Italy

ABSTRACT. VLA measurements of the 6 cm flux of Cyg OB2 No. 5 reveal it has increased by a factor of 4 between 1980 and 1984. From 1983 July to 1984 September the 6 cm flux has slowly increased from 5.8 to 7.3 mJy. During 1984 September the 6 and 2 cm fluxes were measured in 7 occasions. During this month no variability was detected at the 10 percent level and the spectral index was 0.1 ± 0.1. The 6.6-day variability found at 2.2 μm originates from the eclipsing of the contact binary system. There is no evidence of IR variability from the emission originating in the ionized wind. It is yet unclear if the radio emission has a thermal or a nonthermal nature. A model in terms of the ejection of a dense shell (superposed on a steady stellar wind) is discussed.

1. INTRODUCTION

The excess radio and IR emission observed in some hot stars has been interpreted until recently as free-free radiation from a steady-state ionized stellar wind (Wright and Barlow 1975). However, in the last few years several studies have indicated that this simple model is unable to account for the observations. Castor and Simon (1983) and Abbott, Telesco and Wolff (1984) found that a few OB stars have radio emission far in excess of that expected from the IR flux in the framework of the steady-state, ionized wind model. White and Becker (1983) determined for Cyg OB2 No. 9 a brightness temperature considerably larger than the 10^4 K value expected for optically-thick free-free emission. Abbott, Bieging and Churchwell (1984) observed radio variability in Cyg OB2 No. 9 and 9 Sgr. Furthermore, these stars

R. M. Hjellming and D. M. Gibson (eds.), Radio Stars, 127–130.

did not have the spectral index of 0.6 predicted by the simple ionized wind model.

In this contribution we present our radio and IR observations of Cyg OB2 No. 5 (BD+40°4220; V729 Cyg), an evolved contact binary system composed of two Of supergiant stars (Bohannan and Conti 1976).

2. OBSERVATIONS

The radio observations have been made at the Very Large Array of the National Radio Astronomy Observatory. We observed Cyg OB2 No. 5 first in 1983 July as part of a program including stars with large IR excesses. A 6 cm flux of 5.8$\pm$0.7 mJy was obtained. This flux was about 3 times larger than the value of 1.8$\pm$0.3 mJy found by Abbott, Bieging and Churchwell (1981) in 1980 May and indicated strong radio variability. We observed the system again in 1984 January and September, finding a slow increase from 6.3 to 7.3 mJy at 6 cm (Persi et al. 1984). During 1984 September we observed Cyg OB2 No. 5 at 6 and 2 cm in 7 epochs, with different time separations ranging from 2 to 24 days.

The flux at both frequencies remained very stable (with any possible variability smaller than 10 percent). The average flux values over 1984 September were S(6 cm)= 7.3$\pm$0.4 mJy and S(2cm)= 8.4$\pm$0.7 mJy. The spectral index was 0.1$\pm$0.1.

The infrared observations have been made in several occasions between 1980 and 1984 at the Mexican National Astronomical Observatory and the Gornergrat Observatory. At the K magnitude there is a 6.6-day periodicity, which coincides with that observed in the visible (Hall 1974). We attribute this variability to the eclipsing in the contact binary system. There is an absence of variability in the 10 μm excess emission. This result rules out a simple model in terms of an increase in the mass loss rate for the variability in the radio flux. In this model the radio flux behaviour should mimic (with a time lag of months) that of the 10 μm excess emission.

3. DISCUSSION

We have tried to model the time-variable, excess radio emission from Cyg OB2 No. 5 in terms of the following model. The central star is assumed to have a steady-state wind. At a point in time the mass loss rate is assumed to suddenly increase and remain in this higher value for a period of several days. The mass loss rate is then assumed to return,also suddenly, to its original value. Both the wind and the resulting superposed shell are taken to be fully ionized and to expand at the wind's terminal velocity. Under these assumptions it is possible to predict the time behaviour of the radio and infrared fluxes, once initial conditions are adopted. It is interesting to discuss qualitatively the characteristics of the flux variability predicted by this model. At any given frequency in the radio and IR one expects the flux to increase from the steady- state value, reach a maximum, returning later to the steady-state value. The radio spectral index will go from the stellar wind value of 0.6 to ∿2 (as the optically-thick emission from the shell dominates the emission). Then the spectral index will flatten (reaching a value of -0.1 if the shell is massive enough), since the shell will become optically-thin with expansion. As the shell expands further, its contribution to the emission will become negligible and the spectral index will return to the 0.6 value of the stellar wind. The dependence of the flux increase with observing wavelength is as follows. For longer wavelengths, the increase will appear later in time, will last longer, and will be less marked with respect to the steady-state flux value. Thus, increases in the IR flux due to the hypothetical shell will be large but short-lived, while the radio ones will be less conspicuous but with a longer duration.

The shell model can account well for the radio spectral index and slow variability observed in Cyg OB2 No. 5.

To test quantitatively the radio variability, one requires monitoring the source over a period of several years. In contrast, to detect the proposed IR variability one requires intensive day-to-day monitoring as frequently as possible. The model can not account for radio stars with fast radio variability or with large negative spectral indices, such as those reported by Abbott, Bieging and Churchwell (1984). Even in the case of Cyg OB2 No. 5 it is not straightforward to explain the relatively large radio brightness temperature found by Persi et al. (1984). The model of synchrotron emission from chaotic stellar winds proposed by White (1985) appears promising and their predictions should be tested observationally.

L.F.R. and J.C. acknowledge partial support from CONACYT (Mexico) grant PCCBBEU020510.

4. REFERENCES

Abbott, D.C., Bieging, J.H., Churchwell, E. 1981, Astrophys. J., 250, 645.

Abbott, D.C., Bieging, J.H., Churchwell, E., 1984, Astrophys. J., 280, 671

Abbott, D.C., Telesco, C.M., Wolff, S.C., 1984, Astrophys. J., 279, 225

Bohannan, B., Conti, P.S., 1976, Astrophys. J., 204, 797.

Castor, J.J., Simon, T., 1983, Astrophys. J., 265, 304.

Hall, D.S., 1974, Acta Astr., 24, 69.

Persi, P., Ferrari-Toniolo, M., Tapia, M., Roth, M., Rodríguez, L.F., 1984, to appear in Astron. Astrophys.

White, R.L., 1985, to appear in Astrophys. J.

White, R.L., Becker, R.H., 1983, Astrophys. J. (Letters), 272, L19.

Wright, A.E., Barlow, M.J., 1975, Monthly Notices Royal Astron. Soc., 170, 41.

RADIO EMISSION FROM Θ^1A ORIONIS

Guido Garay
European Southern Observatory, Garching bei München, FRG

James M. Moran and Mark J. Reid
Harvard-Smithsonian Center for Astrophysics
Cambridge, USA

ABSTRACT. Multi-epoch and multi-frequency VLA observations of the Orion Nebula show that the radio emission from the Trapezium star Θ^1A Orionis undergoes large flux density variations on a time scale of months. Between 1982 April and 1983 February the flux density increased from 3 to 74 mJy, however it remained constant over a period of eight hours at the latter epoch. The radio spectrum is flat ($\alpha = 0.0$) at all observed epochs. In addition, the radio angular size was measured to be 0"21 ($\sim 1.6 \times 10^{15}$ cm at the distance of the Orion Nebula) at the epochs of maximum and minimum flux density. Interpretation in terms of non-thermal radio emission is difficult because of both the flat spectrum and the large source size. We suggest that the radio emission from Θ^1A is the result of recurrent mass loss outbursts from the B0.5 underlying star.

INTRODUCTION

Observations of the Orion Nebula in August 1981 with the Very Large Array (VLA) of the National Radio Astronomy Observatory revealed the presence of ~20 ultracompact radio sources with flux densities greater than 4.0 mJy at 15 GHz (Moran et al., 1982; Garay, 1983). One of those sources is coincident, within 0"2, with the Trapezium star Θ^1A Orionis. Θ^1A (HD37020) is an eclipsing binary system with a 65.43 day period (Baldwin and Mattei, 1977) containing a B0.5V (HI broad) primary (Levato and Abt, 1976). Moran et al. (1982) attributed the radio emission from Θ^1A to free-free radiation arising in a circumstellar envelope excited and ionized by the Lyman continuum photons from the underlying early B type star.

Subsequently we reobserved the Orion region on three occasions to obtain the angular size and the radio spectral index of the compact radio sources in order to determine their physical characteristics. We report here the data for the radio source associated with Θ^1A. The other sources are discussed elsewhere (Garay, Moran and Reid, 1985).

R. M. Hjellming and D. M. Gibson (eds.), Radio Stars, 131–138.

OBSERVATIONS

We observed the Orion Nebula with the Very Large Array on four occasions. The dates, observing frequencies and other observational parameters are given in Table 1. Typical integration time for each epoch at each observed frequency was 90 minutes obtained from 10 minute scans at different hour angles in order to provide a good (u,v) plane coverage. The on-source observations were preceded and followed by observations for three minutes on the calibrator 0539-057. The data was edited and calibrated following the standard procedures described by Hjellming (1982).

Maps were made by Fourier transformation of the interferometer data. Since the Orion Nebula is a strong emitter at the short VLA fringe spacings we applied a minimum baseline cut-off in order to remove the strong large scale structure (>5"). The minimum baseline spacing used to map the region is given in column 8 of Table 1. The noise level per beam resolution is given in column 9.

RESULTS

The optical position of the Trapezium star Θ^1A and the VLA position of the radio source associated with it are coincident within 0."2. The probability of confusion from a random Trapezium source is $\sim 10^{-3}$ and from a background non-thermal source less than 10^{-6}. We conclude that the optical and radio sources are physically associated.

In Table 2 we summarize the observational results. The flux density as a function of time, for the time span of the observations, is plotted in Figure 1. Figure 2 shows the flux density versus time (on time scale of hours) observed in February 1983. During that epoch, in which Θ^1A reached the largest measured flux density, we observed the Orion Nebula for ~10 minutes on twelve different occasions between 22 and 8 IAT hours. The flux density plotted in Fig. 2 corresponds to an average value as the data was binned on four time blocks with 3 scans each. Finally, the flux density as a function of frequency is plotted in Figure 3.

The radio emission from Θ^1A is characterized by dramatic flux density variations, decreasing by a factor of 4 from August 1981 to April 1982 and increasing by a factor of 20 ten months later. No evidence was found for flux density variations in time scale of hours. At the epoch of maximum flux density it remained constant (variations <2%) over a period of 6 hours. Our observations constrain the variability time scale to range from days to a few months. Recently Abbot et al. (1984) and Persi et al. (1984) discovered radio emission from O type stars, that is variable in time scale of months. The large radio luminosities, as compared to typical O stars, and the time variations led them to propose the existence of a new class of radio objects. Θ^1A may be another member of this class of stellar radio emitters.

The radio spectrum of Θ^1A is flat at all observed epochs, independent of flux density variations. The flat spectrum is consistent with a model in which the continuum radio emission is due

TABLE 1
Observational Parameters

Epoch	Configuration	Phase Tracking Center α (1950)	δ	Frequ. MH_z	Band-width MHz	Synthet. HPBW arc sec	Minimum UV Range Kλ	r.m.s. Noise mJy/beam
1981 Aug 9	B	$5^h32^m50^s.0$	$-5°25'30''.0$	4866	12.5	1.13	50	1.1
1981 Aug 9	B	5 32 47.0	-5 24 23.0	15015	50.0	0.47	35	0.6
1982 Apr 25	A	5 32 48.5	-5 25 20.0	4873	25.0	0.43	-	0.7
1983 Feb 17	C	5 32 47.0	-5 24 45.0	14965	50.0	1.57	20	1.0
1983 Feb 17	C	5 32 48.7	-5 25 15.0	22485	50.0	1.08	25	1.3
1983 Aug 28	A out[(a)]	5 32 49.0	-5 25 10.0	1452	25.0	1.15	30	0.3
1983 Aug 28	A in[(b)]	5 32 49.0	-5 25 10.0	4873	25.0	1.44	30	1.4
1983 Aug 28	A in	5 32 49.0	-5 25 10.0	14977	25.0	0.51	-	0.8

(a) Configuration formed with the four outer antennas in each arm.

(b) Configuration formed with the five inner antennas in each arm.

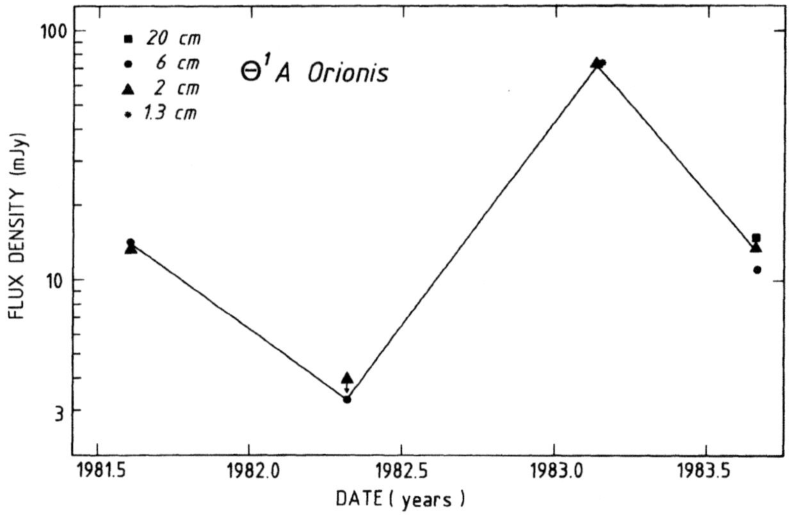

Figure 1: Flux density of Θ^1A Orionis versus time. Squares, circles, triangles and stars refer to flux density values at 20.6, 6.2, 2.0 and 1.3 cm, respectively.

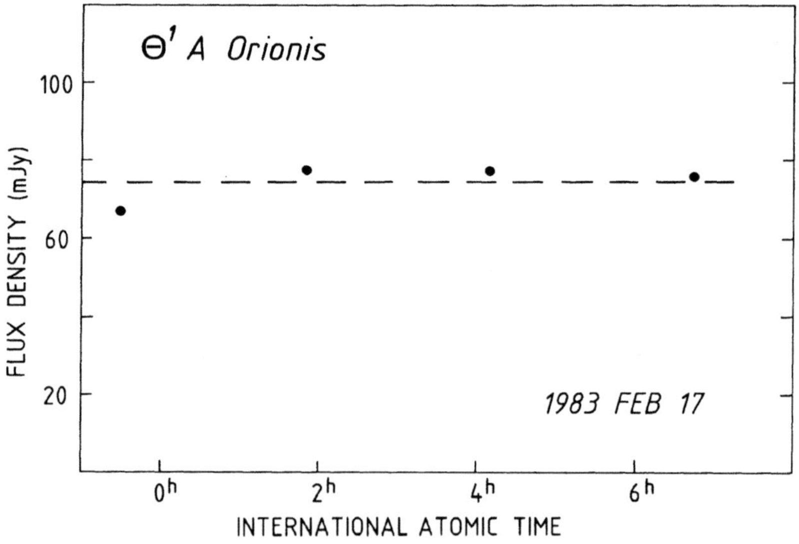

Figure 2: Flux density of Θ^1A Orionis versus time during the February 1983 observations.

to free-free radiation from an optically thin ionized region of constant density. The number of Lyman continuum photons necessary to ionize a constant density HII region is given, in terms of the radio continuum flux density at optically thin wavelengths, by the relation (c.f. Moran, 1983)

$$N_i = 7.7 \times 10^{43} \left[\frac{S_\nu}{\mathrm{mJy}} \right] \left[\frac{T_e}{10^4} \right]^{-0.45} \left[\frac{\nu}{\mathrm{GHZ}} \right]^{0.1} \left[\frac{D}{\mathrm{Kpc}} \right]^2$$

Assuming $T_e = 10^4$ K and $D = 0.5$ Kpc, we obtain $N_i \sim 2 \times 10^{45}$ photon s^{-1} at the epoch of maximum flux density. This Lyman continuum photon rate could be supplied by a B1V star (Panagia, 1973). The flat spectral index is not compatible with a constant velocity stellar wind model, since $\alpha = 0.6$ is then expected. However, a wind model may still be viable if one assumes that the wind is being stopped by its interaction with the surrounding medium.

The angular size of the radio source was measured to be 0".21 ± 0".05 at the epochs of maximum and minimum flux densities. On the other epochs the source was unresolved and smaller than ~0".25 at 2.0 cm. A lower limit for the brightness temperature at 20.6 cm in August 1983, obtained assuming a source angular size of 0".25, is $\sim 2 \times 10^5$K. This lower limit assumes that the source angular size is independent of frequency. Were the angular size to increase as the frequency decreases the brightness temperature could be smaller. However, the flat spectrum observed towards Θ^1A suggests its radio emission is optically thin and therefore that its angular size is the same over the observed range of frequencies.

DISCUSSION

The flat spectrum suggests a thermal origin for the radio emission associated with Θ^1A. However, the high brightness temperature contradicts a simple model in which the ionized region is excited and ionized by the Lyman continuum photons from an early B type star.

We propose that the flux density variations of the radio emission from Θ^1A is the result of recurrent mass loss outbursts from the B0.5 underlying star. The ejected mass is radiatively driven outwards as a strong stellar wind reaching a terminal velocity of ~500 km s^{-1}. The wind interacts with the surrounding circumstellar gas creating a thin dense circumstellar shell outside which the ambient gas is undisturbed (cf. Steigman et al., 1975). The evolution of this wind-driven circumstellar shell, in the case of steady stellar wind (i.e. constant mass loss rate), has been studied by Castor, McCray and Weaver (1975), Weaver et al. (1977) and Shull (1980). In the case of a mass loss outburst the evolution of the wind driven circumstellar shell is conditioned by the transitory nature of the energy source. The shell containing most of the swept up interstellar matter expands as the stellar wind adds energy to the system, but it will only persist for a time scale about the mass loss time scale, after which it dissipates into the surrounding medium.

TABLE 2

Θ^1A Observed Parameters

Date	λ cm	Flux density mJy	α[a]	Θ_s[b] arcsec	T_b K
1981 Aug 9	6.2	13.9	0.0	---	>4700.[c]
	2.0	13.6		---	---
1982 Apr 25	6.2	3.3	$\lesssim$0.0	0.21	5600.
	2.0	<4.0			
1983 Feb 17	2.0	73.9	0.0	0.21	13100.
	1.3	74.3		0.21	5600.
1983 Aug 28	20.6	14.8	0.0	---	>47300.[c]
	6.2	10.6		---	---
	2.0	13.5		---	---

Notes to table:

(a) Spectral index. (b) Deconvolved angular size.
(c) Using an upper limit for Θ_s equal to Θ_{HPBW} at 2.0 cm.

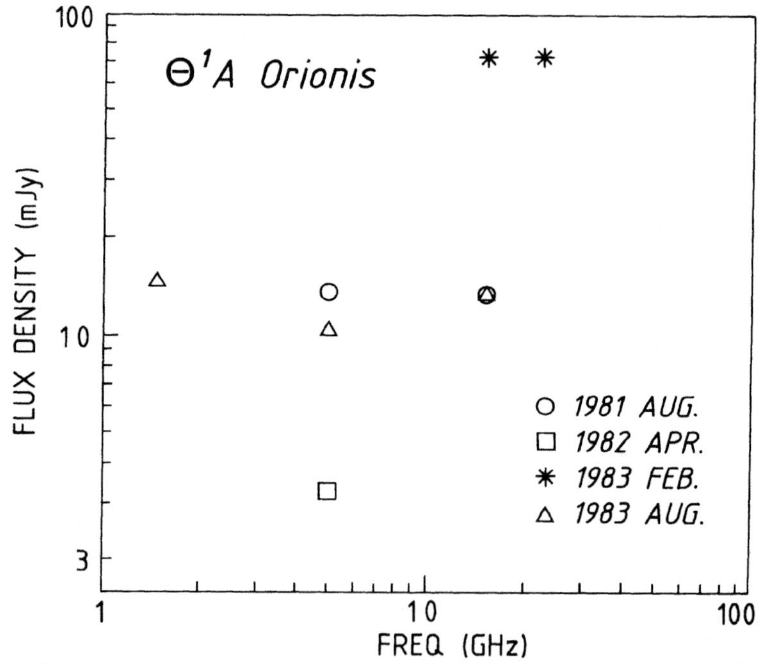

Figure 3:Flux density versus frequency for Θ^1A Orionis. Circles, squares, stars and triangles refer to flux density values in Aug 1981, Apr 1982, Feb 1983 and Aug 1983, respectively.

We suggest that in August 1983 we observed an early stage in the evolution of the wind driven shell. The observed high brightness temperature radio emission arises on the hot (10^6K) dense shocked interstellar gas shell which was recently formed. In February 83, epoch of the largest observed flux density, we suggest that the system was in an intermediate to late stage of its evolution. The radio emission in this case arises in a warm ($T \sim 10^4$K) region of post shocked gas between the dissipating hot shell and the underlying star that has already ended its mass loss outburst.

Abbot et al. (1984) suggested that accretion onto a compact companion is the most plausible explanation for the excess variable radio emission found in O type stars. The binary nature of Θ^1A Orionis could suggest a similar mechanism for the radio emission. Unfortunately, the large uncertainties in the light curve during the eclipse phase and in the radial velocity curve from Θ^1A do not allow an accurate determination of its orbital parameters, much less of the physical parameters of the stars. Lohsen (1976) derived masses of 7.3 and 2.3 $M_\odot$ for the binary members, in discrepancy with the spectroscopic and luminosity mass of ~15 $M_\odot$ inferred for the brightest member (Levato and Abt, 1976). In any case, the collapsed companion hypothesis for Θ^1A is unlikely on evolutionary grounds. The estimated age of the Trapezium cluster of ~10^5 years (Strand, 1958) is more than an order of magnitude smaller than the shortest hydrogen-burning lifetime of very massive stars of $\sim 2 \times 10^6$ years (Hoyle and Fowler, 1963).

ACKNOWLEDGEMENTS

G.G. wishes to thank L.F. Rodriguez for his valuable comments. NRAO is operated by Associated Universities, Inc., under contract with the National Science Foundation.

REFERENCES

Abbot, D.C., Bieging, J.H., Churchwell, E., and Cassinelli, J.P.: 1980, Ap. J. 238, 196.
Abbot, D.C., Bieging, J.H., and Churchwell, E.: 1984, Ap.J. 280, 671.
Baldwin, M.E., and Mattei, J.A.: 1977, R.A.S.C. Jour. 71, 475.
Castor, J., McCray, R., and Weaver, R.: 1975, Ap.J. (Letters) 200, L107.
Garay, G.: 1983, Ph.D. thesis, Harvard University.
Garay, G., Moran, J.M., and Reid, M.J.: 1985, in preparation.
Hjellming, R.M.: 1982, in An Introduction to the VLA NRAO Internal Report, Charlottesville, VA.
Hoyle, F., and Fowler, W.A.: 1963, M.N.R.A.S. 125, 169.
Levato, H., and Abt, H.A.: 1976, P.A.S.P. 88, 712.
Lohsen, E.: 1976, IBVS # 1211.
Moran, J.M.: 1983, Rev. Mex. Astron. Astrof. 7, 95.

Moran, J.M., Garay, G., Reid, M.J., Genzel, R., and Ho, P.T.P.: 1982, Symposium on the Orion Nebula, eds. Glassgold et al., p. 204.
Panagia, N.: 1973, A.J. 78, 929.
Parenago, P.P.: 1954, Works of the Astronomical Institute at Sternberg, p. 25.
Persi, P., Ferrari-Toniolo, M., Tapia, M., Roth, M., and Rodriguez, L.F.: 1984, Astr. Ap., submitted.
Shull, J.M.: 1980, Ap.J. 238, 860.
Steigman, G., Strittmatter, P.A., and Williams, R.E.: 1975, Ap.J. 198, 575.
Strand, K.A.: Ap.J. 128, 14.

HIGH RESOLUTION OBSERVATIONS OF RADIO STARS

Robert H. Becker
Department of Physics
University of California
Davis, CA 95616

Richard L. White
Space Telescope Science Institute
Homewood Campus
Baltimore, MD 21218

ABSTRACT. We have made observations with the VLA at 2, 6, and/or 20 cm of 10 stars: P Cygni, Cyg OB2 No. 12, MWC 349, V1016 Cyg, LkHα 101, HM Sge, Vy 2-2, T Tau, HD 193793, and V410 Tau. Visibility functions have been calculated to study the spatial structure of each source. Most sources appear to be stellar winds; however, some interesting anomalies were discovered. The data and model fits are shown, and the results are discussed.

1. INTRODUCTION

All the observation described in this paper were made at the VLA in August 1983 or in October 1983. The August observations were plagued by bad weather; although we have reduced the data extremely carefully, it is still somewhat suspect. The weather for the October observations was much better.

The data is displayed in the form of visibility functions (correlated flux measured by a pair of antennas as a function of the distance between them). Wind models which are fit to the data have two parameters: the total flux and the temperature of the wind. For each model plotted the flux and temperature are given in the caption at the top. The results for all the stars observed are summarized in Table 1. See White and Becker (1982) for more details on the data analysis techniques.

2. P CYGNI

P Cygni was observed at 2, 6, and 20 cm in August and again at 2 cm in October. Only the 2 cm observations have high signal-to-noise. The August and October 2 cm observations are consistent; P Cyg was observed twice at 2 cm to insure that there were no problems with the weather. The temperature in the 2 cm emitting region is about 6000 K. This is rather surprising, because

R.M. Hjellming and D.M. Gibson (eds.), Radio Stars, 139–146.

TABLE 1
Summary of Observations

Source	*λ (cm)*	*Date*	*Flux (mJy)*	*Temperature (K)*
P Cygni	2	Aug 83	16.8±1.4	6500±1500
	2	Oct 83	17.5±1.3	5400±1100
	6	Aug	7±1	15000±10000
	20	Aug	2.3±0.3	9000±6000
Cyg OB2 No. 12	2	Aug	12±5	4000±3000
	6	Aug	4.0±0.6	11000±7000
	20	Aug	2.1±0.3	3500±2000
MWC 349	2	Oct	316±8	6500±300
	6	Aug	157±5	7400±400
V1016 Cyg	6	Oct	61±6	15000±1500
LkHα 101	2	Oct	70±3	6900±700
HM Sge[a]	6	Oct	51±3	~7000
Vy 2-2[a]	6	Oct	37±3	~5500
T Tau S	2	Oct	5.0±1.5	4000±3000
T Tau N[b]	2	Oct	1.4±0.5	—
HD 193793[c]	6	Oct	24±1	$> 5 \times 10^5$
V410 Tau[d]	2	Oct	< 1	—

NOTES:
(a) Temperature is uncertain because wind model fits poorly.
(b) Source is too weak to fit model.
(c) Source is unresolved.
(d) Source was not detected.

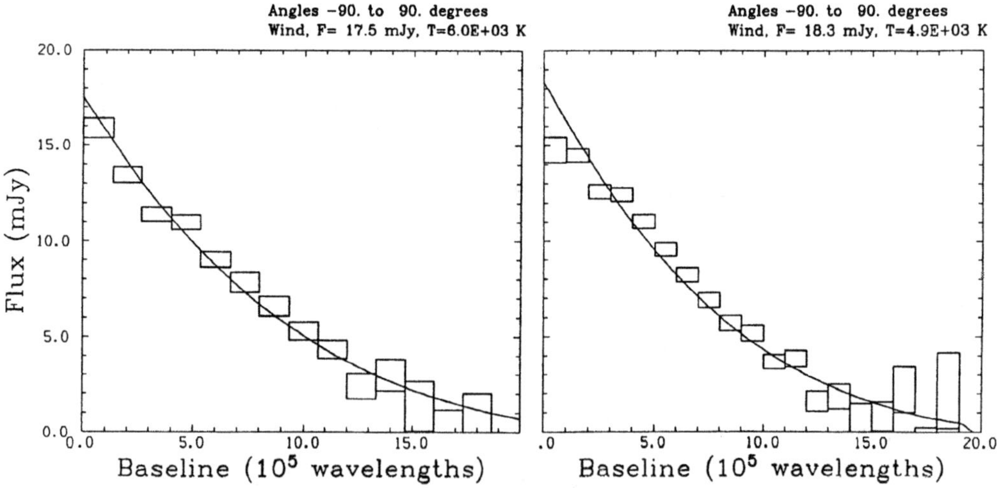

(a) August 1983 (b) October 1983

Figure 1. P Cygni: 2 cm visibility function

it is *cooler* than the temperature in the 6 cm region (~ 20000 K, White and Becker 1982). This means either that P Cygni's wind is heated as it flows away from the star or that the 6 cm region has cooled since it was observed in March 1981 by White and Becker.

The 6 cm observations shown here are consistent with the temperature determined by White and Becker (1982), although the flux is slightly smaller (7 mJy instead of 10 mJy). However, both the 6 cm and 20 cm data admit either high or low temperature fits.

The fluxes at the three wavelengths are consistent with the $\nu^{0.6}$ spectrum expected for a stellar wind.

3. CYG OB2 NO. 12

Cyg OB2 No. 12 was observed in August at 2, 6, and 20 cm. The fluxes are consistent with those expected for a stellar wind. The temperatures are not very well determined, but appear to fall with increasing wavelength between 6 cm and 20 cm (see also the 6 cm observations of White and Becker 1983), as is expected if the wind is cooling adiabatically.

4. MWC 349

MWC 349 is probably the brightest stellar wind source in the sky. It was observed in August at 6 cm and in October at 2 cm. It is easily resolved by the VLA and at both 2 and 6 cm the visibility function looks exactly like a stellar wind. However, at 2 cm MWC 349 is far from circularly symmetric, as can be seen from the 2 cm contour map. Instead, it has a remarkable two-lobed appearance.

This asymmetry also manifests itself in the 6 cm data, although it is not obvious in the 6 cm map. Comparison of the imaginary part of the correlated flux for baselines oriented N-E to that for baselines oriented S-W shows a strong difference.

5. V1016 CYG AND LICK Hα 101

In October, V1016 Cyg was observed at 6 cm and LkHα 101 at 2 cm. Both of these sources are relatively strong and have visibility functions which are fit very well by stellar wind models. Their wind temperatures are quite different, however; V1016 Cyg has a hot wind with $T = 15000$ K, while LkHα 101 has a wind temperature of only 7000 K.

6. T TAU

T Tauri was observed at 2 cm in October. It is a double source with a separation of about 0.5 arcsec. The weaker, northern component coincides with the optical star; the stronger, southern component coincides with T Tau's infrared companion (see Schwartz *et al.* 1984). We find that the visibility

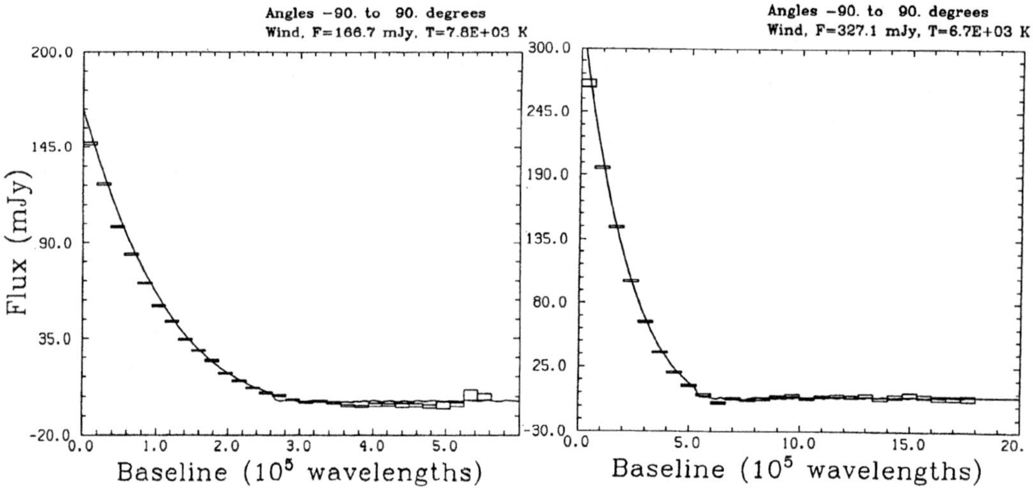

(a) 6 cm (b) 2 cm

Figure 2. MWC 349 visibility functions

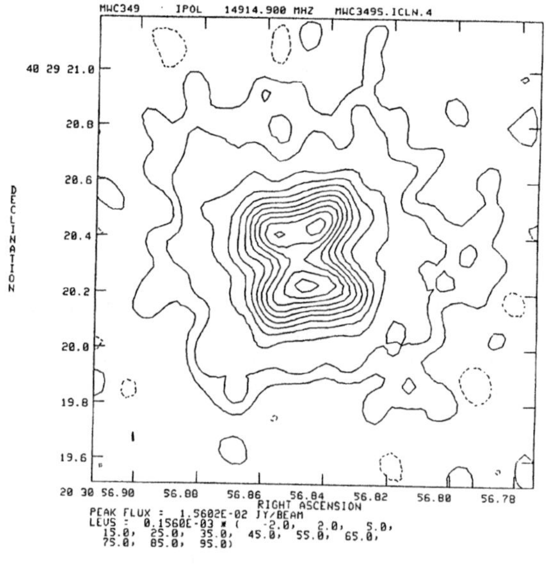

Figure 3. 2 cm map of MWC 349

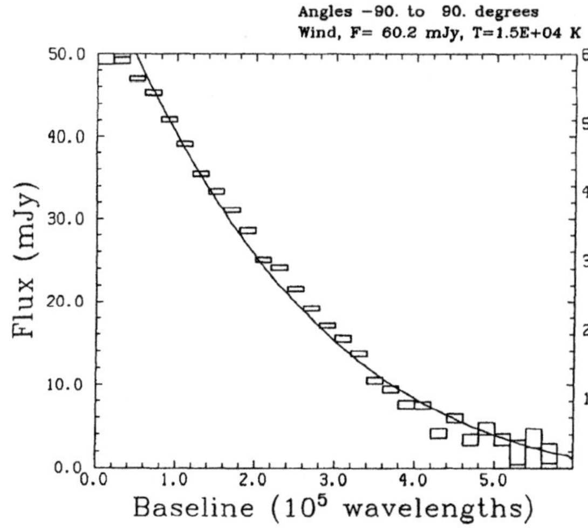

Figure 4. V1016 Cyg: 6 cm visibility function

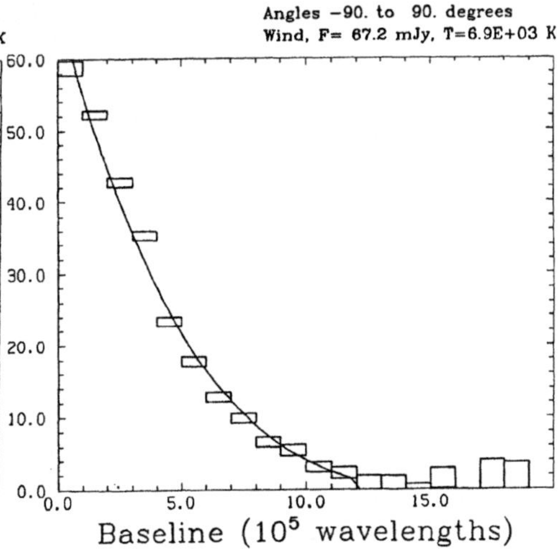

Figure 5. LkHα 101: 2 cm visibility function

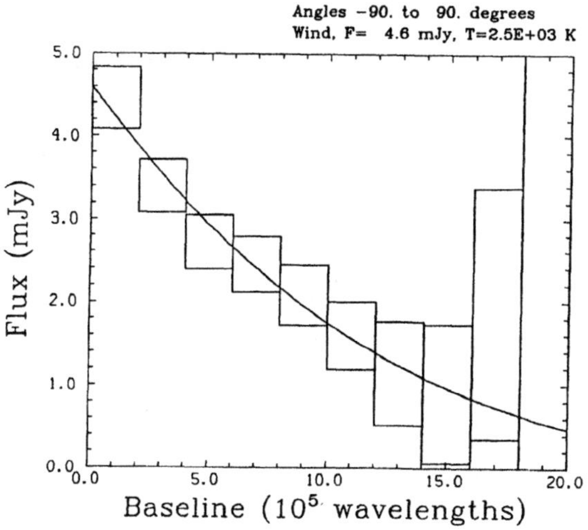

(a) Visibility function

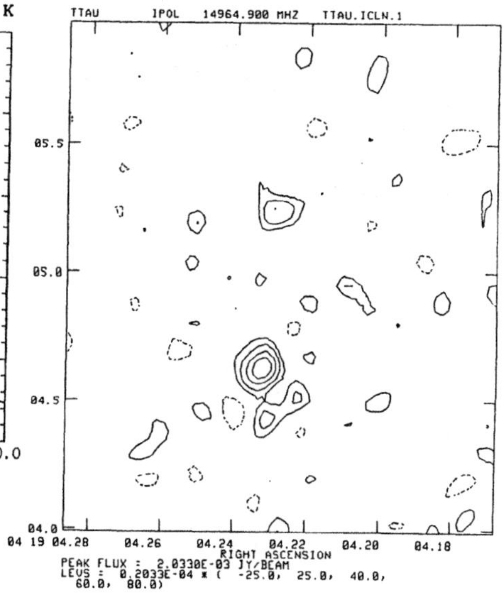

(b) Map

Figure 6. 2 cm observations of T Tau

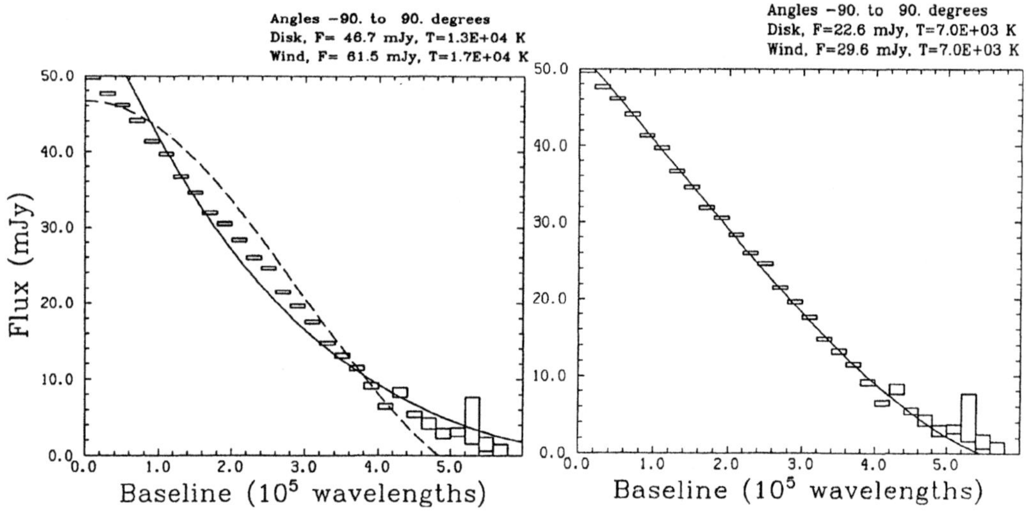

(a) Wind, Disk (dashed) models (b) Sum of wind and disk

Figure 7. HM Sge: 6 cm visibility function

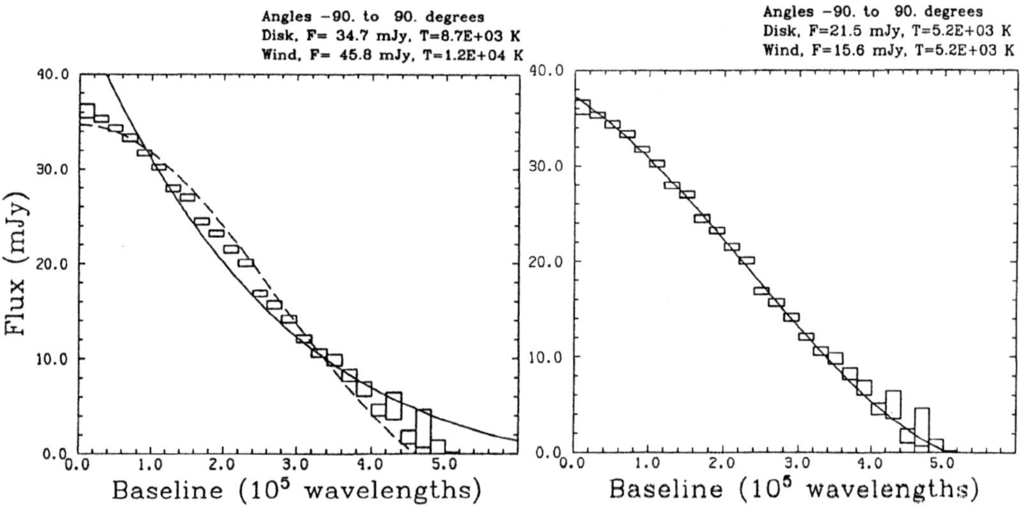

(a) Wind, Disk (dashed) models (b) Sum of wind and disk

Figure 8. Vy 2-2: 6 cm visibility function

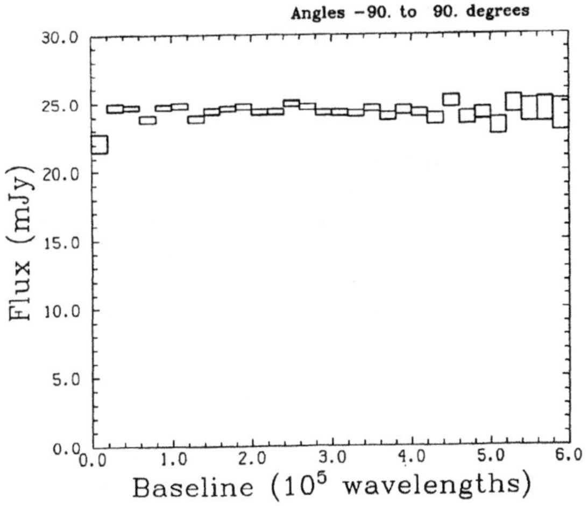

Figure 9. HD 193793: 6 cm visibility function

function of the southern component can be fit with a stellar wind model; however, it could also be fit by an accretion flow model with $n(r) \propto r^{-3/2}$.

The nearby companion makes calculation of the visibility function somewhat complicated; the shortest baseline point is probably somewhat high because of contamination by the northern component.

The 2 cm flux of the northern component is 1.4 ± 0.5 mJy. It was not detected at 2 cm by Schwartz *et al.* (1984), but our (marginal) detection is consistent with their upper limit of 1 mJy. Our 2 cm flux for the southern component is ~4–5 mJy, considerably smaller than the 9 mJy quoted by Schwartz *et al.* (1983); however, they do say that their fluxes might be too high because they self-calibrated their data. It seems premature to argue that the radio emission from T Tau (S) is variable.

7. HM SGE AND VY 2-2

Both of these sources were observed in October at 6 cm. In contrast to the above sources, wind models do not appear to fit the data very well. The figures on the left show the best fit models for both a wind and a disk with a uniform surface brightness. Neither one fits very well. The figures on the right show the best fitting model which is a sum of a wind and a disk with the same brightness temperature. Obviously these are not unique models for these sources, but they do indicate the kinds of modifications which must be made to make wind models fit. These sources are not as extended as ordinary stellar winds, so they must either have compact cores or the wings of the wind emission must be cut off.

If Kwok, Bignell, and Purton's (1984) model for HM Sge is correct, it will be very interesting to monitor the visibility function of HM Sge as time passes.

8. HD 193793 AND V410 TAU

In October, we observed HD 193793 at 6 cm and V410 Tau at 2 cm. The radio emission from HD 193793 is almost certainly non-thermal; we find that it is completely unresolved by the VLA, implying a very high brightness temperature for the emission. V410 Tau may also be a non-thermal source. V410 Tau is highly variable; our upper limit for its 2 cm flux is $<$ 1 mJy, whereas Cohen, Bieging, and Schwartz (1982) observed it to have a 6 cm flux of 19 mJy in March 1981.

REFERENCES

Cohen, M., Bieging, J. H., and Schwartz, P. R. 1982, *Ap. J.*, **253**, 707.
Kwok, S., Bignell, R. C., and Purton, C. R. 1984, *Ap. J.*, **279**, 188.
Schwartz, P. R., Simon, T., Zuckerman, B., and Howell, R. R. 1984, *Ap. J. (Letters)*, **280**, L23.
White, R. L., and Becker, R. H. 1982, *Ap. J.*, **262**, 657.
White, R. L., and Becker, R. H. 1983, *Ap. J. (Letters)*, **272**, L19.

RADIO EMISSION FROM SYMBIOTIC STARS: A BINARY MODEL

A.R. Taylor and E.R. Seaquist
Department of Astronomy, University of Toronto
Toronto, Ontario CANADA

ABSTRACT. We examine a binary model for symbiotic stars to account for their radio properties. The system is comprised of a cool, mass-losing star and a hot companion. Radio emission arises in the portion of the stellar wind photo-ionized by the hot star. Computer simulations for the case of uniform mass loss at constant velocity show that when less than half the wind is ionized, optically thick spectral indices greater than +0.6 are produced. Model fits to radio spectra allow the binary separation, wind density and ionizing photon luminosity to be calculated. We apply the model to the symbiotic star H1-36.

1. INTRODUCTION

The class of symbiotic stars is defined by the basic characteristic of an optical spectrum containing both high excitation emission lines (eg. He II) and absorption features of a cool, late-type star. Extensive infrared photometry of these stars has shown that the late-type spectral features arise from a red giant star, usually of spectral type M, but in some cases as early as K or G. Many also contain hot ($\sim 10^5$ K) continuum sources seen in the far UV. A few symbiotic stars have been known for some time to be radio sources. More recently, a sensitive survey (Seaquist et al. 1984) has shown that more than 25% of known symbiotic stars are radio sources at flux densities of ~ 1 mJy.

Single star ionized wind models (eg. Seaquist and Gregory 1973; Wright and Barlow 1975) have been applied to symbiotic stars. However, such models predict a radio spectral index of +0.6, at odds with the spectral index distribution for symbiotic stars which peaks instead at $\alpha = +1.0$ (Seaquist et al. 1984). In this paper, we present a new model for the radio properties in light of the mounting evidence for the binary nature of symbiotic stars.

2. THE MODEL

We consider a cool star undergoing uniform spherically symmetric mass loss at a rate $\dot{M}$ with wind velocity v. Embedded in the wind at a radial distance, a, is a hot star emitting L_{ph} hydrogen ionizing photons

R.M. Hjellming and D. M. Gibson (eds.), Radio Stars, 147–150.

per second. A schematic of the system is shown in figure 1.

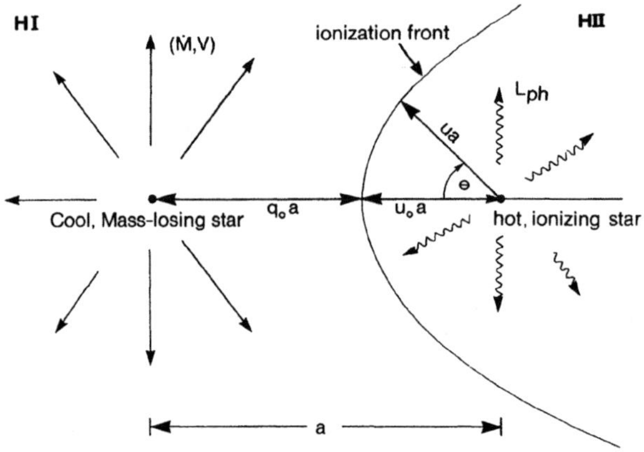

Figure 1. A schematic of the model

Ignoring the very small effect of the wind flow, the location of the surface of the ionized nebula is determined by the balance of recombination and ionization rates inside the nebula, and is given by an expression of the form $f(u,\theta) = X$ where,

$$X = \frac{4\pi\mu^2 m_H^2}{\alpha} a \, L_{ph} \left(\frac{\dot{M}}{v}\right)^{-2} \tag{1}$$

and α is the recombination coefficient to all but the ground state of hydrogen. The function $f(u,\theta)$ is given by Taylor and Seaquist (1984). The size and shape of the ionized portion of the wind depends only upon the single parameter X.

For all values of X there exists a finite inner or outer boundary to the ionized zone. Consequently, at frequencies above some critical value, the nebula will be totally optically thin. The frequency at which $\tau = 1$ along the axis $\theta = 0$ through the ionized zone is given by

$$\nu_t^{2.1} = 8\times10^{66} \; T_e^{-1.35} (q_o a)^{-3} \left(\frac{\dot{M}}{v}\right)^2 \quad \text{GHz}, \tag{2}$$

where a is in cm, $\dot{M}$ is in $M_{\odot}\text{-yr}^{-1}$ and v is in km-s^{-1}. The spectrum turns over to a spectral index of −0.1 at a frequency of a few times ν_t, and the flux density at a frequency ν in the optically thin regime is

$$S_\nu = 1.6\times10^{32} \; T_e^{-1/2} \, D^{-2} \, g_{ff}(\nu, T_e) \left(\frac{\dot{M}}{v}\right)^2 a^{-1} Q(X) \quad \text{mJy}, \tag{3}$$

where D is in units of kpc. Q(X) is a volume integral, of order unity, that may be calculated numerically for a specific value of X.

In the optically thick regime, the shape of the spectrum is highly dependent upon the geometry of the ionized region, and is thus a strong function of X. For $X > \pi/4$, the optically thick spectral index is +0.6.

In this case, the majority of the wind is ionized, and the single star wind model is a good approximation. For $X < \pi/4$, the ionized zone is primarily ionization bounded and the spectral index is >+0.6. For these cases, knowledge of both ν_t and the optically thick spectral index provides a measure of the value of X.

Equations (1), (2) and (3) constitute a set of three independent equations for the unknowns a, $(\dot{M}/v)$ and L_{ph}. Modelling of the radio spectra of symbiotic stars with $\alpha > 0.6$ can thus, in principle, provide a unique solution for the binary separation the wind density and the ionizing photon luminosity. In practice, the derived values of ν_t and X also depend upon the viewing angle of the system, thus fits to the radio spectra alone will yield a range of possible solutions.

3. The Symbiotic Star H1-36

Allen (1983) has carried out extensive optical and infrared spectrophotometry of H1-36 and provides conclusive evidence for the presence of an M giant and a circumstellar dust shell. From fits to the IR continuum, Allen measures A_k=1.5 mag, implying a visual extinction of ~20 mag. Analysis of the emission lines, however, yields A_v=2.2 mag. Allen argues that the high excitation ionization is produced by a hot binary companion exterior to the dust shell.

H1-36 is among the strongest radio sources associated with a symbiotic star. Multi-frequency radio measurements, over the period 1974-1977 were reported by Purton et al. (1977). The spectrum has a low-frequency spectral index of ~1.0 and a high-frequency turn-over at ~10 GHz. Because of its well measured radio spectrum and the evidence for binarity, H1-36 is a prime candidate for our model. We have fit model spectra to the measurements of Purton et al. plus a recent measurement by us at 1.6 GHz. Very nearly equally good fits were found for viewing angles of 0° and 90°. Figure 2 shows the observed spectrum and the model spectra for both viewing angles. The model provides an excellent fit to the observations over the entire frequency range.

Using the best fit model parameters and $T_e = 1.5 \times 10^4$ K (Allen 1983), we have derived values for a, $(\dot{M}/v)$ and L_{ph} as a function of the distance to the star. The solution for each viewing angle is listed in table 1. The physical conditions corresponding to the two solutions agree to within a factor of four.

Table 1. Model Parameters for H1-36

Viewing Angle	X	a (cm)	$(\dot{M}/v)$	L_{ph} (phot-s^{-1})
0°	0.73	3.3×10^{15}D	$1.3\times10^{-7}\,D^{3/2}$	$9.7\times10^{45}D^2$
90°	0.14	1.4×10^{16}D	$5.2\times10^{-7}\,D^{3/2}$	$7.8\times10^{45}D^2$

Adopting a mean value for the model solutions yields, for a distance of 4.5 kpc (Allen 1983), $a = 4\times10^{16}$ cm, $\dot{M} = 3\times10^{-6} v\ M_{\odot}$-yr^{-1} and $L_{ph} = 2\times10^{47}$ photons-sec^{-1}. For a wind velocity of 10 km-s^{-1} (typical, though slightly low, for red giant winds), the mass-loss rate is $3\times10^{-5} M_{\odot}$-yr^{-1}.

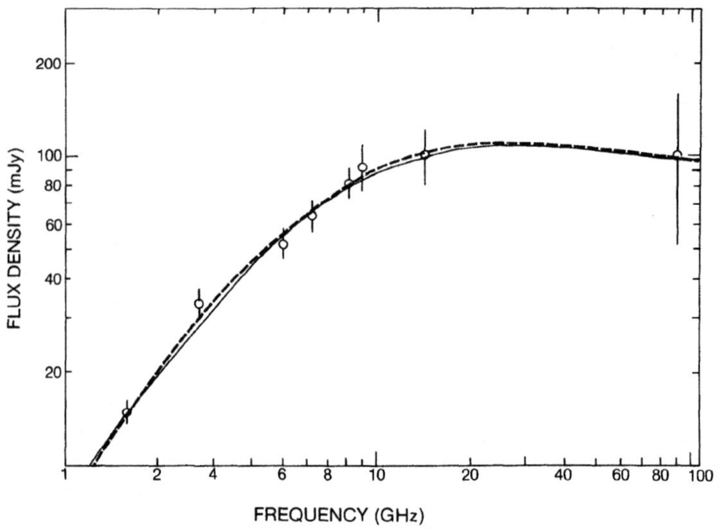

Figure 2. The radio spectrum of H1-36. The solid and dashed lines are model spectra for viewing angles of 0° and 90°.

4. DISCUSSION

We have shown that the radio spectrum of H1-36 can be reproduced by a simple, steady-state model in which a portion of the stellar wind of a cool giant is photo-ionized by a hot companion. Evidence that many, if not all, symbiotic stars are binaries suggests that the high spectral indices of other symbiotics (eg. V2416 Sgr, α =1.01; BF Cyg, α =0.98) may be explained in the same manner. Within this framework sources which exhibit the canonical spectral index of +0.6 are those in which the red giant wind is nearly completely ionized.

The fact that >25% of symbiotic stars are detectable at radio wavelengths, demonstrates that radio emission is a fundamental property of the symbiotic phenomenon. Moreover, we have shown that, through fitting to radio spectra, our model provides a powerful tool for measuring the properties of symbiotic star systems. Coupled with observations at other wavelength regimes, radio observations can provide valuable insights into the nature of symbiotic stars and help fit these unusual systems into an evolutionary scenario.

REFERENCES

Allen, D.A. 1983, Mon. Not. R. Astr. Soc., 204, 113.

Purton, C.R., Allen, D.A., Feldman, P.A., and Wright, A.E. 1977, Mon. Not. R. Astr. Soc., 180, 97p.

Seaquist, E.R., and Gregory, P.C. 1973, Nature Phys. Sci., 245, 85.

Seaquist, E.R., Taylor, A.R., and Button, S. 1984, Astrophys. J., 284, 202.

Taylor, A.R., and Seaquist, E.R. 1984, Astrophys. J., 286, in press.

Wright, A.E., and Barlow, M.J., 1975, Mon. Not. R. Astr. Soc., 170, 41.

THE RADIO EMISSION OF VV CEPHEI-TYPE BINARIES

R. M. Hjellming
National Radio Astronomy Observatory*
Socorro, NM 87801 U.S.A.

ABSTRACT. Seven binary star systems of the VV Cephei type, in which a B star is companion to a cool supergiant star, are known radio sources. These sources are caused by the free-free emission from an ionized subregion in the supergiant's stellar wind. Most of these radio-emitting subregions can be considered a new class of HII region because of their high densities and high temperatures.

1. THE α SCO (ANTARES) NEBULOSITY

The α Sco binary system, a B2.5V star and a M1.5 Iab star (Antares) with a separation of 560 AU (3"), contains the proto-type radio source first mapped by Hjellming and Newell (1983). As seen in Figure 1 this type of object has a resolved nebulosity around the B star inside the cool stellar wind of the supergiant, which may itself exhibit radio emission. Hjellming and Newell (1983) showed that a simple model of an HII region maintained by the B2.5V star explains the characteristics of the α Sco nebulosity. A portion of the cool stellar wind passes through the ionization front maintained between the two stars and remains ionized until it passes through another ionization front on the leeward side of the B star.

2. SIX OTHER BINARIES OF THE VV CEP TYPE

The essential characteristic of the α Sco binary that produces the extended nebulosity is the existence of the B2.5V companion inside the cool stellar wind of Antares; this is also the primary characteristic of the VV Cephei-type binary system as defined by Cowley (1969). Following this clue Hjellming (1985) carried out a VLA search for radio emission from binaries of this kind and found six other systems of this type,

* The National Radio Astronomy Observatory is operated by Associated Universities, Inc., under contract with the National Science Foundation.

R. M. Hjellming and D. M. Gibson (eds.), Radio Stars, 151–154.

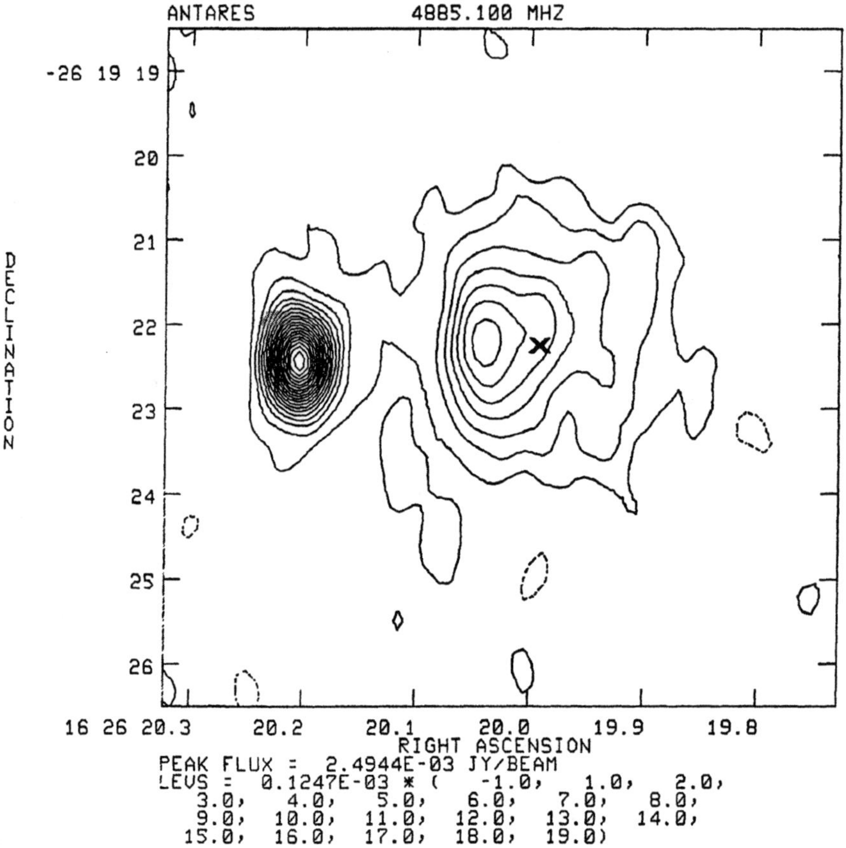

Figure 1. A 4885 MHz VLA map of the α Sco binary system shows a point source on the left on the position of Antares and an extended, ionized nebulosity surrounding the B2.5V star (X).

including VV Cephei. The only VV Cep binaries not detected as radio sources were at distances ≥ 2.5 kpc, whereas the detected systems were all at distances ≤ 2.5 kpc.

A summary of the properties of the VV Cep radio sources is given in Table 1. The main difference between the α Sco nebulosity and the VV Cep, KQ Pup, HR 8164, FR Sct, WY Gem, and HD 237006 nebulosities is the fact that the latter binaries have periods of the order of 20-40 years, making the binary separation more than an order of magnitude smaller than the separation in α Sco. The main effect this has on the ionized nebulosites is that the gases in the HII region around each B star have densities that are a few to several orders of magnitude larger. Therefore, these HII regions are very compact, optically thick at radio wavelengths, and hotter than normal HII regions because of collisional de-excitation of most of the "usual" coolants in the gas. In Table 1 HR 8164 is listed with two different sets of radio fluxes because different

TABLE 1 - Summary of Properties of VV Cep-type Radio Sources

Name	Spectral Types	S(1.465GHz) [mJy]	S(4.9GHz) [mJy]	S(15GHz) [mJy]
VV Cep	M2ep + B1.5V*		0.6 ± 0.1	1.4 ± 0.2
α Sco	M1.5 Iab + B2.5V	8.0 ± 1.0	7.0 ± 1.0	6.0 ± 1.0
KQ Pup	M2ep Iab + B2	0.5 ± 0.1	1.7 ± 0.2	4.9 ± 0.5
HR 8164	M1ep Ib + B2pe + B3V*	2.4 ± 0.4 1.2 ± 0.2	4.4 ± 0.4 3.5 ± 0.2	5.9 ± 0.8 3.8 ± 0.5
FR Sct	M3ep Ia: + O9.5V*	3.3 ± 0.5	14.0 ± 2.0	40.0 ± 5.0
WY Gem	M2ep Iab + B5V*		0.4 ± 0.15	1.4 ± 0.2
HD237006	M1ep Ib + B3:		0.7 ± 0.1	0.84 ± 0.1

*Estimates of spectral type based upon photometric data of Wawrukiewicz and Lee (1973).

results were obtained for observations separated by a few months.

Only α Sco has an optically thin nebulosity; the other six binaries have optically thick radio sources. The VV Cep, FR Sct, WY Gem, and KQ Pup systems have spectral indices from 0.8 to 1.2; HR 8164 and HD 237006 have spectra similar to those of HII regions that are making the optically thick to thin transition between 1.4 and 10 GHz, and are becoming optically thin at about 15 GHz. Models of these nebulosities, similar to that for α Sco (Hjellming and Newell 1983), have been computed to fit these characteristics (Hjellming 1985). In some cases the HII region is a finite, ionized subregion, but in other cases the leeward ionization front effectively extends to infinity. In the latter cases all of the supergiant stellar wind is ionized, with the exception of a zone that begins with an ionization front between the two stars and extends to infinity on the supergiant side of the binary system. The shapes of the ionized region for a range of models (Hjellming 1985) are shown in Figure 2.

The VV Cep radio sources are observable aspects of ionized sub-regions that are previously unrecognized phenomena that must exist in all VV Cep stars. In some cases this indicates that the supergiant wind is divided, with respect to a point half way between the two stars, into solid angles which are ionized and solid angles which are neutral. Thus some solid angles of the source will be like normal internally ionized stellar winds, and others will be largely undisturbed, cool, and un-ionized supergiant winds. In other cases, similar to α Sco, there is

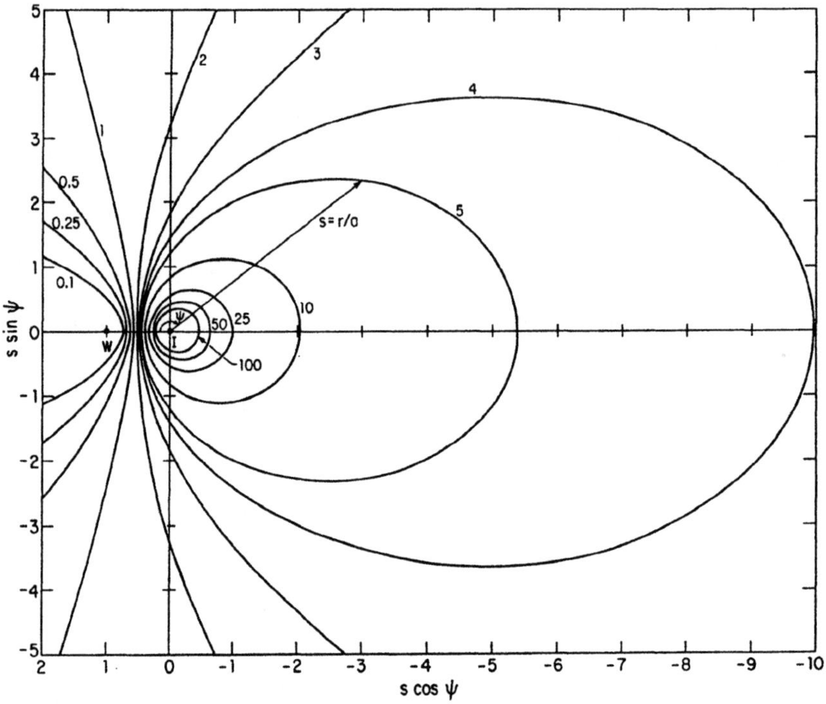

Figure 2. The location of the ionization fronts in various models of ionized regions with an ionizing star (I) at distance, a, from the star (W) with a cool stellar wind. Coordinate (r,ψ) is a point on the ionization front in the plane through the two stars.

a finite ionized subregion inside an otherwise cool, un-ionized wind. The latter wind has three different physical states: a normal un-ionized supergiant wind; an ionized subregion with the stellar wind flowing in one side and out the other; and a portion of un-ionized stellar wind that has passed through the HII region, so that the chemical evolution of molecules and dust may be very different from that of normal supergiant winds.

REFERENCES

Cowley, A.P. 1969, P.A.S.P., **81**, 297.
Hjellming, R.M. 1985, in preparation.
Hjellming, R.M., and Newell, R.T. 1983, Ap.J., **275**, 704.
Wawrukiewicz, A.S., and Lee, T.A. 1973, B.A.A.S., **5**, 346.

MASS-LOSS DURING THE STAR FORMING PROCESS IN CEP A.

V.A. Hughes
Astronomy Group, Dept. of Physics,
Queen's University,
Kingston, Ontario, K7L 3N6.

ABSTRACT. The star forming region in Cep A is at present unique. It consists of two strings of the most compact HII regions yet detected, each region appearing to be produced by a B3 star, and all are contained inside a diameter of 40", or 0.1 pc for the distance of 725 pc. OH and H_2O masers are associated with the central regions, and these normally indicate star formation, but the linear dimensions of the regions of 200 to 800 au suggested initially that they were no more than 1,000 years old. Measurment of the spectral indices of the regions are consistent with mass-loss due to stellar winds, which may account for their small size, and give some indication of how B3 protostars evolve.

A map at λ6 cm of the Cep A region, made using the Very Large Array (VLA) of the National Radio Astronomy Observatory* in the "B" configuration, is shown in Figure 1. The angular resolution is 1", corresponding to a linear distance of 725 au at the distance to Cep A of 725 pc. It shows the two strings of HII regions, a total of about 14, contained in a diameter of 0.1 pc. The average flux density per region is 3.6 mJy, with most regions having flux densities between 2.4 and 5.5 mJy. Assuming that the HII regions are optically thin, and spherical, then it can be shown that in this case the excitation parameter, U (pc cm^{-2}) is given by $U = 1.14\ S^{1/3}$ where S(mJy) is the flux density. Thus a typical value of $U = 1.8$ pc cm^{-2} applies to each region, such that if they are produced by stars, then using the stellar models of Panagia (1973), the exciting stars are of spectral type B3, with luminosity of $2.0 \times 10^3 L_o$ and mass of $7.8\ M_o$. The total luminosity expected from the 14 stars, of $2.8 \times 10^4 L_o$, is approximately equal to that for the infrared luminosity of the region of $2.4 \times 10^4 L_o$ (Evans et al. 1981), which gives some confidence to the assumption that it is stars that are exciting the HII regions. The total radiation which escapes is absorbed in the surrounding medium and radiates in the infrared.

The above interpretation has experienced some difficulties, not only because all the stars appear to be of the same spectral type and because of their high spatial density, but also the fact that the HII regions are so small, and appear to be very young. We expect the

R. M. Hjellming and D. M. Gibson (eds.), Radio Stars, 155–158.

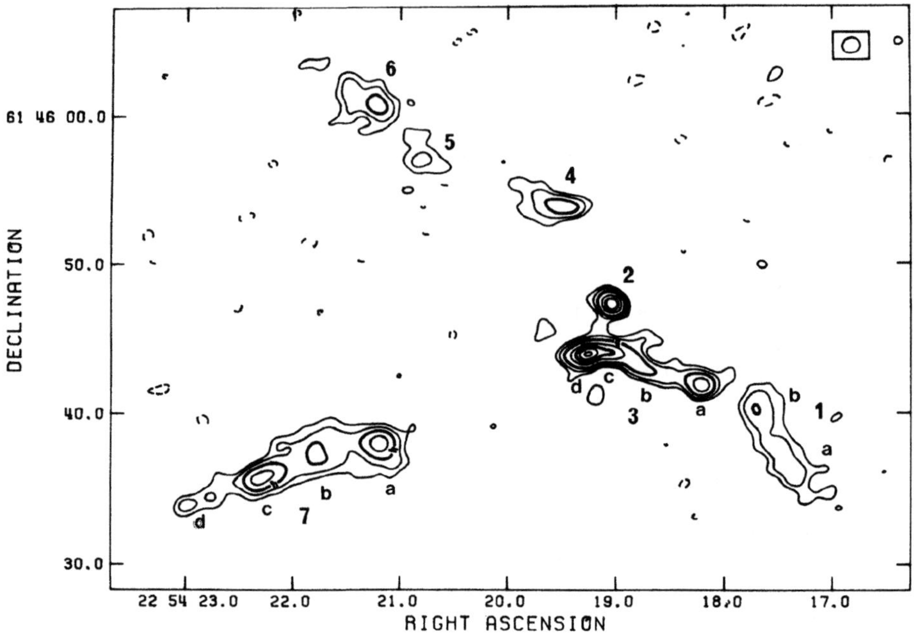

FIGURE 1. VLA map of Cep A at λ6 cm. The resolution is 1". (Hughes and Wouterloot 1984.)

initial HII region to be of radius, $s_o = U n^{-2/3}$, where n is the electron density, and for the ionization front to expand at a speed ~ 10^4 km s^{-1}. If $n = 10^5$ cm^{-3}, then $s_o = 170$ au, which size would be reached in less than 1 year. In fact, the HII region would grow as the star turned on its Lyα. Even at a speed of 1 km s^{-1}, a size of 500 au would be reached after 2,400 years, compared with the estimated time for the star to evolve onto the Main Sequence of 2×10^5 years.

There are two clues as to what is going on in the region, and both come from additional VLA observations in the "A" configuation at both λ6 cm and λ20 cm. The first comes from measurements of the spectral index of the individual components. These have been obtained by comparing the observations at λ6 cm in the "B" configuration with those at λ20 cm in the "A" configuration, which have the same angular resolution. Most of the components have peak flux densities that are not greater than 1 mJy, so that the errors in determining the total flux density can be somewhat large, but in general the larger size components away from the center have values of $\alpha = 0.0 \pm 0.1$, (where we define α by $S \propto \lambda^{\alpha}$, where S is the flux density and λ is the wavelength), which is consistent with optically thin HII regions, while sources 2 and 3 have values of $\alpha >$ 0.1. In particular, source 2 has $\alpha = 0.8$ and sources 3c and 3e have values of 0.2 and 0.5; all these appear to be very compact sources. The only other source with large α is source 5, with $\alpha = 0.5$, but this is comparatively extended with only a small peak flux density. These

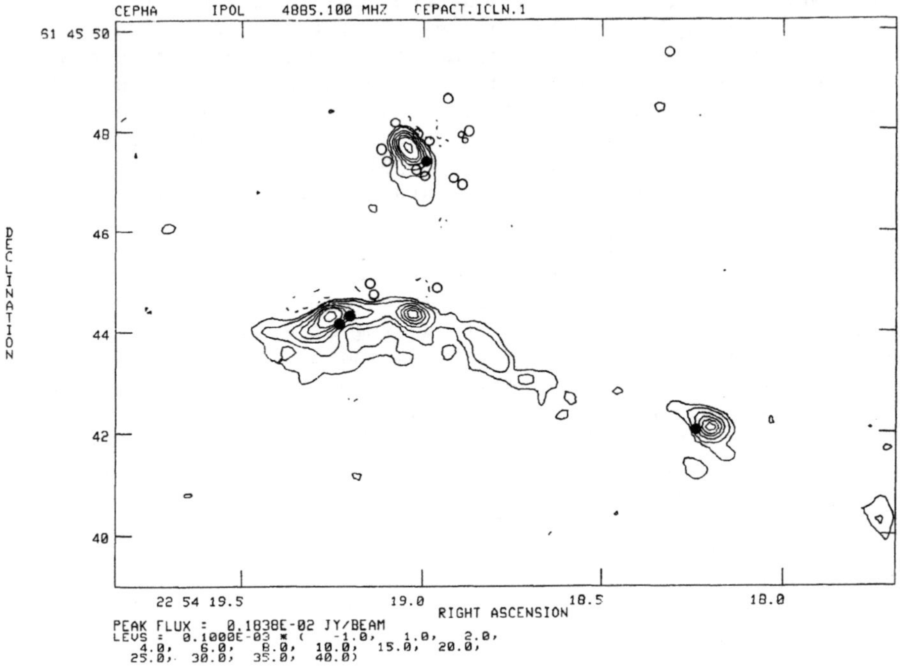

FIGURE 2. VLA map of the central region of Cep A at λ6 cm. The resolution is 0''3. Open circles show position of OH masers, closed circles that of H_2O masers, as determined by Cohen et al. (1984).

large values for α imply either mass-loss from the exciting star, or mass accretion.

Wright and Barlow (1975) have shown that for a star undergoing mass-loss so that the electron density, n, varies as $n \propto r^{-2}$, then $\alpha = 0.6$, and in fact if recombination occurs in the stellar wind so that there is a sudden cut-off in the ionization density, then the flux density can approximate that from a black-body, so values in the range $0.6 < \alpha < 2.0$ are possible. Thus, though the integrated flux from all the regions suggests that they are thermal in origin, a few, and in particular those that have associated masers, show strong indication that they could be losing mass. If we use the analysis by Wright and Barlow (1975), then the rate of mass-loss is given by $\dot{M} = (3-5) \times 10^{-9}$ v $M_\odot$ yr^{-1}, or an estimated rate of $\sim 10^{-7}$ $M_\odot$ yr^{-1}. It appears that we could be observing in most of the components, a combination of stellar wind and ionization producted by Lyα radiation from the central stars.

The second clue comes from the high resolution maps of the central region. Figure 2 shows a map of sources 2 and 3 with angular resolution of 0''3, obtained by combining λ6 cm data from VLA observations in both the "A" and "B" configuations, on which has been plotted the positions of OH and H_2O maser sources obtained by Cohen et al. (1984) using MERLIN. It can be seen that the OH masers are situated outside the HII region, while the H_2O masers are at the edge of the HII region.

The regions appear to be the result of the "turning-on" of $8M_\odot$ stars. As the protostar evolves, and the temperature increases, there will be very little effect on the surrounding gas and grains (of radius $\sim$0.1μm) until the temperature reaches $\sim$4800K. At this point, H_2O will have been evaporated off the grains and radiation pressure will cause them to be accelerated away. At some point, radiation pressure will balance the pressure due to matter infalling as the result of gravitational forces, and at this point a cocoon will form. This is delineated in Figure 2 by the position of the OH masers. As the temperature reaches $\sim$15,000K, an ionization front will be initiated, the star finally settling down at a temperature $\sim$19,000K. Mass-loss keeps the HII region small by effectively adding mass, but when the rate of mass-loss falls, the HII region will expand until it overtakes and evaporates the cocoon. For earlier-type, hotter stars, the ionization front will overlap the dust cocoon, the H_2O inside will be dissociated and the star will have OH masers surrounding the HII regions. For the cooler, later-type stars, the cocoon will be close in with OH being produced effectively in the stellar atmosphere.

Thus, in this case we appear to be detecting the combined effects of a stellar wind from B-type stars, together with the HII region produced by Lyα from the stars. The mass-loss seems to occur near to the time that the stars reach the Main Sequence, and could be activity analogous to that of the later type T Tauri stars, but in this case the stars are still optically obscured.

REFERENCES.

Cohen, R.J., Rowland, P.R., and Blair, M. 1984, M.N.R.A.S., 210, 425.

Evans, N.J., Becklin, E.E., Beichman, C., Gatley, I., Hildebrand, R.H., Keene, J., Slovak, M.H., Werner, M.W., and Whitcomb, S.E. 1981, Ap.J., 244, 115.

Hughes, V.A., and Wouterloot, J.G.A. 1984, Ap.J., 276, 204.

Wright, A.E., and Barlow, F. 1975, M.N.R.A.S., 170, 41.

* The National Radio Astronomy Observatory is operated by Associated Universities Inc., under contract with the National Science Foundation.

DISCUSSION AFTER ABBOTT'S TALK IN PART II

KUIJPERS: When you say the emission is "nonthermal", do you mean that the emission can not be explained by bremsstrahlung?

ABBOTT: Yes.

KUIJPERS: I would like to emphasize to everyone that it is an inappropriate word choice to describe bremsstrahlung as "thermal" emission. Thermal means that the particle distribution function is a Maxwellian characterized by a single temperature parameter. Consequently "thermal" emission does not describe the emission process. For a "one-temperature" gas one can still have free-free, gyrosynchrotron, synchrotron, or any other radiative process. The same is true for "nonthermal". Most of the time speakers who use the phrase "thermal" emission are describing free-free emission from a one-temperature gas.

THOMAS: Dave, I think you should be more explicit in your categorical statements. When you say "the OB stars" show such and such, you should be very explicit about which OB stars, because your conclusions do not hold for all of them. I gather that they are restricted to the OB supergiants, not the main sequence, and not the Oe and Be stars. Correct?

ABBOTT: Yes, and in particular only to those stars detectable at radio wavelengths.

THOMAS: Second, we have considerable evidence for many stars, notably the Be stars, that the "asymptotic velocity" you use to derive the mass flux from the radio measurements is highly variable (1) for a given star on a time scale of weeks, and (2) at a given place in the atmosphere. The far UV (IUE) measured velocities are, for many stars, much larger, as much as a factor of ten, in the regions where the far UV spectrum is formed, compared to the regions where the visual (e.g. Hα) and radio originate. Thus, when you say that the radio-inferred mass loss is the most reliable measurement, you must qualify that statement for those stars showing strong variability, of either type (1) or (2) above. Myself, I think the radio data, as the Hα data, give us information about the density distribution in the atmosphere. You need to combine that with simultaneous velocity measurements from another source to infer mass loss rates. Departure of the radio spectrum from an $\alpha = 0.6$ spectral index need not be nonthermal emission - it can simply be a non-constant velocity distribution, resulting in a departure from an inverse square law density distribution (with constant mass flow), or it can also mean a variable mass loss rate.

ABBOTT: I am not aware that any of my stars exhibit significantly variable winds.

UNDERHILL: I do not understand your use of nebular recombination theory

R. M. Hjellming and D. M. Gibson (eds.), Radio Stars, 159–170.

to estimate the emission measure for WR atmospheres which have densities of the order of 10^{11} cm^{-3}, which is what is found for the values of dM/dt used by you for WR stars. Would you be kind enough to explain what you did, and which lines you analyzed. I should think that collisional excitation and de-excitation would be important when the particle densities are of this order or larger.

ABBOTT: Recombination radiation is formed over a wide range of densities. I used optical data derived from high-level lines where the use of the theory is most appropriate. The theory is certainly adequate for a log-log plot.

WHITE: Would you care to suggest an interpretation for the correlation of dM/dt with M for WR stars?

ABBOTT: Mass loss is correlated with luminosity and, thus, mass in the same way as it is for the OB stars.

THOMAS: You keep telling us the OB stars that you have studied give no evidence of variability. Hence you think your results confirm the radiative-pressure origin and acceleration of the mass loss. First, we know that many stars have strong variability in the far UV spectrum, in magnitude of velocity, in strength of the super-ionized lines, etc., so for these stars you can hardly claim the radiative-pressure origin suffices. Second, if the accumulation of data shows that a star is variable in mass flow velocity, mass loss rate, etc., does this not prove the inadequacy of your radiative-acceleration approach to explain these phenomena wholly by itself?.

ABBOTT: I hesitate to answer yes or no. The question addresses problems for some OB stars but not for the ones I am talking about. The theory I have presented seems to work well even if the mass loss is clumpy.

MUTEL: What are the typical upper limits for circular polarization in the radio?

ABBOTT: 10 - 20%.

MUTEL: Then nonthermal emission mechanisms are possible.

WHITE: The rotation measure through the wind is enormous, so you never expect to see linear polarization for any source of radio emission in the wind.

MULLAN: Are you equating radio variability with nonthermal emission?

ABBOTT: Yes, it has been shown to be the case for all but two stars, P Cyg and Cyg OB2 No. 5. In the former case, the wind may be variable because it is ionization limited. In the latter case the situation is unclear.

DISCUSSION FOLLOWING KWOK'S TALK IN PART II

MUTEL: A comment. The first few VLBA antennas, e.g. the Pie Town, New Mexico antenna correlated with the VLA, will just be able to resolve some of these objects (see Mutel, this volume).

A question. The scatter in the visibilities of many of the objects that you have shown seems large. Can this mean the objects are asymmetric?

KWOK: Possibly.

BIEGING: Your model visibility curves for HM Sge were for a truncated wind model. The data appeared to me to be equally well fit by a simple inverse square law wind model extending to arbitrarily large radii. What is the reason for using a truncated wind?

KWOK: This was necessary to reconcile the light curve and the visibility data. In fact, there are still some discrepancies between the data and the model.

DISCUSSION RELATED TO PART II

LAMERS: The poster session of this afternoon contains papers on three different classes of objects: radio emission from winds of early-type stars; radio emission from pre-main sequence stars; and symbiotic stars.

The most interesting result for the early-type stars, in my opinion, is the possibility to resolve the winds spatially and then to derive the density distribution and the electron temperatures far out in the winds ($\geq$ 100 stellar radii). Several authors now seem to find that these temperatures are very much lower than the effective temperatures, which indicates efficient cooling. Some stars have variable radio fluxes, and possibly also variable temperatures, on time scales which are shorter than are expected from variations in the mass loss.

The observations of radio emission from pre-main sequence stars shows spatially resolved structures, indicating jets or flattened structure. Some of these structures may be due to the binary characteristics of the object, or to rotation, or to the structure of the surrounding material.

It is clear that the very high spatial resolution obtained with the VLA provides important clues on the physical processes occuring in the immediate vicinity of pre-main-sequence stars.

I would like to organize the discussion into three sections: 1) winds from early-type stars, 2) winds from pre-main-sequence stars, and 3) winds from symbiotic stars.

CHURCHWELL: It appears that the temperatures of stellar winds that have been resolved are substantially lower than the photospheric temperatures. Is the wind cooling consistent with simple expansion, or is some other mechanism, such as over-abundant coolants, or blocking of stellar ionizing radiation, required to explain the inferred wind temperatures?

WHITE: For P Cygni the temperature actually appears to increase as the wind flows away from the star. We found T ~ 7000 K at 2 cm (Aug.-Oct. 1983) and T ~ 18000 K at 6 cm (March 1981). Either the wind is heating up with radius or the wind has cooled off since 1981.

LAMERS: Again, as pointed out by Dave Abbott, the variability of P Cygni occurs on time scales that are short compared with the dynamical time scale. This may be due to variations in the degree of ionization on time scales of 100 days. One way to get this recombination variability is by having the star eject thick (in the UV) shells which temporarily block the ionizing flux in the outer portions of the radio emitting region.

UNDERHILL: How do you know the variable emission doesn't arise close to the photosphere of the star?

LAMERS: If the star's emission is resolved, you know where it comes from. It's far from the star.

UNDERHILL: A comment on the short time scale variations in P Cygni. Attributing all of the radio flux of P Cygni to free-free emission leads to an estimate of $(dM/dt)v_\infty/(L/c) >> 1$. This is unsatisfactory because you have no source for much of the momentum in the wind. It seems desirable to attribute part of the radio flux to processes which occur in a low-beta plasma close to the photosphere. It is easy to account for rapid changes in this part of the radio flux. The variability observed in the radio flux from P Cygni is one more observation pointing toward the need to include both free-free and gyro-radiation as a result of processes in a low-beta plasma in models of the mantles of hot stars.

THOMAS: A comment. There is lots of evidence for a drop in temperature and velocity in the winds of Be stars and the like and now we must come up with a mechanism. What's exciting is that the radio observations may not measure mass loss rates but rather the density distribution and, perhaps, provide another way of getting at the velocity distribution far from the star, which you don't get from UV observations. So the radio emission gives density information and has nothing to do with mass loss rates!

LAMERS: Is there a rebuttal to that?

ABBOTT: Dick, how do you get this evidence for slow-down?

THOMAS: Observationally. We see it in many Be stars.

Another question. Has anyone looked at Be or Bp stars with the VLA? If so, at what level were they detected, or what are the upper limits?

DRAKE: I, Abbott, and Linsky have observed six Bp stars (with kilogauss magnetic fields) with the VLA at 2 cm, following the serendipitous discovery of σ Ori E (a prototypical helium-strong star) as a 3.6 mJy source at 6cm by Abbott, Bieging and Churchwell. We have detected one additional star (HD 37017). In addition we determined that the spectral index between 2 and 6 cm is -0.15 for σ Ori E (cf. Drake et al. paper in this volume).

THOMAS: A review paper by Vera Doazan (B Stars With and Without Emission Lines, NASA SP-456) lists some Be stars as radio sources. Why aren't we discussing them?

ABBOTT: I challenge Dick Thomas's statement that Be or B main sequence stars are radio sources other than the poster paper by Drake et al. concerning the magnetic B stars. I know of no other B star detectionss.

HJELLMING: I agree. I know of no detections of B or Be stars other than those just mentioned by Steve Drake.

THOMAS: What about LSI +61°303?

HJELLMING: That's a binary X-ray source.

LAMERS: So we must conclude that there are no observations of radio emission from Be stars!

Editors' Note: The above discussion points out one of the basic difficulties in this field. There has been a general tendency for those not directly involved in radio observations to draw conclusions based on papers which would be regarded by those active in the field as either incorrect or outdated. The field of stellar radio emission is a fast moving one in which it has been necessary to abandon old interpretations when new data become available. The review mentioned by R. Thomas was discussing objects, which exhibit radio emission, like MWC349, V1016 Cyg, etc., which are complex binaries and not single Be stars.

HJELLMING: Let me mention another category of wind where the degree of ionization is a few percent or less, and highly variable throughout the wind. Recent VLA observations of α Ori (Betelgeuse) have resulted in a preliminary result whereby the radio disk has been resolved at 15 GHz, with an angular diameter of 0.08 arcseconds. If you combine this result with the measured radio spectrum, which is $0.24(\nu_{GHz})^{1.32}$ mJy (Newell and Hjellming, Ap.J.Lett., **265**, L85, 1982), one finds that the electron concentration is determined to be $3.7 \times 10^{7} (R/R_*)^{-3.6}$ cm^{-3} for radii from 1.6 R_* outward. This leads to mismatches of factors of 3 - 10 in the 1.6 to 2.5 stellar radii region when compared with the Alfven-wave driven models of Hartmann and Avrett (Ap.J., **284**, 238, 1984) for the Betelgeuse atmosphere, in the sense that more ionization (and hence energization of the wind) is necessary in these regions. The main point is that this is the first case where radio measurements (possible for α Ori, α Sco, and α Her) are beginning to give us direct measurements of the electron concentration distribution in the critical regions of their winds where mechanical energy in some form is driving the winds.

UNDERHILL: Don't you think the discrepancy in the Alfven-wave driven models compared to your observations should make one re-think the physics of the acceleration region?

HJELLMING: Maybe, but my current opinion is just the opposite. I'm surprised that in their first attempt, without the benefit of the most recent radio observations, they get as close as they do. I'm encouraged that they are on the right track.

UNDERHILL: The ultraviolet spectrum of α Ori shows only emission lines from the low-temperature ions, I believe. Therefore the temperature probably does not exceed 10000 K or thereabouts in the mantle of the star.

HJELLMING: The Hartmann and Avrett model, which matches the above-mentioned "model" very well from 2.5 stellar radii outwards, indicates the temperature is rising roughly linearly from 6000 K at 1.1 stellar radii to about 8000 K at 4 stellar radii, after which it declines precipitously.

MOLNAR: Might the excess radio emission at 2 cm and above be interpreted in terms of other radiation mechanisms, besides free-free emission, being involved?

HJELLMING: Quite possibly. I point out that the problem exists not only at 2 cm but also at 1.3 cm and 9 mm as well. We have considered only free-free emission so far in our simple minded model. You could use another emission mechanism to alter the spectrum, but my bias would be to produce such a spectral change by altering the structure of the source. That's one of the problems in this field, there are lots of ways to alter the spectrum; one needs to provide adequate justification (perhaps through observations in other bands) to back up his choice.

MUTEL: A question to Anne Underhill. What level of brightness temperatures do you expect to see in the regions where you propose that gyroresonance emission arises?

UNDERHILL: 10^7 K.

MUTEL: This should be testable by short-wavelength VLBI measurement. If the stars are detected, this would prove your assertion that the emission arises close to the surface; if not, the emission arises in the wind. If we work hard, I believe we can make detections down to the level of 5mJy.

UNDERHILL: I would like to see the mass loss rates from WR stars and P Cygni reduced by a factor of the order of 10. There would be no difficulty in interpreting the optical spectrum with a reduced dM/dt. For O stars a small reduction would be useful, but it is not demanded.

The observations of OB supergiants do not demand a change in interpretation of the radio flux, but one could be accomodated. It should be noted that adopting a normal cosmic chemical composition for WR stars results in a lower inferred dM/dt, by a factor of four, compared to what one gets assuming pure He, an assumption which is unproven.

CONTI: Instead of reducing the mass loss rates for WR stars by a factor of ten, which would make me unhappy, I would rather increase their effective temperatures by a factor of two - thus greatly increasing their luminosities.

VAN BUREN: What is the relationship between the presence of X-ray emission from a wind and the nature of the radio continuum emission seen from the wind?

WHITE: The x-ray and nonthermal radio emission can be related if the x-rays come from shocks - as in the model of Lucy (1982) and Lucy and White (1980). The same shocks (that produce the x-ray emission) can accelerate particles to high energies; these particles then coast out in the wind and emit synchrotron radio emission (see paper by White in this volume).

VAN BUREN: Is there an observational answer to my question?

WHITE: No.

LAMERS: If the x-rays are produced by instabilities in the wind, and the temperature of the hot regions become of order 10^8 K, this would hardly change the radio flux.

LINSKY: Don't forget that there is a transport problem if you want to produce the relativistic particles near the surface and then have them radiate in the outer parts of the wind. What about the magnetic fields and energy losses in your chaotic wind model?

WHITE: The magnetic fields that are required are only about 1-10 Gauss at the stellar surface, and lower out in the wind, so synchrotron losses for the particles are small. Adiabatic losses are important, but do not prevent the particle from retaining enough energy to produce the observed quantities of nonthermal emission.

ABBOTT: I would comment that the difficulty with the gyroresonance explanation in hot stars is multiplied by the fact that the radio photosphere is large in hot stars, even if you think a fraction of the radio flux is not free-free emission. Thus you need 300 Gauss fields, not at the surface, but at 100 stellar radii.

KUNDU: With regard to Underhill's comment on solar loops extending out beyond 20 stellar radii, I know of only one case of a solar loop extending out this far. Stone and his group observed a "U" burst at very low frequencies (100 kHz). This observation was associated with a type III burst producing (10-100 keV) electrons which were guided by magnetic field lines delineating a loop. The NRL P78-1 satellite coronagraph (white light) goes out only to 10 solar radii and the HAO coronograph on SMM only goes out to 8 solar radii.

GIBSON: Gyroresonance emission from a low-beta plasma, which Underhill has been arguing is important, cannot be a factor in the radio emission from early-type mass loss stars, since the spectral peak would be at low frequencies. If you have, as you suggest, a 500 G field to produce gyroemission, the local plasma $\beta \sim 6 \times 10^{-14} n_e$. Clearly the B-field would dominate the gas dynamics for any reasonable estimate of the local n_e. However, this does no appear to be the case.

MULLAN: Friend and MacGregor have studied the loss of angular momentum during the evolution of OB stars. They use observed mass loss rates, and assume mono-polar (open) magnetic fields of B_0 on the stellar surface. Magnetic braking is found to be so strong as to be inconsistent with observed rotation velocities if $B_0 \sim 100$ Gauss. However, this argument says nothing about the upper limits on magnetic field strengths in closed field regions on the surface of the OB stars.

KWOK: For the moment let us restrict the discussion to the two stars for which we have good visibility curves: P Cygni and γ^2Vel. There is no doubt that in these two stars the mass loss rate is very high and the optically thick sphere at radio frequencies is very large. If there are any nonthermal processes near the stars, their contributions cannot be seen (because of optical depth effects). I am more concerned about the low temperatures derived for γ^2Vel, 5000 K, and other winds observed by White and Becker. Either they are seriously wrong in fitting the visibility curves by isothermal wind models, or we do not know enough about heating and cooling mechanisms in winds.

MULLAN: What do we know about circular symmetry in γ^2Vel?

HOGG: The VLA data at 6cm are consistent with the star being circularly symmetric in the plane of the sky. However, the data are limited, since the array is foreshortened by a factor of six in the North-South direction, but within that limitation there is no evidence for a change of visibility with position angle.

MULLAN: Speckle data suggest departures from spherical symmetry in α Ori (Roddier and Roddier, Ap.J.Lett., **270**, L23, 1983). Do radio data suggest departures from spherical symmetry in other stars?

HJELLMING: The α Ori radio data I mentioned earlier indicates a size scale of 0.08" at 2 cm, but the source is too weak to indicate asymmetries such as those seen by Roddier and Roddier. However, since you ask about any star, let me show you the following map of V1016 Cyg obtained at 1.3 cm by R.T. Newell with the VLA (Ph.D. thesis, also published in Science, **216**, 1283, 1982). It is salutary to remember that, in terms of its spectrum, this object has always been interpreted as a spherically symmetric stellar wind object.

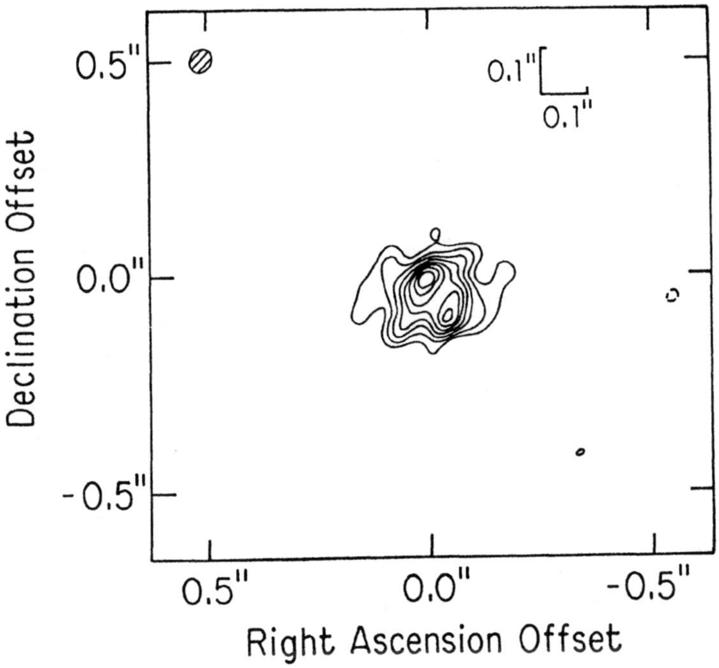

ABBOTT: In response to Kwok's question, I ask what is disturbing about a temperature of 6000 K in a hot star wind? It seems like a very probable outcome of a radiative equilibrium calculation at large distances from the star.

BOCHKAREV: UV data also show temperature decreasing with radius in stellar envelopes. Eclipsing binary systems are very good probes of the inner parts these envelopes. Comparison of IR, optical, near UV, and far UV (IUE) observations of several WR + OB binaries show that temperatures of WR envelopes are only 20000 to 25000 K, much less than the temperatures of WR cores. UV data do not show as low temperatures (6000-7000 K), as seen sometimes from radio data, but clearly indicate, for several WR stars (e.g. V444 Cyg), a decrease of temperature with radius for R = 5 to 50 solar radii. This corresponds to one or several effective radii of WR stars. Thus in several ways we can see negative gradients of temperatures in WR envelopes.

WHITE: It is okay to have a wind which is cool around an O star; however, it is hard to have a constant cool temperature, as D. Hogg's γ^2 Vel observations show because, by the time the gas can cool radiatively, it generally can also cool adiabatically. Thus the temperature should drop with increasing radius.

RODRIGUEZ: In the beginning the radio emission from OB stars was interpreted as thermal. Now we know that some are nonthermal. I believe that most cases of radio emission from pre-main-sequence stars are being interpreted as free-free emission. Is there evidence of nonthermal activity in them?

BIEGING: Many of the radio-detected T Tauri stars are probably not thermal radio sources, e.g. V 410 Tau. Our survey of Taurus-Auriga shows that 100% of the T Tauri stars associated with Herbig-Haro objects are detected at 6 cm, but only about 2% of T Tauris not associated with H-H objects are detectable at 6 cm. How many of these are thermal sources is at present unclear.

CAMPBELL: Just how steep was the spectrum? How did you determine at first glance that it was nonthermal?

BIEGING: In the low state, that is at about 1 mJy, the spectrum was essentially flat.

CAMPBELL: Can you guess at a crossing time for the ionized region?

BIEGING: It is unresolved, and a K star, so we cannot estimate a size.

COHEN: Responding to Luis Rodriguez's question about nonthermal emission in T Tauri stars, our survey of the ρ Oph cloud showed one star, DoAr21, which increased by a factor of ten or more in a day, with a rise from 35 to 48 mJy in 2.5 hours. Though we are trying an eruptive wind model, a nonthermal flare interpretation is possible, and probably unavoidable.

MULLAN: Do T Tauri stars look like RS CVn stars in the radio? Does the position of the star in the HR diagram control the radio emission in a unique way?

COHEN: Yes, the event in DoAr21 looks just like several RS CVn events. A crucial issue is whether V410 Tau and DoAr21 are rare objects - perhaps due to rapid rotation (like RS CVn's) - or whether most pre-main sequence stars are flaring with a low duty cycle. Two pieces of evidence point tentatively toward the latter possibility: 1) DoAr21 was seen to flare only on one out of seven days it was observed; and 2) most pre-main sequence stars in the ρ Oph cloud show continuous flaring at x-ray wavelengths.

GIBSON: A star in a particular part of the HR diagram need not be detectable, but I am very tempted to claim that detected emitters will occur in only special parts of the HR diagram (see paper by Gibson, this volume). On another point just discussed, to detect more flares on T Tauri stars one should go to 2 cm and beat the "cocoon" problem you probably have at 21 cm.

MULLAN: Do T Tauri stars show pronounced departures from spherical symmetry in their mass loss?

RODRIGUEZ: We have made calculations for anisotropic winds, and surprisingly, the mass loss rates that you derive are not very different from the isotropic case.

BIEGING: The optical and radio images of DG Tau illustrate the non-spherical mass loss of some T Tau stars. From optical polarimetry one finds that there must be a circumstellar dust disk which is orthogonal to the flow directions - as deduced from the position of the associated H-H object. In Feb. 1982 DG Tau looked biconical in 5 GHz maps, with elongation along the same flow direction. By Nov. 1983 it was unresolved at 5 GHz. If one interprets this as evidence of an episode of mass loss, then it requires flow velocities of the order of 130 km/sec to evacuate the biconical nebula. This is remarkably close to the radial velocity of the H-H object.

CAMPBELL: I have observed a similar object, an O6 or O7 star (NGC 7538 IRS1) where the double-lobed structure is also seen. The two lobes are exactly bisected by a thin IR disk which is unresolved in thickness. Both 5 and 15 GHz emission are unresolved at VLA A-array resolution, indicating sizes less than 100 AU. We interpret the two lobes as two radio-thick lobes of an outflowing wind, and not as a binary system.

KWOK: How does this object compare with MWC349?

CAMPBELL: The composite spectral slope is very steep, with spectral index of the order of 1.6, over roughly 1.5 to 24 GHz. I have not been able to de-convolve a 2/3 spectral index component from either an optically thick or an optically thin free-free component. We are tentatively assuming a steeper density gradient, like $1/r^3$, using the Wright and Barlow formula.

GIBSON: Is DG Tau's shape caused by a disk-type absorption of a spherical object, or is it really bipolar?

COHEN: It really has to be bipolar to be consistent with the observed outflow.

BIEGING: MWC349 is a binary star system whose radio images have "horns" at about 90°, and there is a radio "hot spot" on the line between the two stars in the system. The radio morphology strongly suggests an interaction of winds from the two binary components.

KWOK: MWC349 has a very well observed wind-like spectrum, and its visibility curve can be fitted by an isothermal wind, although high resolution observations show that is is bipolar on a small scale. Nevertheless, from the global structure a very high (dM/dt)/v is derived. If v is anywhere close to the escape velocity of an early-type star, dM/dt is unacceptably high. We have no clue whatever about the nature of this star, although it is well observed in the radio.

HJELLMING: One of the lessons that we have learned from the combination of radio and optical data about nova shell ejecta is that many nova shells have separate "ring" and "polar cap" structures (determined from optical velocity data) which behave in the aggregate as density distributions that are power laws. This is clearly because expanding segments of shells still obey geometric mass conservation principles, so even the most complex structures can counterfeit density power laws as seen in radio spectra. Structures in winds or wind-related ejecta could show similar effects that we would not know about because of lack of data.

PART III

ACTIVE BINARIES AND FLARE STARS

NON-THERMAL RADIO EMISSION FROM FLARE STARS AND RS CVn SYSTEMS

D. J. Mullan
Bartol Research Foundation
University of Delaware
Newark, DE 19716

ABSTRACT. We review the observations of continuum radio emission from flares on red dwarf stars and from outbursts on RS CVn Systems. In the RS CVn systems, the emission appears to be mainly incoherent, whereas in the flare stars, particularly at the lower frequencies, a coherent mechanism must be at work. In the RS CVn systems, it appears that during a radio outburst, a large fraction of the coronal electrons become highly energetic. We interpret this in the context of deposition of mechanical energy in the corona, followed by a turbulent cascade, and find that the electric fields induced in the corona of an RS CVn star may be large enough to give rise to electron runaway in certain large magnetic loops.

1. INTRODUCTION

Bursts of continuum radio emission have been observed on many occasions from two classes of stars. One class is the flare stars, main sequence dwarf stars of spectral type M or late K. These stars have effective temperatures of typically 4000 K and gravities of typically 3-7 times the solar value. The processes which occur during a flare on such a star are generally believed to be analogous to those which occur during a solar flare, except that the total energies involved are larger, and the rates of energy release are greater (cf. Mullan, 1984a).

The second class of stars which shows outbursts of radio emission are the RS CVn systems. These are binaries with periods of 1-20 days, in which chromospheric activity is strong (as evidenced by strong emission in the calcium K line), and in which one the components is a subgiant of spectral class K (or late G). It is usually the K subgiant which is the more active member of the system, and it is to that star that the outbursts of energy are generally attributed. The effective temperatures of the active components are typically 5000 K, while the gravities are less than solar by factors of 5-10. Because the effective temperature of the active star is not much less than solar, and because the surfaces of many RS CVn stars show cool starspots, there is a widespread belief that the radio outbursts on RS

R. M. Hjellming and D. M. Gibson (eds.), Radio Stars, 173–184.

CVn stars are also analogous to radio outbursts which occur during solar flares. From this point of view, therefore, the radio outbursts on both flare stars and RS CVn stars essentially involve identical physics, and both groups of stars can be discussed within the context of the models which have been developed for explaining energy release in solar flares.

In this paper, we examine the data for both classes of stars, and we argue that a rather different point of view appears to be consistent with the data, namely, that whereas radio bursts in dMe stars may indeed be analogs of solar bursts, the bursts in RS CVn systems need not be due to flare activity in the conventional sense, but instead to a form of coronal heating. Our approach here will be to consider a flare process as one which involves impulsive release of energy, while coronal heating involves the release of energy in a gradual way over a prolonged period of time.

2. RS CVn RADIO BURSTS

In Table I we list the largest outbursts which have been reported in several RS CVn systems. We list the quiescent luminosity of the star (in units of 10^{15} ergs/sec/Hz), and the maximum luminosity observed during an outburst at wavelength λ (cm). The brightness temperature T_B has been derived by assuming that the scale size of the source is comparable to the size of the star, i.e. about 3 times the solar radius. Such scale sizes have been confirmed recently by VLBI

TABLE I. RADIO OUTBURSTS IN RS CVn STARS

Star	Qu.Lu.	Flare Lu.	λ	T_B	kT_B/m_ec^2
AR Lac	< 10	1650	3	3×10^{10}	5
UX Ari	~ 20	920	3	2×10^{10}	3
RT Lac	≤100	4000	6	30×10^{10}	50
HR1099	~10–20	1600	3	3×10^{10}	5
SZ Psc	< 50	1300	3	3×10^{10}	3
HD224085	< 10	260	3	0.6×10^{10}	1.2

Notes:

Luminosities (both Quiescent and Flare) are given in units of Lu = 10^{15} ergs/sec/Hz

λ in units of cm

T_B in units of °K (Source radius assumed to be $3R_{sun}$)

observations of several systems (Lestrade et al, 1984). It is remarkable that apart from the case of RT Lac, the values of T_B are in all cases of order a few times 10^{10} K. Mutel (1984) has stated that in his extensive observing of RS CVn systems, the upper limit on T_B is indeed observed to be a few times 10^{10} K. Now, the maximum possible brightness temperature from electrons with Lorentz factor γ is given by $kT_B = \gamma m_e c^2 f(s)$, where s is the energy spectral index of the fast electrons. With a typical value of s=2, $f(s) \approx 0.28$. Hence, the observed values of T_B in RS CVn systems correspond to electrons with γ=3-15 (except for RT Lac).

Circular polarization is observed in the emission, with typically 10% amplitude. If the emission is incoherent gyrosynchrotron (Owen et al, 1976), a degree of polarization of π_C corresponds to an electron Lorentz factor of order $1/\pi_C$, i.e. about 10. This is consistent with the values deduced from the observed intensities. Moreover, the linear polarization has never been observed to be greater than the observing noise (of order a few %) in RS CVn systems, and this is consistent with a large rotation measure due to the fast electrons themselves (Owen et al, 1976). Thus, it appears that interpreting the radio emission from RS CVn outbursts in terms of incoherent gyrosynchrotron from electrons with energies of a few MeV is quite self-consistent.

This then leads to an estimate of the magnetic field strength in the radio emitting region. Owen et al (1976) find B≈30 gauss; Spangler (1977) finds 30-125 gauss; and Doiron and Mutel (1983) find 5-80 gauss, in various systems at different observing epochs. Thus, it seems that the fields in the radio emitting regions are typically a few tens of Gauss.

Now that the fields are known, the source luminosities can be converted into a total number of energetic electrons, using the known emissivity of the gyrosynchrotron process (Dulk and Marsh, 1982): in the case of UX Ari and HR 1099 at 1.65 GHZ, Mutel et al (1984) have derived the following range of values for the number density of energetic electrons (i.e. the numbers of electrons in excess of a lower cut-off in the power law spectrum; in this case, the lower cutoff is assumed to be ε_0=10 KeV):

$$2 \times 10^6 < N_\varepsilon < 4 \times 10^8 \text{ cm}^{-3}.$$

At higher frequencies (8.4 GHz), Lestrade, et al (1984) have observed HR1099 and found the following results:

$$1 \times 10^7 < N_\varepsilon < 4 \times 10^8 \text{ cm}^{-3}.$$

The most interesting aspect of these results emerges when we compare them with thermal electron densities N_{th} in the coronae of these stars. Spangler (1977) has discussed the constraints which can be imposed on the thermal electrons, and finds that the most likely range of densities is $N_{th} \approx 10^8 - 10^9$ cm^{-3}. Gibson et al

(1978) also estimate, on the basis of free-free absorption, that $N_{th} \simeq 10^9$ cm^{-3}.

It is interesting that the range of N_ε overlaps with the range of N_{th}. Even in the most pessimistic case, N_ε is almost 1% of N_{th} and data are consistent with $N_\varepsilon \approx N_{th}$, i.e. essentially 100% of the thermal electrons may be involved in the fast electron population during the outbursts. This appears to be an important property of the outbursts of RS CVn stars to which sufficient attention has not previously been paid. Note that if the fast electrons have a spectral index s=3, the mean energy of the fast electrons, $\bar{\varepsilon}$, is $2\varepsilon_0 \approx 20$ keV. Thus, even with the maximum values of N_ε, the total energy density in fast electrons ($N_\varepsilon \bar{\varepsilon}$) ($\approx 13$ ergs cm^{-3}) is less than the magnetic energy density (if B>18 gauss). Hence, the field in the radio emitting region dominates the energetic particles, and the particles do not disrupt the field.

We also note that the outbursts in RS CVn systems have a time evolution which is remarkably smooth. The data of Feldman et al (1978) for example, show a series of very large outbursts in HR 1099, consisting of essentially monotonic increases of intensity on time scales of hours, followed by essentially monotonic decreases in intensity, also on time scales of hours. The total duration of an outburst is almost 24 hours. Data taken with high time resolution (down to one minute or so), occasionally shows structure with periodicity of 4-5 minutes (Brown and Crane, 1978), but in other events, no periodicity or structure on such time scales is observed (Lestrade et al, 1984).

Various investigators have attempted to correlate the intensity of RS CVn emission with orbital phase, but have found no such evidence. This has led them to the conclusion that the energetic electrons which produce the emission are not accelerated primarily at favored places in the binary orbit. Instead, the electrons are accelerated in one of the stars (cf. Feldman et al., 1978). There may occasionally be events which are caused by a process which is essentially related to the presence of a binary component, e.g. reconnection of magnetic loops between the two stars (Simon et al, 1980), but such events may be rare. The fact that radio outbursts in Algol (which is generally not classified as an RS CVn system because it contains a hot B star) are identical in many respects to those in an RS CVn system suggests (Gibson, 1985) that the unseen companion of Algol is an RS CVn type star, i.e. a K subgiant. Thus, we should perhaps think in terms of an RS CVn star, rather than an RS CVn system. The presence of a binary companion would then serve merely to keep the RS CVn star rotating rapidly enough (due to tidal forces) that the magnetic flux in the star is maintained at a high (and observationally interesting) level.

We will therefore adopt the view that the source of energy which accelerates the fast electron population in an RS CVn outburst resides in the K subgiant.

In this regard, it is important to note that Owen and Gibson (1978) derived luminosity distributions for moderate (max. lum. = 250 Lu) outbursts on 5 RS CVn stars and found that the low level activity in these stars (when they are classified as "quiescent") can be interpreted as an extension of the "outburst distribution" to low levels. Their conclusion was that even in the "quiescent state", the RS CVn stars are actually experiencing a series of small, superposed "outbursts". Here we point out that the data are equally consistent with the opposite suggestion, i.e. the moderate outbursts are a large-amplitude extension of the processes which are occurring essentially at all times in the coronae of the RS CVn stars. Thus, rather than interpreting the "quiescent" emission as a low level form of "flaring", we might equally well interpret the moderate outbursts as a high level form of "coronal heating". We will return below to reasons for this interpretation. (Note however that some very large outbursts, up to 1400 Lu, may not follow the Owen-Gibson distribution: Feldman, 1983).

3. RADIO FLARES ON dMe STARS

A survey of the literature on flare stars yields the maximum intensities at various frequencies as shown in Table II. There is a striking difference from the case of the RS CVn Stars. In the case of

TABLE II. PEAK INTENSITIES IN dMe FLARES

Freq. (MHz)	Star	Flux(Jy)	Log $T_{B(R*)}$
5000	ATMic	0.02	9.5
1420	UVCet	0.04	10.9
430	YZCMi	0.5	13.0
408	YZCMi	5.7	14.1
318	YZCMi	1.4	13.7
240	YZCMi	36	15.4
196	ADLeo	12.5	15.1
150	UVCet	15	15.7
136	Ori F.S.	220	(20)?
80	UVCet	37	16.4
19.7	V371Ori	23000	>20

Notes: Ori F.S. denotes an unidentified flare star in the Orion aggregate.
T_B in units of °K (Source radius = R*)

the dMe stars, the brightness temperatures are in almost all cases so large ($T_B > 10^{12}$K) that the emission mechanism must be coherent. And even in the cases of somewhat lower T_B (at the highest frequencies), we should point out that our derivation of T_B values assumes that the source dimension is equal to one stellar radius, and that the star is at a distance of 5 parsec. There are certainly cases where the source size is smaller than a stellar radius. For example, Lang et al (1983) observed emission from a dMe flare star which varies on time scales of less than 0.2 sec, implying a maximum source size of less than 0.3 stellar radii. Hence, even the highest frequency emission may require a coherent mechanism.

Suppose, however, for the sake of argument that the high frequency emission is incoherent. The emission is observed to be circularly polarized and so, we may interpret it as gyrosynchrotron in nature. The lack of absorption at the observing frequency (5 GHz) suggests that the magnetic field is not more than 500 gauss. In that case, $N_\varepsilon \approx 10^8$ cm^{-3}. However, in this case, stellar coronal flare plasma is quite dense: $N_{th} \approx 10^{12-13}$ cm^{-3} (Haisch, 1983). Hence, in this case, the energetic electron population represents only one part in 10^{4-5} of the thermal electron population. And if the emission is coherent, the fraction of fast electrons may be even smaller. The conclusion is that in dMe flares, only a small number of electrons are accelerated to the high energies needed to create the radio emission. This is a significant difference from the case of RS CVn outbursts.

Another remarkable difference in the dMe flares is that the radio emission is observed to have a high degree of linear polarization. At a frequency of 430 MHz, the linear polarization in dMe flares can be as high as 60–70% (Spangler, 1975). Now, synchrotron emission becomes linearly polarized in the ultrarelativistic limit, and this suggests that the electrons in dMe radio flares may be much more energetic than in RS CVn flares (where linear polarization has never been detected). Spangler (1975) proposed that a synchrotron maser might be at work in creating the high brightness temperatures which he reported. His calculations required electron spectra peaked around 10–20 MeV i.e. about one order of magnitude more energetic than the electrons required in RS CVn outbursts.

A striking feature of radio flares in dMe stars is their highly impulsive nature. Bursts as short as the observing resolution (0.2 sec) have been reported, implying either that the energetic electrons are being accelerated in short bursts, or else that the conditions for the maser to work are only satisfied very sporadically. (Rather severe constraints must be satisfied for a synchrotron maser to operate; Melrose, 1980). Even apart from the impulsive nature of the events, their overall lifetime is quite short: durations of order several seconds or several tens of seconds are typical at frequencies of 196–430 MHz (Spangler, 1975). This behavior seems quite different from the very gradual (and almost monotonic) behavior in RS CVn stars, where almost an entire day is required for an event to run its course.

As regards flare frequency, there is insufficient data for dMe radio flares. However, on the basis of optical data, Lacy et al (1976) have shown that the frequency of flares of various amplitudes is such that the overall energy release rate from dMe stars in the form of flares is dominated by the largest flare events. (There is no star in their sample which is dominated by small flares, at the one sigma level of confidence.) Thus, it appears that the dMe stars have two modes of energy release, one involving energetic flaring (and being highly impulsive), the other involving low level activity ("coronal heating"), but neither can be considered to be an extension of the other. In this sense, then, we have here again a distinction between the flares in dMe stars and the outbursts in RS CVn stars.

It seems that flares in dMe stars may be a manifestation of a source of energy in the stellar atmosphere which is truly distinct from the source of energy for "coronal heating". The source of flare energy presumably involves the highly irregular, violent, and impulsive release of magnetic energy which is so characteristic of large-scale reconnection.

4. DEPOSITION OF MECHANICAL ENERGY IN THE CORONA

4.1. RMS Velocity Amplitude in a Turbulent Cascade

In an attempt to understand the differences between outbursts in RS CVn stars and in dMe stars, we will discuss in this section the amplitude which one expects to find for the non-thermal velocity field in the corona. We will use the concept of turbulent cascade proposed by Hollweg (1983). We recognize that the question of coronal heating in any star is a very complex one, and that a detailed description of the heating process may eventually require a formalism much more sophisticated than that developed by Hollweg. Nevertheless, in the present context (which is of an exploratory nature), Hollweg's approach appears to yield some interesting results.

Hollweg shows that if a turbulent cascade develops in the corona with a Kolmogoroff spectrum of turbulence, the volumetric heating rate E_H corresponds to an RMS turbulent speed V_{rms} as follows:

$$E_H = \frac{2^{3/2}\, k_o\, \rho\, V^3_{rms}}{(1.5\, C_o)^{3/2}}$$

Here, ρ is the density of the gas on the loop, C_o is a universal constant of the turbulence (C_o = 1.5), and k_o is the wavenumber which corresponds to the maximum wavelength of the turbulence, λ_{max}, $k_o = 2\pi/\lambda_{max}$. Presumably, λ_{max} is related to one of the dimensions of the magnetic loop on which the energy is being deposited. However, the actual value of λ_{max} is uncertain. It is convenient in what follows to discuss the mechanical energy deposition of a flux entering the corona from below, F_H. In that case, a dissipation length scale L_d also enters into the discussion: $E_H = F_H/L_d$, where L_d is also presumably related to one of the dimensions of the loop. The ratio $R = \lambda_{max}/L_d$ is probably of order 0.1-1. Then the RMS turbulent amplitude becomes

$$V_{rms} = 0.5\ (F_H/\rho)^{1/3} R^{1/3}$$

Because of the weak dependence on R, we will neglect the R term in the expression for V_{rms}. Even if R is as small as 0.1, the value of V_{rms} would be reduced by a factor of only about 2, and this will not affect the conclusions we will draw below.

The question now is: What value do we use the mechanical energy flux entering into the coronal loop from below?

4.2. Mechanical Energy Flux from Convection Zone

The maximum mechanical energy flux F_{mech} which is associated with the convection zone in cool stars is of order 1% of the bolometric flux, $F_{bol} = \sigma T^4$, (Mullan, 1984b). The fraction of F_{mech} which couples into a coronal loop is given by an electrodynamic coupling efficiency ε: $F_H = \varepsilon F_{mech}$. According to Ionson (1984), the value of ε under the conditions which apply to the atmospheres of the stars in which we are interested here can be related to two fundamental time scales as follows:

$$\varepsilon = \frac{\tau_c}{\tau_A} \frac{1}{\left[1 + \left(\frac{\tau_c}{\tau_A} - \frac{\tau_A}{\tau_c}\right)^2\right]}$$

The time scale τ_c is the correlation time of the convection. Based on information available in the solar convection zone, it can be argued (Mullan, 1984b) that τ_c is essentially the turnover time of a convection cell, i.e. proportional to H/V, where H is the cell depth and V is the convection velocity. It is generally believed that H is of order the pressure scale height, i.e. $H \propto T/g$ where T is the stellar temperature and g is the acceleration due to gravity. In dMe stars, τ_c has a value of a few tens of seconds (Mullan, 1984b). In RS CVn stars, the convection velocities are not well known. However, results of Bell and Gustaffson (1975) suggest that the values of V are larger than the solar values by perhaps a factor of 2. Hence, in RS CVn stars, τ_c is expected to be larger than the solar values by factors of about 3-5. Using the solar value of 500 seconds (Mullan, 1984B), this leads us to expect $\tau_c \approx 1500$-2500 seconds in the RS CVn stars.

The time scale τ_A is the bounce time of an Alfven wave along a coronal loop: $\tau_A = 2L/V_A$, where L is the loop length and V_A is the Alfven wave speed. In dMe stars, it can be argued that τ_A is on the order of a few tens of seconds (Mullan, 1984b). Thus, it appears that, at least among the early M dwarfs, $\tau_c \approx \tau_A$, ε is of order unity, and $F_H \approx 0.01\ \sigma T^4 \approx 10^8$ ergs/cm^2/sec.

In the RS CVn stars, the observational data yield information on the source scale size (presumably related to L, the loop length), and on the magnetic field strength plus electron density. The latter than yield V_A, and we can deduce a value of τ_A in these stars. With L = 4 x 10^{11} cm, B = 50-100 gauss, and thermal electron densities of 10^8 cm^{-3}, we find $\tau_A \approx$6-12 minutes. (We note that Brown and Crane (1978) reported periodicities of 4-7 minutes in a burst in HR 1099; thus, our estimates of the timescales are reasonable if the oscillations observed by Brown and Crane are associated with Alfven waves bouncing on a loop, and creating a radio burst each time they encounter either end of the loop, i.e. at time intervals of 0.5 τ_A). In the RS CVn stars, then, we find τ_C/τ_A may be as large as 10. Hence, the value of ϵ may be as small as 0.1. As a result, the fraction of the bolometric luminosity which enters the corona is ~0.1%. This is consistent with the maximum observed X-ray luminosities of RS CVn systems: the maximum X-ray flux is between 0.1 and 0.3% of the bolometric flux (Dupree, 1983). Thus, we expect F_H in RS CVn stars to be of order 10^8 ergs/cm^2/sec.

4.3. Coronal Velocities and Induced Electric Fields

We can now estimate the non-thermal velocities in the coronae of the two classes of stars.

In the RS CVn stars, where the coronal densities are of order 10^8 cm^{-3}, we find V_{rms} of order 400 km/sec. In the presence of magnetic fields of order 50-100 gauss, the induced electric fields ($E_i \approx V_{rms}$ B/c) associated with the above turbulent velocities are of order 0.1-0.2 statvolts/cm. In the turbulent state of the corona of an RS CVn star, different cells of the turbulence will in all likelihood come into contact with one another with magnetic field vectors which are sheared with respect to one another. Hence, reconnection at the boundaries or turbulent cells seems to be likely. In those particular cells where the neighboring fields find themselves directed at essentially 180 degrees with respect to each other, the formation of a neutral sheet at the cell boundary is expected. In that case, electric fields of the above order are expected to become available to accelerate individual particles as they execute non-adiabatic motion near the neutral sheet. The question is: are electric fields of the above order sufficiently large as to give rise to interesting effects in the corona? We will address that question in the next section.

In the dMe stars, the coronal densities are high: Haisch (1983) has derived densities of 10^{12-13} cm^{-3} in flare plasma. The plasma in coronal loops may not be quite as dense as that, but a coronal density of 10^{11} cm^{-3} on a loop in a dMe star is not unreasonable: this represents an enhancement of ~10^2 above densities in active solar loops, and density enhancements of 10-100 in dMe star chromospheres/coronae appear to be justified (Mullan, 1977). This leads to V_{rms} of order 40 km/sec. Hence, if the coronal field is B gauss, the induced electric fields will be of order 10^{-4} B statvolts/cm.

5. ELECTRON RUNAWAY

If an electric field imparts a larger increment of velocity to a charged particle between collisions than the particle can lose in the next collision, the particle velocity will accelerate indefinitely. In an electron gas where the velocity distribution has a finite spread, there are always some electrons which experience this effect. However, the effect becomes of interest only if the electric field is large enough to induce the bulk of the distribution to accelerate. This condition is referred to as electron runaway (cf. e.g. Delcroix, 1960).

The critical electric field for runaway can be derived from the requirement

$$\frac{e}{m} E_{crit} \tau_{mfp} \approx V_{th}$$

where e and m are the charge and mass of the electron, τ_{mfp} is the mean free time between collisions, and V_{th} is the thermal velocity. Usually, τ_{mfp} is evaluated in terms of binary collisions between isolated charged particles. In such a case, the critical electric field is referred to as the Dreicer field. In the present case, however, where the coronal densities are quite low, it is more likely that the mean free time between collisions is determined not by particle collisions, but by interactions between particles and turbulence. In the latter case, the collision frequency is expected to become comparable to the plasma frequency, and we therefore find $\tau_{mfp} \approx 10^{-4}/N^{0.5}$ seconds, where N is the electron density. Then the critical electric field for runaway becomes

$$E_{crit} \approx 10^{-8} (NT)^{1/2} \text{ stat volts/cm},$$

where T is the electron temperature (K).

In the RS CVn stars, $T \approx 10^{6.6-7.6}$ (Swank and White, 1980), and hence, using $N=10^8$ cm^{-3}, we find $E_{crit} \approx 0.2$–0.6 statvolts/cm. In the dMe stars, T is of order $10^{6.7}$ (Golub, 1983), while $N \approx 10^{11}$ cm^{-3}. Therefore, in the dMe stars, we find $E_{crit} \approx 10$ statvolts/cm. Referring to the dMe stars first, we see that unless the coronal magnetic fields are as large as 100 kilogauss or more, the induced electric fields in the coronae will fall short of the runaway electric field. In fact, since in the coronae of flare stars, fields of typically a few hundred gauss have been reported (Gary and Linsky, 1981), it seems that the induced electric fields will fall short of critical by factors of almost one thousand. This leads us to conclude that electron runaway is not a likely phenomenon on a large scale in dMe star coronae. The coronal electric fields induced by turbulence appear to be such that electric field acceleration will be rapidly redistributed as heat. Thus, coronal heating, although efficient in these stars, does not lead to "spectacular" effects. In order to produce the violent and impulsive behavior observed in flares in dMe stars, something else, over and above coronal heating, must take place.

On the other hand, in the RS CVn stars, we consider it interesting that our estimates of the induced electric fields E_i are remarkably close to the estimated range of E_{crit}. Now, we have pointed out above that during a large outburst in an RS CVn star, a large fraction, perhaps 100%, of the thermal electrons become accelerated to high energies (mean energies of 20 keV, i.e. an order of magnitude larger than thermal). This is a feature of electron runaway, and the fact that our estimates of the induced turbulent electric fields in the corona are also close to runaway conditions suggests that this interpretation of the data is not inconsistent.

6. RADIO OUTBURSTS IN RS CVn STARS

We suggest that a radio outburst in an RS CVn star involves merely an episode in coronal heating as a magnetic loop emerges from the stellar interior into the atmosphere. When the loop first emerges, its Alfven wave crossing time scale, τ_A, is very short compared with the convection time scale, τ_c. Therefore, the electrodynamic coupling efficiency ε is small, and the loop cannot "tap" significant mechanical energy from the underlying convection zone. However, as the loop expands, and the Alfven wave speed decreases (due to weakening field stength), the time scale τ_A becomes longer, and so the "tapping" becomes more efficient. Moreover, the loop is extending to greater heights, where the densities are becoming smaller, with the result that the mechanical energy flux F_H which does reach the loop is being deposited in more and more rarefied material. This leads to larger turbulence velocities, and larger induced electric fields, E_i. The lower density also reduces the critical electric field, E_{crit}, with the result that conditions approach closer and closer to satisfying the constraints of electron runaway (i.e. E_i can exceed E_{crit}). We suggest that when runaway becomes possible, the whole loop "turns on" in the sense of becoming visible in radio emission: the electrons are accelerated to energies of many times the mean thermal energy, and conditions for gyrosynchtrotron emission become optimal.

In this view, then, an outburst in an RS CVn star has a timescale which is controlled by the evolutionary timescale of the entire loop. We suggest that this is a natural explanation fo the gradual (almost monotonic) rise and fall in the profiles of many outbursts in RS CVn stars. The outburst "turns off" when the loop extends so far that either the loop severs its connection with the convection zone, or else the coupling efficiency again falls off and the flux of mechanical energy reaching the corona once again falls to low levels.

Also in this view, it is natural to consider the outbursts as an optimal manifestation of the low level "coronal heating" which is always going on in the atmosphere. Thus, it is not surprising that Owen and Gibson (1978) found the "quiescent" radio emission to be an extension of the moderate "outbursts" luminosity distribution to low levels. An upper limit on the outburst luminosity would occur in events where all of the electrons on the loop had "runaway". The existence of a limiting luminosity in RS CVn outbursts has in fact been reported by Feldman (1983).

REFERENCES

Bell, R. A. and Gustaffson, B. 1975, Astron. Astrophys. 42, 407.
Brown, R. L. and Crane, P. C. 1978, Astron. J. 83, 1504.
Delcroix, J. L. 1960, Introduction to the Theory of Ionized Gases (Interscience: New York), p. 129.
Doiron, D. J. and Mutel, R. 1984, Astron. J. 89, 430.
Dulk, G. and Marsh, K. 1982, Ap. J. 259, 350.
Dupree, A. K. 1983, in: Activity in Red Dwarf Stars, ed. P. Byrne and M. Rodono (Dordrecht, Reidel), p. 447.
Feldman, P. A. 1983, in Activity in Red Dwarf Stars, ed. P. Byrne and M. Rodono, p. 434.
Feldman, P. A. et al, 1978, Astron. J. 83, 1471.
Gary, D. L. and Linsky, J. L. 1981, Ap. J. 250, 284.
Gibson, D. M. 1985, this conference.
Gibson, D. M., Hicks, P. D., and Owen, F. N. 1978, Astron. J. 83, 1495.
Golub, L. 1983, in: Activity in Red Dwarf Stars, ed. P. Byrne and M. Rodono (Dordrecht, Reidel), p. 83
Haisch, B. M. 1983, in: Activity in Red Dwarf Stars, ed. P. Byrne and M. Rodono (Dordrecht, Reidel), p. 255
Hollweg, J. V. 1983, in Solar Wind Five, ed. M. Neugebauer (NASA CP-2280),p. 5.
Ionson, J. A. 1984, Ap. J. 276, 357.
Lacy, C. et al. 1976, Ap. J. Suppl. 30, 85.
Lang, K. et al. 1983, Ap. J. Letters 272, L15.
Lestrade, F. et al. 1984, preprint.
Melrose, D. B. 1980, Plasma Astrophysics, Vol. I (Gordon and Breach), pp. 215-217.
Mullan, D. J. 1977, Solar Phys. 54, 183.
Mullan, D. J. 1985, in Proc. IAU Symp. No. 107, ed. M. Kundu and G. Holman (Dordrecht, Reidel).
Mullan, D. J. 1984b, Ap. J. 282, 603.
Mutel, R. L. 1985, this conference.
Mutel, R. et al, 1984, Ap. J. 278, 220..
Owen F. N. and Gibson, D. M. 1978, Astron. J. 83, 1488.
Owen, F., Jones, T. and Gibson, D. M. 1976, Ap. J. Letters 210, L27.
Simon, T. et al. 1980, Ap. J. 239, 911.
Spangler, S. R. 1975, PhD. dissertation, Univ. of Iowa.
Spangler, S. R. 1977, Astron. J. 82, 169.
Swank, J. H. and White, N. E. 1980, Smithsonian Ap. Obs. Spec. Rep. 389 (ed. A. K. Dupree), p. 47.

QUIESCENT STELLAR MICROWAVE EMISSION

D. E. Gary
Solar Astronomy, 264-33
Caltech
Pasadena, CA 91125
USA

ABSTRACT. The quiescent microwave flux we expect from hot stellar coronae, as predicted from analogy with the Sun and from X-ray derived temperatures and densities, is about 10 times smaller than the observed fluxes for many dMe stars observed with the VLA. The implications of these high observed fluxes to the coronal parameters of magnetic field strength, temperature, and source size are discussed, and observations are suggested that potentially allow a choice among the several possibilities.

1. INTRODUCTION

For the purposes of this paper, the term "quiescent" microwave emission will be used to define the emission, observed to be associated with some late-type dwarf stars, that is nearly always present and shows only slow variations with time. Thermal stellar wind and accretion disk sources are excluded, and are discussed elsewhere in these proceedings.

The observation of a quiescent component of stellar microwave emission was impossible before the completion of the VLA due to the low fluxes involved. After an unsuccessful attempt by Bowers and Kundu (1980) to detect quiescent emission from a number of nearby stars with the partially completed VLA, Gary and Linsky (1981) first reported the detection of steady 6 cm emission from χ^1 Ori (G0 V) and UV Cet (M5.6e V). Although it now appears possible that the emission from χ^1 Ori is due to a previously unknown red dwarf companion (Linsky and Gary 1983), the quiescent emission from UV Cet and other dMe stars has been well established (Topka and Marsh 1982; Linsky and Gary 1983; Fisher and Gibson 1982; Kundu and Shevgaonkar 1984). Study of this emission can tell us something about the quiescent (or quasi-steady) conditions in the coronae of these stars when the precise mechanism for the emission is known. Possible emission mechanisms are Bremsstrahlung (free-free emission), plasma emission (due to nonthermal electrons), gyroresonance emission (due to thermal electrons--$T_b < 10^8$ K), or gyrosynchrotron emission (due to high temperature or nonthermal electrons).

R. M. Hjellming and D. M. Gibson (eds.), Radio Stars, 185–196.

We discuss in Section 2 what radio emission we expect from stellar coronae in analogy with the Sun, using coronal temperatures and densities derived from X-ray observations and magnetic field strengths estimated from photospheric observations. In Section 3 we summarize results from radio observations with the VLA and compare what we observe with what we expect. The chief result is that the observed fluxes for dMe stars are about ten times greater than we would predict. In Section 4, the emission mechanisms are briefly reexamined in an effort to understand the discrepancy between observed flux and expected flux. We conclude in Section 5 with suggestions of future observations that can choose among several possibilities of source conditions.

2. WHAT WE EXPECT

2.1. Analogy with the Sun

The microwave brightness temperature (T_b) of the solar corona is low over much of the Sun's disk. Above active regions, however, the brightness temperature reaches the coronal temperature of ~ 10^6 K, as depicted in Fig. 1 from Dulk and Gary (1983). The combined area of the high T_b 1.4 GHz sources in Fig. 1 is about 20% of the Sun's disk, and is smaller at higher frequencies. At the distance of a typical nearby star, say 3 pc, the 1.4 GHz flux would be only ~ 2×10^{-3} mJy, two orders of magnitude lower than is observable with the VLA. The flux from other stars may be greater than the solar case due to two effects: (1) The brightness temperature may be greater than for the Sun. Since the emission in Fig. 1 is optically thick, this can be accomplished only

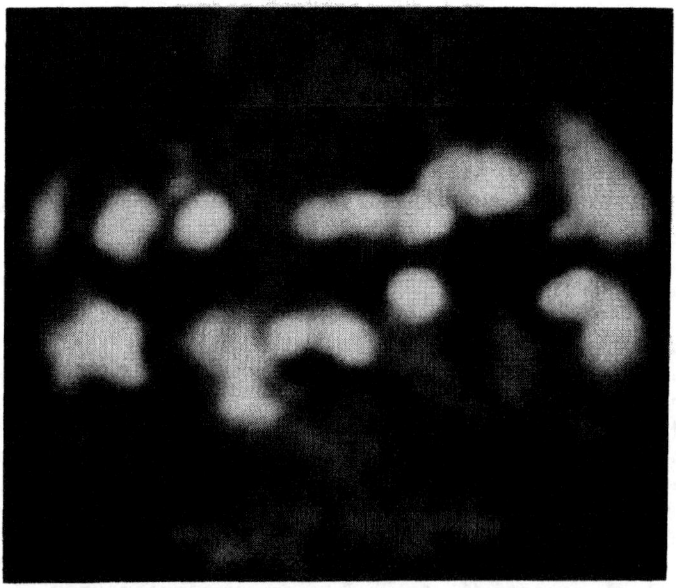

Fig. 1. The Sun at 1.4 GHz on 1981 Sep 26. Shading from black to white indicates brightness temperature ranging from 0 to 2.2×10^6 K. From Dulk and Gary (1983).

by an increase in electron temperature. (2) The stellar source area may be greater, due either to larger or more numerous active regions. In order to properly scale up the solar microwave emission to match what we expect from other stars, we must first determine likely stellar values for electron temperature and density, magnetic field strengths, and source area.

2.2. X-Ray Observations

Observations with HEAO-1 A2 (Walter et al. 1980) and the Einstein (HEAO-2) IPC and HRI imaging instruments (Vaiana et al. 1981; Johnson 1981) have revealed that many stellar types are represented as coronal X-ray sources, including main sequence late-type stars of interest here. An excellent review of X-ray emission from quiescent stellar coronae can be found in Golub (1982), where 35 dM star detections are enumerated. To obtain coronal temperatures from X-rays, spectral observations are required. Golub (1982) summarizes results of temperature measurements for 16 M-dwarfs.

The spectra for AD Leo (Swank et al. 1981) and Wolf 630 AB (Swank and Johnson 1982) are best fit by two components, a low temperature one at $\sim 7 \times 10^6$ K (several times the solar corona temperature) and a higher temperature component at a few times 10^7 K. Swank and Johnson (1982) note that the similar mix of temperatures for the two systems suggests this type of spectrum is common to dMe stars with high X-ray luminosity. The spectrum for Wolf 630 AB indicates that the lower temperature component is dominant, with an EM that is consistent with loops of length 1.6×10^{10} cm covering the surfaces of both stars at ~ 6 dyne cm^{-2}. Golub (1982) links the higher temperature component seen in IPC spectra, at least for YZ CMi and Prox Cen, with flaring activity. However, the Wolf 630 AB result given above was obtained in a quiet period with no evident flares.

We can estimate the flux expected from Wolf 630 AB and AD Leo due to free-free emission using these values of temperature and emission measure. At 5 GHz, assuming a source covering the star, the low temperature component has optical depth only slightly less than unity, while the high temperature component is optically thin. Even assuming optical depth $\tau > 1$ at $T = 7 \times 10^6$ K over the entire star, the fluxes expected are 0.05 mJy for Wolf 630 AB and 0.02 mJy for AD Leo. These fluxes cannot be increased by postulating higher temperatures, since this reduces the optical depth. Nor can the assumed source sizes be much increased since this would be inconsistent with the X-ray EM. Similar remarks apply to other late-type stars with measured X-ray fluxes, indicating that free-free emission is not important.

2.3. Stellar Magnetic Fields

Free-free emission is associated with active regions at low frequencies as shown in Fig. 1. At higher frequencies, where the free-free component has become optically thin, the corona remains optically thick over sunspots due to gyroresonance opacity (Alissandrakis, Kundu, and Lantos 1980; Pallavacini, Sakurai, and Vaiana 1981; see also Hurford, Gary, and Garrett 1984 in this proceedings). This and other related gyromagnetic emission mechanisms are associated with high magnetic field strengths, and so are limited in the solar case to small regions directly above sunspots. We now review the evidence for strong magnetic fields in late-type dwarf stars.

Effects of stellar magnetic fields can be observed by several techniques. Broadband photometric observations of dark "starspots" have been observed for RS CVn systems (cf. Eaton and Hall 1979) and BY Dra stars (cf. Vogt 1981), and have been shown to be correlated with chromospheric line fluxes as well (Baliunas and Dupree 1982). From rotational modulation it can be determined that the scale of such starspots is much larger than for sunspots, covering tens of degrees in longitude on the star. The radiative flux in chromospheric lines, which can be measured over a much wider range of spectral type, is also an indirect indicator of magnetic activity. Using this diagnostic, it is clear that many stars of otherwise solar type exhibit much greater activity than the Sun, implying more extensive active regions (Kelch, Linsky, and Worden 1979).

Direct evidence for stellar magnetic fields is more difficult to obtain. Efforts to measure circular polarization induced by the Zeeman effect in photospheric magnetically sensitive lines have yielded null results to an upper limit of ~ 10-20 gauss (Brown and Landstreet 1981). The circular polarization due to the Zeeman effect from bipolar magnetic regions tends to cancel, however, making this technique unreliable for stars with complex magnetic topology. Recently an idea revived by Robinson (1980) has received a great deal of interest and some success--that broadening of spectral lines due to the Zeeman effect can be detected in integrated starlight when compared with an appropriate magnetically insensitive line. At least three groups have reported positive detections using this technique--Robinson, Worden, and Harvey (1980), Timothy, Joseph, and Linsky (1981), and Marcy (1983).

Summarizing the results of these detections, which have been made only for stars of spectral type K5 or earlier due to blending of lines in later types, magnetic field strengths of ~ 500 to 3000 gauss have been detected covering 10 to 40% of the stellar surface on Xi Boo A, 70 Oph A, Epsilon Eri and 61 Cyg. Thus, at least some solar-type stars show evidence for much greater magnetic flux than occurs on the Sun.

The importance, to microwave observations, of high field strengths over much of the stellar surface is that under these conditions gyroresonance opacity can make the high temperature stellar coronae optically thick. Thus, where free-free emission was determined to be undetectable as a quiescent source of emission for solar-type stars, gyroresonance emission may perhaps be a viable alternative. We have already seen that optically thick emission at a temperature of $\sim 7 \times 10^6$ K is not enough for even nearby dMe stars to be detectable with the VLA. But with optically thick gyroresonance emission at the higher temperature of the two X-ray components, and assuming a source covering the entire star, quiescent emission from nearby stars may reach ~ 0.3 mJy and thus be marginally detectable. On the basis of the fore-going arguments, then, under the most favorable conditions we might expect some nearby stars to exhibit quiescent microwave emission at the < 0.5 mJy level, especially earlier type (G and K) stars with large stellar radii.

3. WHAT WE OBSERVE

The number of published obervations of quiescent microwave emission from solar-type stars is small. Primarily the reports are of detection of individual (Topka and Marsh 1982; Cox and Gibson 1984--this proceedings) or small numbers of objects (Gary and Linsky 1981; Kundu and Shevgaonkar 1984). The most comprehensive report of quiescent observations to date was by Linsky and Gary (1983) who gathered observations from several observing runs. A synthesis of the results is given below. A useful parameter for this discussion is

$$(R/R_*)^2 T_b = 2.23 \times 10^4 \, S(\mathrm{mJy}) \, d^2(\mathrm{pc}) \, \lambda^2(\mathrm{cm}) \, (R_*/R_{sun})^{-2} \quad , \qquad (1)$$

where R is the source radius, R_* is the stellar photospheric radius, T_b is the brightness temperature of the source, S(mJy) is the observed flux or upper-limit, d is the stellar distance in parsecs, and λ is the observing wavelength in cm.

3.1. F-K Stars

Of 11 such objects observed by Gary and Linsky (1981) and Linsky and Gary (1983), only one, χ^1 Ori, was detected. Subsequent observations by myself and others have failed to detect it again, however, and it now appears likely that the emission was due to a previously unknown red dwarf companion reported by McCarthy (1983). A few of the upper-limits for the other, undetected stars are important, however. Most notably, the stars for which magnetic fields have been measured--Xi Boo A, 70 Oph A, Epsilon Eri, and 61 Cyg A--have extremely low 6 cm flux upper-limits < 0.2-0.3 mJy (3σ). Using equation (1), the values of $(R/R_*)^2 T_b$ for these stars are as listed in Table 1. These results are consistent with our expectations from the previous section; if the size of the optically thick source is $R/R_* \approx 1$ for any of these objects, the coronal temperature must not be much higher than that for the solar corona. Higher temperatures are allowed if the optically thick source is only a fraction of the stellar size.

3.2. dMe Stars

The situation is not so clear for dMe stars. Other things being equal, their smaller sizes should make them less easy to detect than F-K stars,

Table 1

Quiescent Flux Upper-Limits for Stars With Detected Fields

Source	Spectral Type	6 cm Flux upper-limit	$(R/R_*)^2\ T_b$ (K)
ξ Boo AB	G8V+K4V	< 0.2 mJy	< 8.1 (6)
70 Oph AB	K0V+K5V	< 0.2 mJy	< 5.0 (6)
ε Eri	K2V	< 0.3 mJy	< 4.0 (6)
61 Cyg AB	K5V+K7V	< 0.2 mJy	< 1.4 (6)

Table 2

Quiescent Emission from dMe Stars

Source	Spectral Type	Dist. (pc)	Band (cm)	Flux (mJy)	Ref.	Quiescent?
EQ Peg A	dM4e	6.5	6	0.7	1	Probable
EQ Peg B	dM6e	6.5	6	0.4	1	Probable
UV Cet	dM5.6e	2.6	20	1	2,4	Yes
			6	1 - 2	2,3,4	
L726-8A	dM5.5e	2.6	20	< 0.3	2	No
			20	2 - 3	4	
			6	< 0.2	2,3	No
			6	0.3 - 0.6	4	
YY Gem	dM1e+dM1e	14.6	20	0.5 - 2	5	Yes
			6	0.5 - 2	5	
			2	< 0.3 - 2	5	
Wolf 630 AB	dM4e+dM4e	6.4	6	< 0.2 - 2	3,5	Yes
AU Mic	dM0e	8.3	20	1.5 - 2.0	6	Yes
			6	0.7	6	
			2	1	6	
YZ CMi	dM4.5e	6.1	20	0.5 - 3	2,4	Probable
			6	0.5 - 1	2,4	
			6	< 0.2	5	
AD Leo	dM4.5e	4.9	20,6	< 0.2	5	No
CN Leo	dM6.5e	2.3	20,6	< 0.2	2	No
Wolf 424	dM2e	4.3	20,6	< 0.2	2	No
BD+16 2708	dM2e	10.4	20,6	< 0.2	2	No

1: Topka and Marsh (1981); 2: Fisher (1982); 3: Linsky and Gary (1983);
4: Kundu and Shevgaonkar (1984); 5: This Paper; 6: Cox and Gibson (1984)

yet many are quiescent sources at the 1 to 2 mJy level. A list of potential quiescent sources is presented in Table 2. In interpreting Table 2, one must keep in mind that each of these sources is variable on some timescale, and all are capable of flaring. Thus, on some occasions an object may show quasi-steady emission that is due decay of a long-lived flare; alternatively the source may not be detected even though it exhibits quiescent emission on other occasions. The assignment of 'Yes', 'Probable', or 'No' in the last column is subjective and Table 2 is biased to my view. Those objects listed with 'No' in the last column have not been detected outside of flares, to my knowledge. The L726-8 primary was reported by Kundu and Shevgaonkar (1984) as a quiescent source, but its variability and high polarization lead me to believe that this is low level flare emission. Numerous observations by Fisher (1982), Gary and Linsky (1981), Gary, Linsky, and Dulk (1982), and Linsky and Gary (1983) have failed to detect it as a quiescent source. Those sources listed with 'Yes' in the last column of Table 2 are observed on many occasions as quiescent sources at about the same flux level, and the degree of polarization is small except during short time scale brightenings. Polarization measurements at such low flux levels are uncertain, however, and are likely to contain contamination from small flares, as is apparent for UV Cet in Fig. 2.

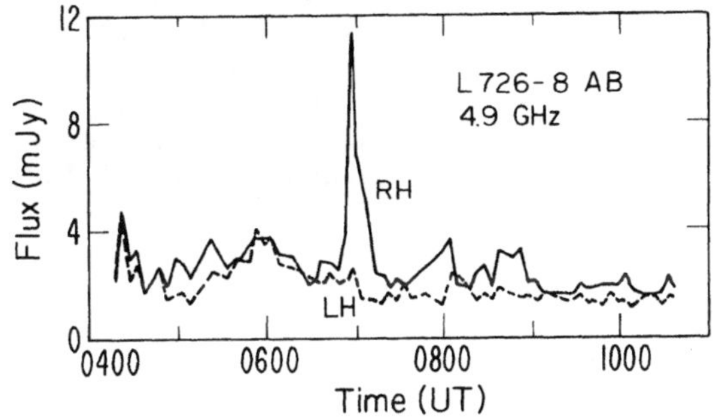

Fig. 2. The 6 cm (4.9 GHz) flux of L726-8 in right-hand (RH) and left-hand (LH) circular polarization at a time resolution of 5 minutes. Most of the emission is from UV Ceti, but the flare from 0650 to 0730 UT is mainly from the primary star, L726-8 A. Typical 1σ uncertainty is ±0.22 mJy.

3.2.1. Variability. Most published observations of quiescent emission are single flux measurements integrated over a relatively short integration time. The fact that these flux measurements change from one observation to the next indicates variability that could be due to rotational modulation, constant low level flaring, or growth and decay of active regions. It is extremely important to obtain flux measurements at sufficient time resolution and over a sufficiently long time that these various kinds of variability can be separated. Fig. 2 shows an example of flux variations on a 5 min timescale that is perhaps most consistent with low level flaring. Note, however, that the flux never goes below about 1.5 mJy, and UV Cet has never failed (to my knowledge) to be detected at a flux level of ~ 1 to 2 mJy. It does not seem likely that low level flaring can account for this steady component.

Rotational modulation of flux from dMe stars has not been demonstrated, primarily because observing times have been too short. However, new observations shown in Fig. 3 suggest that emission from the short period (20 hours) eclipsing binary system YY Gem is rotationally modulated. Unfortunately, the 14 hours of observation is only ~ 3/4 of the rotation period, but a clear variation at 20, 6, and 2 cm is apparent. VLA observations over three consecutive nights is planned for November 1984 that should definitively establish rotational modulation for YY Gem if it exists. Rotational modulation of flux and polarization is perhaps one way to determine the size and location of the quiescent source(s), since the amplitude and abruptness of the variations give limits to these parameters.

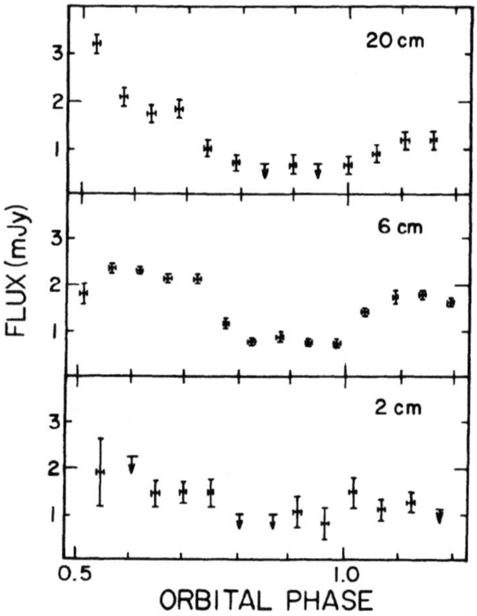

Fig. 3. Variation with orbital phase of total flux from the eclipsing binary YY Gem at 20, 6, and 2 cm. The vertical error bars denote 1σ uncertainties, and downward arrows denote 3σ levels when no source was visible. Each frequency shows a clear variation with orbital phase. Photospheric eclipses occur near phases 0.5 and 1.0.

3.2.2. Flux Levels. The striking thing about the flux levels shown in Table 2 is that they are at least a factor of 10 higher than we expected from Section 2. For many of these sources, the parameter $(R/R_*)^2T_b \sim 10^8$ K. Apparently distance is not the controlling factor in which stars are quiescent sources, since UV Cet is a source while its companion L 726-8A is not. Nor is measured coronal X-ray temperature or flux, since Wolf 630 AB and AD Leo have similar parameters. In the following section, we briefly examine the possible emission mechanisms in an effort to understand where our estimate of expected flux may have gone wrong.

4. EMISSION MECHANISMS

We have already seen that bremsstrahlung is insufficient to account for the flux from dMe stars, since the required source size and density is inconsistent with X-ray emission measures. Linsky and Gary (1983) briefly examined another emission mechanism, plasma emission, and suggested that the free-free optical depth of the overlying material would obscure any source due to that mechanism, at least at 6 cm. It is not impossible that some emission at 20 cm could escape, especially if the emission occurs at the second harmonic of the plasma frequency, but the problem of the too high fluxes remains to be explained at 2 and 6 cm. The only remaining alternative emission mechanism is some form of gyro-emission, and at least two forms are possible--gyroresonance and gyrosynchrotron emission.

Since the measured parameter $(R/R_*)^2T_b$ is a factor of 10 greater than we had previously predicted, the problem can be stated simply: either (1) the source is 3 times larger than the stellar size, or (2) the brightness temperature of the emission (and therefore the effective electron temperature) is up to 10 times greater than the already generous 7×10^6 K we used in Section 2. The first possibility perhaps may be met through the gyroresonance emission mechanism (emission at the first few harmonics $s = f/f_B$ of the gyrofrequency f_B), while the latter requires emission at higher harmonics of the gyrofrequency, implying gyrosynchrotron emission.

4.1. Gyroresonance Emission

Gyroresonance emission is the favored mechanism for high frequency sources on the Sun, and we have already appealed to a scaled up version of this mechanism in section 2.3 to predict the flux we should observe. Gary and Linsky (1981) examined how this mechanism could be scaled up still further to account for source sizes larger than the stellar radius. Basically, by postulating an arbirarily high magnetic field strength covering essentially the entire visible stellar surface, the size of the shells of opacity given by the gyroresonance mechanism can be made arbitrarily large. The exercise becomes one of determining the highest optically thick harmonic for an atmospheric model, and choosing the base field strength that gives the source size, for that harmonic, indicated by the observations. In fact, the formula for optical depth used by Gary and Linsky (1981) erroneously accounted twice for the gyroresonance line-shape and the resulting opacities are about 100 times too small. When the correct formula is used, the highest optically thick harmonic increases to about $s = 10$, at which point the gyroresonance formula breaks down and formulas for gyrosynchrotron emission must be used. Although a correct analysis requires more work, it is safe to say that gyroresonance emission can lead to sources of greater than the stellar size, and could be as large as the required $R = 3\ R_*$ if the magnetic field strength and coverage are large enough. We reserve the gyroresonance emission mechanism as a possible explanation of the large fluxes measured for dMe stars.

4.2. Gyrosynchrotron Emission

The other alternative to account for the large fluxes is to increase the electron temperature to ~ 10^8 K. This may be a true temperature (that is, the electron distribution may be maxwellian at this temperature) or it may be the effective temperature of a non-thermal electron distribution. In either case, emission from such high energy electrons is regarded as gyrosynchrotron emission, and new formulas must be used.

Linsky and Gary (1983) discussed the thermal case and found that a relatively few electrons at the higher temperature of a two temperature distribution (as is found in X-ray spectra) could be optically thick at this higher temperature. Thus, if only about 10^5 electrons cm^{-3} exist at a temperature of 10^8 K in a magnetic field strength of ~ 100 G, the fluxes from dMe stars are accounted for with a source size $R \approx R_*$.

The fluxes can also be explained if the electron distribution has a nonthermal tail, perhaps populated by quasi-steady, low level flaring. A nonthermal model was invoked by Kundu and Shevgaonkar (1984) to account for the quiescent emission from UV Cet at both 20 and 6 cm. They found that although the 6 cm flux could be explained with the thermal model of Linsky and Gary (1983) the 20 cm flux was too high unless the source at 20 cm is ~ 3 times the size of the 6 cm source. With a nonthermal model, in which brightness temperature is a function of frequency, the fluxes at both wavelengths could be fit if both sources originate from the same physical volume of radius ~ 2.7 R_*. This study points out clearly that multi-frequency observations are necessary to constrain the possible models.

5. FUTURE OBSERVATIONS

The lack of detections of F-K dwarfs is consistent with our expectations of Section 2. Because of variation in activity on these stars, it will remain of interest to continue monitoring the nearest and most active of these stars (Epsilon Eri and Xi Boo A, perhaps), especially when chromospheric activity is reported to be high. A firm detection of one of these stars would be of great interest, but until then only modest observing time will probably be expended.

We have enumerated at least three possible source models that can account for the high fluxes measured for dMe stars. Determination of which of these is the correct one depends on (1) elucidation of the time behavior, both long and short term, of the quiescent emission, (2) measurements at as many frequencies as possible of flux and polarization, and (3) a clear understanding of which stars are quiescent sources, and which are not.

5.1. Continuous Monitoring

To determine if rotational modulation is important, it is very worthwhile to observe some of the short period rotators over as much of a rotational cycle as possible. This is being done at present with the eclipsing binary YY Gem, as shown in Fig. 3. Longer term variations unfortunately can be detected only with long observing times, such as the program of Cox and Gibson (1984) with AU Mic. Comparison with dipole field models such as Hurford, Gary, and Garrett (1984) present, including rotation, may indicate how large the sources must be.

Even for long term observations, it is of the utmost importance for the workers to make short term flux measurements as well, to guard against contamination of the quiescent fluxes by flares. Also, if the quiescent emission is due to constant, low level flaring, the flux level should bear a simple relationship to detectable flares. As shown in Fig. 2, very small flares are observable, and the quiescent flux level should decrease steadily in their absence, the life-time of the fast electrons being only 6 hours or so, depending on the magnetic field strength (Linsky and Gary 1983).

5.2. Multi-Frequency Observations

As we saw in Section 4.3, a source model that fits the flux at one frequency may be completely unable to account for the flux at another frequency. Although low frequency emission may be partly due to plasma emission, and other effects can mislead us when dealing with a poorly determined (2 point?) spectrum, the mismatch of a model at some frequencies mitigates against it.

5.3. Complete Samples

From X-ray observations, it is clear that dM stars are ubiquitous emitters. If the quiescent microwave emission is due to thermal electrons, whose properties presumably do not greatly depend on the flare activity, most dMe stars also should have roughly the same microwave flux. From Table 2, this apparently is not the case. Observations of a complete (perhaps distance limited) sample of dMe stars would be useful in defining the class of stars that show quiescent emission, especially comparing the microwave flux or upper-limit to the X-ray flux.

The study of quiescent microwave emission from late-type stars has come far since its birth only four years ago. For the first time we are able to directly observe the coronae of stars other than the Sun from the ground. Such observations promise to become as integral a part of stellar atmospheric studies as they are for the Sun.

Acknowledgement: This work was supported in part by NSF grant ATM-8309955 to the California Institute of Technology.

REFERENCES

Alissandrakis, C.E., Kundu, M.R., and Lantos, P.: 1980, Astron. Astrophys. 82, 30.
Baliunas, S.L., and Dupree, A.K.: 1982, Ap.J. 252, 668.
Bowers, P.F., and Kundu, M.R.: 1981, Astron.J. 86, 569.
Brown, D.N., and Landstreet, J.D.: 1981, Ap.J. 246, 899.
Dulk, G.A., and Gary, D.E.: 1983, Astron.Astrophys. 124, 103.
Eaton, J.A., and Hall, D.S.: 1979, Ap.J. 227, 907.
Fisher, P.L.: 1982, Masters Thesis, New Mexico Inst. of Tech.
Fisher, P.L., and Gibson, D.M.: 1982, in Second Cambridge Workshop on Cool Stars, Stellar Systems, and the Sun (eds. M.S. Giampapa and L. Golub), (SAO: Cambridge), Vol. II, 109.
Gary, D.E., and Linsky, J.L.: 1981, Ap.J. 250, 284.
Gary, D.E., Linsky, J.L., and Dulk, G.A.: 1982, Ap.J.(Letters) 263, L79.
Cox, J., and Gibson, D.M.: 1984, in Proceedings of the Workshop on Stellar Continuum Radio Astronomy.
Golub, L.: 1983, in Activity in Red Dwarf Stars (IAU Colloq. No. 71), eds. P.B. Byrne and M. Rodono (Dordrecht: Reidel), p. 273.
Hurford, G.J., Gary, D.E., and Garrett, H. 1984: in Proceedings of the Workshop on Stellar Continuum Radio Astronomy.
Johnson, H.M.: 1981, Ap.J. 243, 234.
Kelch, W.L., Linsky, J.L., and Worden, S.P.: 1979, Ap.J. 229, 700.
Kundu, M.R., and Shevgaonkar, R.K.: 1984, submitted to Ap.J.
Linsky, J.L., and Gary, D.E.: 1983, Ap.J. 274, 776.
Marcy, G.W.: 1983, in Solar and Stellar Magnetic Fields: Origins and Coronal Effects (IAU Symp. No. 102), ed. J.O. Stenflo (Dordrecht: Reidel), p.3.
McCarthy, D.W., Jr.: 1983, in IAU Colloquium 76, The Nearby Stars and the Stellar Luminosity Function
Pallavicini, R., Sakurai, T., and Vaiana, G.S.: 1981, Astron. Astrophys. 98, 316.
Robinson, R.D.: 1980, Ap.J. 239, 961.
Robinson, R.D., Worden, S.P., and Harvey, J.W.: 1980, Ap.J.(Letters) 236, L155.
Swank, J.H., Boldt, E.A., Holt, S.S., Marshall, F.E., and Tsikoudi, V.: 1981, Bull. Am. Phys. Soc. 26, 570.
Swank, J.H., and Johnson, H.M.: 1982, Ap.J.(Letters) 259, L67.
Timothy, J.G., Joseph, C.L., and Linsky, J.L. 1981, Bull. Amer. Astron. Soc. 13, 828.
Topka, K, and Marsh, K.A.: 1982, Ap.J. 254, 641.
Vaiana, et al. 1981:, Ap.J. 245, 163.
Vogt, S.S.: 1981, Ap.J. 250, 327.
Walter, F.M., Cash, W., Charles, P.A., and Bowyer, C.S.: 1980, Ap.J. 236, 212.

THE SOLAR-STELLAR CONNECTION

R. Pallavicini
Arcetri Astrophysical Observatory
Largo E. Fermi 5
50125 Florence, Italy

ABSTRACT. This review summarizes the many contact points between solar and stellar physics stressing the similarities and differences between phenomena observed on the Sun and on stars with outer convective mantles. Topics discussed include: 1) the relationship between convection, rotation and magnetic fields as a source of magnetic activity in late-type stars; 2) the observation of solar-type phenomena (spots, plages, activity cycles etc.) on other stars; 3) the structuring and heating of high temperature coronae; 4) the observation and interpretation of radio emission from the Sun and late-type stars.

1. INTRODUCTION

The Sun is a normal star of spectral type G2V. It is to be expected that phenomena analogous to those observed on the Sun (spots, plages, activity cycles, X-ray coronae and winds) may be present in other stars of similar spectral type. In spite of this, for a long time the fields of solar and stellar astronomy have developed quite independently, each with its own methods and virtually no connection in between. Today this situation has changed, mainly as a consequence of high-sensitivity X-ray and UV observations from space, which have demonstrated that activity phenomena of the solar type are not restricted to somewhat exotic objects such as RS CVn binaries or BY Dra variables, but are common to virtually all stars of spectral type later than F. These stars have an internal structure which is totally different from that of stars of earlier spectral types, and is characterized by a radiative core and an outer subphotospheric convection zone.

The cross-fertilization of solar and stellar physics is mutually beneficial. However, some caution should be used in extrapolating solar analogies to all late-type stars. One must realize that physical conditions change substantially for stars of spectral types from early F to late M, and from dwarf stars to giants. Two parameters, in particular, change dramatically over this range. One is the depth of the convection zone which varies from zero at spectral type ~ F0 to values comparable with the stellar radius - i.e. fully convective stars - at

R.M. Hjellming and D.M. Gibson (eds.), Radio Stars, 197–211.

spectral type ~ M5. The other is the average rotational velocity which drops precipitously at about spectral type ~ F5 from values of the order of ~100 km s^{-1} to very low rotation rates, comparable with the equatorial rotational velocity of the Sun (~ 2 km s^{-1}). Surface gravity does not change very much along the main sequence, but it changes by orders of magnitude from main-sequence stars to giants and supergiants. We may expect, therefore, that giants and supergiants have atmospheric properties quite distinct from those of the Sun.

The Sun is a single star, but most stars are not. Components of widely separated binaries may retain the same properties as single stars, but it is unlikely that the same occurs for close binaries as detached RS CVn systems, semi-detached Algol-type systems with mass transfer in between, or contact W UMa binaries. The observations indicate that binarity may have important effects, either directly or indirectly, in determining the activity level of the component stars. Finally, with an age of 4.5×10^9 years, the Sun is an intermediate age star. Both younger and older stars exist. It is unclear how far back in time we can extrapolate solar properties. For instance, T-Tauri stars and other pre-main sequence objects have many properties similar to those of other late-type stars, but they have additional features, such as dense circumstellar envelopes, which have no solar counterpart.

With these caveats in mind, I shall summarize in the following sections some of the contact points between solar and stellar physics. In particular, I shall discuss in some detail the problem of generation of magnetic fields in late-type stars, the inferred presence of solar-type activity phenomena on other stars, and the heating and structuring of high temperature coronae. In the last section, I shall discuss the relationship between solar and stellar radio emission as expecially relevant to the subject of this Workshop.

2. CONVECTION, ROTATION AND DYNAMO ACTION

All activity phenomena on the Sun are magnetohydrodynamic and result from the emergence of magnetic fields at the solar surface. The magnetic field is unlikely to be primordial in the Sun and other late-type stars, and a mechanism is needed to maintain and regenerate continuously the field against resistive losses. The mechanism which is generally assumed is dynamo action, which results from induction effects in the rotating conducting fluid body of the star (Parker 1979, Cowling 1981, Gilman 1981, Belvedere 1983). Most dynamo theories developed in the astrophysical context treat kinematic dynamos, in which the fluid motions are assumed rather than explicitly computed. A subclass of these kinematic dynamos are the so-called α-ω dynamos, to which the discussion below will be restricted.

An α-ω dynamo originates from two different mechanisms: a) the ω-effect which produces a toroidal field from a seed poloidal field by virtue of differential rotation (which in turn originates from the interaction of rotation and convection); b) the α-effect which regenerates a poloidal field of opposite polarity under the action of cyclonic turbulence produced by Coriolis forces acting upon convective motions.

In the case of the Sun, kinematic α-ω dynamos have been able to reproduce quite successfully the observed properties of the 22-year magnetic cycle (Yoshimura 1975, 1978 a,b). However, the agreement between theory and observations is more apparent than real, since the kinematic dynamos developed so far are heavily parameterized theories and the many free parameters are chosen ad hoc so as to reproduce the observations. This procedure, although not correct, may be somewhat justified in the case of the Sun, for which the observations tell us what to expect. However, the same procedure is certainly not applicable in the case of other stars, for which the observational constraints are too scanty. Fully hydrodynamic calculations are necessary; unfortunately they are still in their infancy, even for the Sun (Gilman 1981, 1983).

Recently, a number of authors have tried to extend α-ω dynamo calculations to stars other than the Sun. For instance, Belvedere et al. (1980 a,b, 1981, 1982) have computed differential rotation models and dynamos for stars in the spectral range F5 to M0, and have made "predictions" with regard to differential rotation, average magnetic field, length of activity cycles, level of coronal X-ray emission etc. Although suggestive, the predictions of these calculations must be viewed with caution. These models are based on the assumption that the dynamo process works in all stars as in the Sun (which is far from having been proved), and they are all calibrated by the solar case. Even more important, the predictions of these models depend critically on the assumed free parameters, and different choices of these produce quite different answers (Moos and Vilhu 1983).

As an example of this state of affairs, I shall refer again to the calculations of Belvedere et al. (1981, 1982). Their model predicts a maximum of differential rotation at spectral type F5, which is not confirmed by Fourier analysis of line profiles (Gray 1982). The observations indicate a degree of differential rotation for F5 stars substantially less than for the Sun (see also Belvedere and Paternò 1983). Moreover, the calculations of the linear theory of Belvedere et al. predict longer activity cycles for stars of later spectral types, which is exactly the opposite of what is predicted by the non linear α-ω dynamo calculations of Durney and Robinson (1982). Neither of these predictions is apparently confirmed by the observations of Vaughan (1983). In conclusion, α-ω dynamo calculations for late-type stars appear at present to have little, if any, prediction power. In this field, observations must lead the theory, to provide constraints and guidance for the development of theoretical models.

What do the observations tell us about the role of rotation, convection and magnetic fields in late-type stars? It was shown in the sixties by Wilson (1963), Kraft (1967) and Skumanich (1972) that CaII chromospheric emission and rotation both decline with age in late-type main-sequence stars. In addition, stellar chromospheres first appear at early spectral type F, i.e. at about the same spectral type at which stars begin to have appreciable outer convection zones. At about the same spectral type, the average rotation rate of stars decreases sharply.

The ensemble of these observations was interpreted by a simple scenario which has remained valid up to now, with only some minor modifications tending to increase the role of magnetic fields in the whole process. According to it, stars which possess outer convection zones can develop - by acoustic heating and/or magnetic effects - chromospheres, coronae and thermally-driven winds. The outflowing of wind in the presence of magnetic fields produces magnetic braking (Weber and Davis 1967), thus explaining the decrease of rotation rate with age. On the other hand, the decrease of chromospheric emission with age was attributed to the decay of stellar magnetic fields, as suggested by the solar analogy, which shows the existence on the Sun of a one to one correspondence between regions of enhanced CaII emission and regions of enhanced magnetic fields (Skumanich et al. 1975). These observations indicate that an intimate, and possibly causal, relationship must exist between magnetic fields and rotation, as implied by the dynamo mechanism.

Recent observational work at X-ray, UV and optical wavelengths has substantially confirmed this picture, while strongly supporting the concept that rotation per se, rather than age is the crucial factor in determining the activity level of late-type stars. For instance, Pallavicini et al. (1981a, 1982) found a quadratic dependence of coronal X-ray emission on stellar rotation, and somewhat similar correlations have been found by other investigators for chromospheric and coronal emission (Walter 1981, 1982, Middelkoop 1982, Catalano and Marilli 1983, Vilhu and Rucinski 1983, Noyes et al. 1984, Hartmann et al. 1984). In addition, evidence has been produced that the high degree of activity of RS CVn binaries and BY Dra variables is due to enhanced rotation, either in young single stars (as in some BY Dra variables) or in close binary systems (RS CVn and Algol-type binaries), where rapid rotation is enforced by tidal interaction (Bopp and Espenak 1977, Bopp and Fekel 1977).

All these findings are strong observational support of the notion that magnetic activity in late-type stars is produced by dynamo action and that the degree of dynamo efficiency increases with stellar rotation rate. Unfortunately, at present the agreement between theory and observations does not go beyond the qualitative stage. The rotation-activity connection has been firmly established by the observations, but the exact functional dependence of chromospheric and coronal emission on rotation, and the relevance of additional parameters such as convection zone depth, are still not well determined owing to the large scatter in the data. More X-ray observations of stars from future space missions, as well as new accurate determinations of rotation rates are needed, before the observations can be used to constraint dynamo models effectively.

3. SPOTS, SPOT CYCLES AND RELATED PHENOMENA

Sunspots are the best known and most characteristic activity phenomena on the Sun. They are dark areas cooler than the surrounding photosphere and with high magnetic fields of up to a few thousand Gauss. Their statistical properties define a recurrent pattern of activity known as the 11-year sunspot cycle. Sunspots cover only a very small fraction of the solar surface (~10^{-4}- 10^{-5} of the visible hemisphere) and their presence would go undetectable in integrated sunlight. Only recently, space observations from the Solar Maximum Mission have allowed the detection of variations of the solar constant (at the level of ~0.1%) due to the presence of large spots on the Sun (Willson et al. 1981).

The presence of dark areas resembling spots has been inferred on other stars, notably in RS CVn and BY Dra variables (Hall 1976, Vogt 1983). Their presence has been established from photometric variations observed in their light curves. Differently from the solar case, all properties of these dark areas (size, shape, distribution on the stellar surface, migration rates etc.) must be derived from the light curve, a procedure which generally has no unique solution. The dark areas are cooler than the surrounding photosphere as demonstrated by the strengthening of TiO bands near the photometric minimum (Ramsey and Nations 1980). Their magnetic nature, however, has still to be proved, since no measurement of magnetic fields in starspots has been possible so far.

In spite of the many uncertainties with regard to the properties of starspots, two facts seem well established. First, starspots occupy a much larger fraction of the star surface (up to 30 - 40%) than in the Sun; and second, their distribution in longitude must be fairly inhomogeneous. The latter, however, may be a selection effect, since spots distributed uniformely in longitude would not produce appreciable variations in the light curve. In RS CVn stars, spots are concentrated on one hemisphere (Eaton and Hall 1979), a fact which is reminiscent of active longitude belts on the Sun, although on an extreme level. At any rate, the large fractional area covered by spots in RS CVn binaries and BY Dra variables suggests a degree of perturbation of the normal stellar atmosphere and of the subphotospheric convection flow far in excess than for the Sun.

Chromospheric plages are usually observed on the Sun in association with sunspots. Plages are best seen in the CaII lines, and are accompanied by enhanced transition region and coronal emission at X-ray, UV and radio wavelengths. The best evidence for the presence of plages on other late-type stars is the observation of rotational modulation of the CaII lines (Vaughan et al. 1981). In the solar analogy, we would also expect an anticorrelation between the presence of plages and the minimum of the photometric light curve. Search for this effect has produced rather contradictory results, which might indicate that stellar plages are more uniformly distributed than starspots, as they are on the Sun. However, Baliunas and Dupree (1982) found that chromospheric emission is enhanced at the photometric minimum in the long-period RS CVn star λ And. Furthermore, collaborative work at Colorado, Catania and Armagh (Marstad et al. 1982) has shown that chromospheric and transition region lines observed by IUE were enhanced at the photometric minimum in the star II Peg, which is a border case between RS CVn and BY Dra stars.

The existence of stellar activity cycles similar to the 11-year sunspot cycle was first proved by O. Wilson in a paper published in 1978 and summarizing the results of a 10-year effort in monitoring stellar variations in the CaII lines (Wilson 1978). Photometric variations in integrated photospheric light cannot be easily determined for stars of spectral types earlier than late K and M. CaII chromospheric emission, on the contrary, has a much higher contrast and in the Sun it varies by as much as 40% during the solar cycle (Sheeley 1967, White and Livingston 1981). Wilson's observations showed the existence of cycles in many solar-type stars with periods ranging from ~7 to $\gtrsim$10 years and amplitude variations of ~10% to ~35%. The Sun, observed with the same technique, gives a cycle with a relative amplitude of ~10% (Wilson 1978). This program has been continued by Vaughan and collaborators, and the observations extend now over ~15 years. The detected cycles have periods ranging from ~6 to ~14 years with no apparent dependence on spectral type (Vaughan 1983). Interesting enough, only old stars with relatively low chromospheric activity (i.e. stars below the so-called Vaughan-Preston "gap") have been found so far to have cycles (Vaughan 1980). Whether this is related to different modes of dynamo action (e.g. Durney et al. 1981, Knobloch et al. 1981), or more simply to the fact that cycles in young active stars are more easily masked by the high level of short-term variations, is unclear at present.

More uncertain is the situation with regard cycles in RS CVn binaries and BY Dra variables. Phillips and Hartmann (1978) have found long-term, possibly cyclic variations of the mean level of photospheric emission in BY Dra and CC Eri with periods of ~50 years. Analyses of light curve amplitudes and wave migration rates of RS CVn stars have produced "periods" ranging from 5 to more than 30 years (Catalano 1983). It is uncertain whether these variations represent true activity cycles. Observations over much longer time spans are necessary before the cyclic nature of these variations can be established with some confidence.

In spite of the large body of indirect observational evidence which suggests that activity phenomena on stars are of magnetic origin, our knowledge of magnetic fields in late-type stars is still extremely scanty. Only recently direct measurements have become possible by determining the subtle differences between the profiles of a magnetic sensitive and of a magnetic insensitive line (Robinson et al. 1980, Marcy 1984). In principle, the method allows the determination of both field strength and area coverage factor. Detected magnetic fields range in strength from ~750 Gauss (the minimum detectable value) to ~3000 Gauss, and the area coverage factors from ~20% to ~80% (Marcy 1984). For comparison, the average magnetic field on the Sun is ~1500 Gauss (i.e. the typical field strength in magnetic flux knots) and the area coverage factor is ~1% - 2%. It is important to notice that the method does not provide the field strength in individual starspots or active regions, but only an average value over the entire stellar surface. Moreover, although promising, the method does not appear yet to provide releable quantitative results, and physical implications derived from these observations should be viewed with caution.

4. X-RAY CORONAE

The topic of stellar coronae is a good example of how the interplay of solar and stellar physics has allowed substantial progress to be made in the understanding of the physical processes responsible for stellar activity. Prior to the advent of the EINSTEIN Observatory, there was a general expectation that, among single stars, only those with both outer convection zones and substantial convective velocities - i.e. stars of spectral type F and G - should possess high temperature coronae, similar to the solar corona (Mewe 1979). These expectations were a consequence of the universally accepted theory of coronal heating by acoustic waves, with magnetic fields relegated to play only a secondary role (Kuperus 1969, Ulmschneider 1979).

The observations from the EINSTEIN and IUE satellites (Vaiana et al. 1981, Stern 1983, Linsky 1983) have demonstrated, instead, that transition regions and coronae are fairly common among stars of all spectral types and luminosity classes. The only exception is apparently constituted by late-type giants and supergiants, for which the absence of appreciable material at $T \gtrsim 2 \times 10^4$ K is accompanied by the presence of mass losses far in excess than for the Sun (Stencel and Mullan 1980). Even more importantly, the observations from space have shown that chromospheric and coronal emission of late-type stars scale with effective temperature and gravity in a way totally different from that predicted by the acoustic theory (Basri and Linsky 1979, Vaiana et al. 1981). A broad range of chromospheric and coronal emission levels exists for stars with the same effective temperature and gravity - and hence with the same convective flow pattern. Moreover, chromospheric and coronal activity appear to depend on stellar rotation rate (Pallavicini et al. 1981a, Noyes et al. 1984). These results indicate the inadequacy of the acoustic theory and point at dissipation of dynamo-generated magnetic fields as a more likely heating mechanism.

Similar conclusions have been reached by high spatial and spectral resolution studies of the Sun. X-ray and UV observations from Skylab, OSO-8 and the Solar Maximum Mission (Withbroe and Noyes 1977, Vaiana and Rosner 1978) have revealed that the solar transition region and corona are highly structured and variable, and that the observed spatial and temporal inhomogeneities are intimately connected with the presence and intensity of magnetic fields. They have also shown that the energy flux carried by acoustic waves at transition region and coronal levels is insufficient to provide for coronal heating (Athay and White 1978, Bruner 1978). Alternative heating mechanisms, such as dissipation of MHD waves and/or DC currents, are required in the upper atmospheric layers (Rosner et al. 1978, Ionson 1978, Wentzel 1981).

The combined analysis of solar and stellar observations has proved to be extremely fruitful in tackling the fundamental problem of coronal heating and has stimulated the development of new heating theories (for a detailed discussion, see e.g. Pallavicini 1984). At the same time, the spatial information on coronal structures provided by solar observations is essential for interpreting spatially unresolved observations of stellar coronae.

In the solar corona, two topologically distinct types of magnetic structures exist: a) closed arch-shaped structures (loops), responsible for most of the observed X-ray emission; b) open field regions (coronal holes), from which high velocity wind streams originate. Loops apparently trace magnetic field lines in the corona, where the plasma β factor - i.e. the ratio of gas pressure to magnetic pressure - is $\ll 1$

(Poletto et al. 1975). In first approximation, loops can be considered in energy balance and hydrostatic equilibrium (Rosner et al. 1978, Landini and Monsignori-Fossi 1981, Pallavicini et al. 1981b). Under these restrictive hypotheses, simple orders of magnitude estimates of the energy losses and gains allow the derivation of scaling-laws between global loop parameters, such as maximum coronal temperature T_{max}, pressure p and loop semi-length L. In the formulation of Rosner et al. (1978) the scaling-law reads:

$$T_{max} \sim 10^3 \, (p \, L)^{1/3} \qquad \text{c.g.s.}$$

This scaling law has been proved to be in good agreement with observations.

In the stellar case, it is usually not possible to determine the three parameters T_{max}, p and L directly from observations. In this case, scaling-laws can be quite useful to infer coronal pressures and/or loop sizes (or equivalently area coverage factors), once the other parameters have been determined from observations (or estimated otherwise). Although this is probably the best we can do, the importance of these simplifying relationships should not be overestimated. In the stellar models developed so far (Walter et al. 1980, Landini et al. 1984b), the assumption is usually made that all loops are equal, an assumption which is certainly not true for the Sun, whose coronal emission originates from at least two distinct families of loops (large low-pressure loops associated with quiet regions, and small high-pressure loops in active centers). Moreover, in the stellar case it is usually not possible to derive separately the loop pressure p and the fractional area A_f of the star covered by loops, unless further constraints are imposed, often quite arbitrarely. Only the product ($A_f \cdot p$), upon which the level of coronal emission depends, can be determined. In spite of these drawbacks, the application of solar-type loop models to X-ray observations of stars is a quite promising approach which allows us to obtain at least a first glimpse of the physical conditions in stellar coronae.

Coronal temperatures derived from EINSTEIN and EXOSAT observations are usually in the range ~3-4 x 10^6 K for active solar type stars, while the emission measure may be up to two orders of magnitude higher (Schrijver et al. 1984, Landini et al. 1984a). High pressure loops and large filling factors are derived for active stars and RS CVn systems. For the latter, two-temperature coronae (at $T \sim 7 \times 10^6$ K and $T \gtrsim 20 \times 10^6$ K) have been derived from observations with the Solid State Spectrometer on EINSTEIN (Swank and White 1981). The two temperature components may represent spatially distinct coronal regions in RS CVn systems (Walter et al. 1983).

5. RADIO EMISSION FROM THE SUN AND LATE-TYPE STARS

The Sun emits radio waves by a variety of mechanisms and in a variety of different physical conditions (Kundu 1965, 1982). The thermal free-free emission from the quiet Sun is of no relevance for the stellar case, because emission of this type is several orders of magnitude below the detection threshold of the VLA even for the nearest stars. The situation may be somewhat better for the slowly varying component associated with active centers, although even in this case expectations based on solar analogies do not look very promising for other stars.

On the Sun, the slowly varying component at centimetric wavelengths consists of diffuse (halo) emission associated with plage areas, and of compact high brightness temperature components usually - but not exclusively - associated with sunspots (Kundu et al. 1977, Pallavicini et al. 1979). The low brightness temperature components associated with plages ($T_B \lesssim 10^5$ K) can be easily interpreted as due to optically thin thermal free-free emission. On the contrary, the high brightness temperature components associated with sunspots ($T_B \geqq 2 \times 10^6$ K) require an additional opacity source, which has been attributed to optically-thick thermal gyroresonance emission (cyclotron emission) in the intense magnetic fields of sunspots (Kakinuma and Swarup 1962, Zheleznyakov 1970). At longer wavelengths ($\lambda \gtrsim 20$ cm) the corona above active regions becomes optically thick by free-free absorption, and loop-like structures similar to X-ray loops are observed at radio wavelengths (Velusamy and Kundu 1981, Lang et al. 1982).

Recent detailed models based on high spatial resolution observations of solar active regions at X-ray, UV and radio wavelengths have given strong support to the gyroresonance interpretation of sunspot associated components (Alissandrakis et al. 1980, Pallavicini et al. 1981c, Schmahl et al. 1982). The absorption levels which are relevant for cyclotron emission in the solar case correspond usually to the second and third harmonics of the gyrofrequency, thus implying magnetic fields of several hundred Gauss in the lower corona. The models are able to reproduce quite satisfactorily the observed brightness and polarization distributions, including such detailed features as ring (or horse-shoe) structures observed with the VLA around sunspot umbrae (Alissandrakis and Kundu 1982, Lang and Willson 1982).

An alternative explanation for the high brightness temperature components is non-thermal emission. For instance, these sources could be produced by gyrosynchrotron emission by a small population of non-thermal electrons, analogously with solar microwave bursts (Pallavicini et al. 1979, Chiuderi-Drago and Melozzi 1984). The main problem in this case is the short life-time (~ 1 hour) of non-thermal electrons against collisional and radiative losses, which implies a continuous acceleration process. Although this is unlikely in the solar case, it may be a good explanation for quiescent microwave emission from dMe stars, owing to the high degree of activity and frequent flaring of these stars.

What are the microwave fluxes expected from nearby stars if the emission mechanism is similar to the Sun's slowly varying component? Thermal free-free emission from active centers can be excluded on the basis of temperatures and emission measures derived from X-ray observations of stellar coronae. The predicted free-free fluxes are at least one order of magnitude lower that the minimum detectable fluxes at the VLA (~ 0.1 mJy). Somewhat higher fluxes can be expected for gyroresonance emission prodided the source size is sufficiently extended. The microwave flux at λ = 6 cm in the optically thick case is given by:

$$S(6\text{ cm}) = 1.2\times10^{-6} \left(\frac{R_s}{R_*}\right)^2 \left(\frac{R_*}{R_\odot}\right)^2 T_B \left(\frac{1}{d_{pc}}\right)^2 \text{ mJy}$$

where R_* is the star radius and R_s is the radius of the optically-thick source at 6 cm. The brightness temperature T_B is equal in this case to the coronal electron temperature T_{cor} determined by X-ray observations ($T \sim 4\times10^6$ K). For typical nearby stars at distances of $\approx$ 10 pc, radio emission will be detectable with the VLA at a level $\gtrsim$ 0.5 mJy only if the source size is several times the stellar radius (the source size must be $\approx$ 3 times the stellar radius for solar-type stars and $\gtrsim$ 6 times the stellar radius for dwarf M stars). Magnetic fields of several hundred Gauss must exist at such large distances from the stellar surface if the gyroresonance interpretation is correct. This situation is quite different from the solar case, where gyroresonance absorption occurs at low heights in localized regions overlying sunspots. It is not surprising, therefore, that most radio surveys of solar type stars have failed so far to produce detections even for stars with active chromospheres and coronae (Bowers and Kundu 1981, Linsky and Gary 1983, Lang et al. 1985).

However, quiescent microwave emission has been observed in a number of dMe flare stars and has been attributed to thermal cyclotron emission (Gary and Linsky 1981, Topka and Marsh 1982, Fisher and Gibson 1982, Linsky and Gary 1983). As discussed above and elsewhere in this volume (Gary 1985), the thermal gyroresonance interpretation requires an extended source much larger than the stellar radius. The high magnetic fields required by the gyroresonance process imply either that the magnetic fields at photospheric level are much larger than on the Sun and of the order of $\simeq 10^4$ Gauss, or that the magnetic field decreases with distance much more slowly than in the potential case. Both alternatives are difficult to accept. It seems more plausible that quiescent microwave emission from dMe stars originates by a non-thermal process (e.g. gyrosynchrotron emission) involving a population of non-thermal particles accelerated more or less continuously in stellar active regions (Kundu and Shevgaonkar 1985, Lang et al. 1985). It is worth noticing that this kind of "quiescent" emission has been detected so far only in dMe stars with a high frequency of flare events. The detection of quiescent microwave emission from the GOV star χ^1 Ori (Gary and Linsky 1981) has not been confirmed by subsequent VLA observations (e.g. Lang et al. 1985). The detected emission was likely produced by a flare-like event in a previously unknown dMe companion (Linsky and Gary 1983).

On the Sun, microwave bursts are a well-known and characteristic feature of solar radio emission (Kundu and Vlahos 1982). They are usually associated with the impulsive phase of flares, and coincide in time with hard X-ray bursts. Both emissions indicate the acceleration of particles to supra-thermal energies and provide crucial information on the primary energy release mechanism. Microwave bursts are produced by gyrosynchrotron emission of mildly relativistic electrons spiralling in coronal magnetic fields. High spatial resolution observations with the VLA have shown that solar microwave bursts occur in magnetic loops and that the emission region is localized near the loop top (Marsh et al. 1979, Marsh and Hurtford 1980).

Flaring activity at centimetric wavelengths has been detected in a number of late-type stars, including UV Ceti flare stars, RS CVn binary systems and Algol-type systems. Flares on dMe stars are strongly reminiscent of microwave bursts on the Sun (Gibson 1983, Mullan 1984), but the energy involved may be many orders of magnitude larger. On the contrary, substantial differences exist between radio flares on RS CVn stars and solar microwave bursts. The time scales associated with flares on RS CVn stars are usually much longer than those characteristic of flares in dMe stars and in the Sun (Feldman et al. 1978, Feldman 1983). Moreover, VLBI observations indicate that radio emission from RS CVn and Algol-type systems involve regions as large as the binary separation (Clark et al. 1976, Lestrade et al. 1984, Mutel et al. 1984a), suggesting the existence of interconnecting magnetic loops as predicted by the model of Uchida and Sakurai (1983). Similar interconnecting loops have also been inferred from UV observations (Simon et al. 1980). Recent VLBI observations have revealed the existence of a core-halo structure in some RS CVn systems, with a compact unpolarized region, possibly associated with the active late-type subgiant, and an extended polarized region comparable in size with the binary separation (Mutel et al. 1984b). Unfortunately, the extreme degree of variability of RS CVn systems at radio wavelengths and the lack of systematic monitoring with adequate sensitivity, makes it difficult to distinguish between flaring activity and low-level, possibly quiescent emission from these systems.

In addition to microwave bursts, the Sun emits also a large variety of bursts at metric wavelengths. They have been classified in a number of different types (I to V) according to their dynamic spectra (Kundu 1965). While centimetric bursts give information on the flare primary energy release site in the transition region and low corona, metric bursts give information on the propagation of disturbances (mass ejections, shock waves etc.) through the outer corona. A variety of emission mechanisms, not all completely understood, are involved in solar metric bursts.

Decimetric and metric bursts have been observed on UV Ceti flare stars for more than two decades (Lovell et al. 1963, Spangler et al. 1974, Spangler and Moffett 1976). In spite of this, our knowledge of metric bursts in late-type stars is still very poor. We do not have anything similar to the classification of solar bursts in different types, and even the temporal association between metric bursts and optical flares is unclear. There are evidences, however, that suggest that stellar metric bursts are produced, as on the Sun, by mechanisms different from those responsible for microwave bursts, and that the emission refers to coronal phenomena which occur later than the impulsive phase of flares. An example is the recent observation by Kahler et al. (1982) of 408 MHz emission during a flare on YZ CMi, which was delayed by 17 min with respect to the impulsive phase at optical and X-ray wavelengths. On the basis of this rather long time lag, Kahler et al. have interpreted the observed radio burst in terms of Type IV metric emission, rather than as a Type II slow-drift burst. Systematic observations of metric bursts on flare stars with high sensitivity and at different frequencies would be extremely useful to elucidate the emission mechanism of stellar metric bursts and to investigate further the connection between solar and stellar phenomena.

REFERENCES

Alissandrakis, C.E., and Kundu, M.R. (1982) Astrophys. J. Letters 253, L49.

Alissandrakis, C.E., and Kundu, M.R. (1981) Astrophys. J. Letters 243, L103.

Athay, R.G., and White, O.R. (1978) Astrophys. J. 226, 1135.

Baliunas, S., and Dupree, A.K. (1982) Astrophys. J. 252, 668.

Basri, G.S., and Linsky, J.L. (1979) Astrophys. J. 234, 1023.

Belvedere, G. (1983) in Activity in Red-dwarf Stars (P.B. Byrne and M. Rodonò ed.), p. 579.

Belvedere, G., Chiuderi, C., and Paternò, L. (1981) Astron. Astrophys. 96, 369.

Belvedere, G., Chiuderi, C., and Paternò, L. (1982) Astron. Astrophys. 105, 133.

Belvedere, G., and Paternò, L. (1983) Astrophys. J. 268, 246.

Belvedere, G., Paternò, L., and Stix, M. (1980a) Astron. Astrophys. 88, 240.

Belvedere, G., Paternò, L., and Stix, M. (1980b) Astron. Astrophys. 91, 328.

Bopp, B.W., and Espenak, F. (1977) Astron. J. 82, 916.

Bopp, B.W., and Fekel, F.Jr. (1977) Astron. J. 82, 490.

Bowers, P.F., and Kundu, M.R. (1981) Astron. J. 86, 569.

Bruner, E.C. Jr. (1978) Astrophys. J. 226, 1140.

Catalano, S. (1983) in Activity in Red-dwarf Stars (P.B. Byrne and M. Rodonò eds.), p. 343.

Catalano, S., and Marilli, E. (1983) Astron. Astrophys. 121, 90.

Chiuderi-Drago, F., and Melozzi, M. (1984) Astron. Astrophys. 131, 103.

Clark, T.A. et al. (1976) Astrophys. J. Letters 206, L107.

Cowling, T.G. (1981) Ann. Rev. Astron. Astrophys. 19, 115.

Durney, B.R., Mihalas, D., and Robinson, R.D. (1981) Publ. Astron. Soc. Pacific 93, 537.

Durney, B.R., and Robinson, R.D. (1982) Astrophys. J. 253, 290.

Eaton, J.A., and Hall, D.S. (1979) Astrophys. J. 227, 907.

Feldman, P.A. (1983) in Activity in Red-dwarf Stars (P.B. Byrne and M. Rodonò eds.), p. 273.

Feldman, P.A., Taylor, A.R., Gregory, P.C., Seaquist, E.R., Balonek, T.J., and Cohen, N.L. (1978) Astron. J. 83, 1471.

Fisher, P.L., and Gibson, D.M. (1982) in Cool Stars, Stellar Systems and the Sun (M.S. Giampapa and L. Golub eds.), p. 109.

Gary, D.E. (1985) this volume.

Gary, D.E., and Linsky, J.L. (1981) Astrophys. J. 250, 284.

Gibson, D.M. (1983) in Activity in Red-dwarf Stars (P.B. Byrne and M. Rodonò eds.), p. 273.

Gilman, P.A. (1981) in The Sun as a Star (S. Jordan ed.), p. 231.
Gilman, P.A. (1983) in Solar and Stellar Magnetic Fields: Origins and Coronal Effects (J.O. Stenflo ed.), p. 247.
Gray, D.F. (1982) Astrophys. J. 258, 201.
Hall, D.S. (1976) in Multiple Periodic Variable Stars (W.S. Fitch ed.), p. 287.
Hartmann, L., Baliunas, S.L., Duncan, D.K., and Noyes, R.W. (1984) Astrophys. J. 279, 778.
Ionson, J.A. (1978) Astrophys. J. 226, 650.
Kahler, S., et al. (1982) Astrophys. J. 252, 239.
Kakinuma, T., and Swarup, G. (1962) Astrophys. J. 136, 975.
Knobloch, E., Rosner, R., and Weiss, N.O. (1981) Mont. Not. Roy. Astron. Soc. 197, 45P.
Kraft, R.P. (1967) Astrophys. J. 150, 551.
Kundu, M.R. (1965) Solar Radio Astronomy, New York, Wiley Interscience.
Kundu, M.R. (1982) Rep. Progress. Phys. 45, 1435.
Kundu, M.R., Alissandrakis, C.E., Bregman, J.D., and Hin, A.C. (1977) Astrophys. J. 213, 278.
Kundu, M.R., and Shevgaonkar, R.K. (1985) this volume.
Kundu, M.R., and Vlahos, L. (1982) Space Science Rev. 32, 405.
Kuperus, M. (1969) Space Sci. Rev. 713, 739.
Lang, K.R., and Willson, R.F. (1982) Astrophys. J. Letters 255, L111.
Lang, K., Willson, R., and Pallavicini, R. (1985) this volume.
Lang, K.R., Willson, R.F., and Rayrole, J. (1982) Astrophys. J. 258, 384.
Landini, M., and Monsignori-Fossi, B.C. (1981) Astron. Astrophys. 102, 391.
Landini, M., Monsignori-Fossi, B.C., and Pallavicini, R. (1984) in X-ray Astronomy '84, Bologna, June 1984, in press.
Landini, M., Monsignori-Fossi, B.C., Paresce, F., and Stern, R. (1984b) Astrophys. J., in press.
Lestrade, J.-F., Mutel, R.L., Preston, R.A., Scheid, J.A., and Phillips, R.B. (1984) Astrophys. J. 279, 184.
Linsky, J.L. (1983) in Solar and Stellar Magnetic Fields: Origins and Coronal Effects (J.O. Stenflo ed.), p. 313.
Linsky, J.L., and Gary, D.E. (1983) Astrophys. J. 250, 284.
Lovell, B., Whipple, F.L., and Solomon, L.H. (1963) Nature 198, 228.
Marcy, G.W. (1984) Astrophys.J. 276, 286.
Marsh, K.A., and Hurford, G.J. (1980) Astrophys. J. Letters 240, L111.
Marsh, K.A., Zirin, H., and Hurford, G.J. (1979) Astrophys. J. 228, 610.
Marstad, N., Linsky, J.L., Simon, T., Rodonò, M., Blanco, C., Catalano, S., Marilli, E., Andrews, A.D., Butler, C.J., and Byrne, P.B. (1982) in Advances in Ultraviolet Astronomy: Four Years of IUE Research, p. 554.
Mewe, R. (1979) Space Science Rev. 24, 101.
Middelkoop, F. (1982) Astron. Astrophys. 107, 31.
Moss, D., and Vilhu, O. (1983) Astron. Astrophys. 119, 115.
Mullan, D.J. (1985) this volume.
Mutel, R.L., Doiron, D.J., Lestrade, J.F., and Phillips, R.B. (1984a) Astrophys. J. 278, 220.
Mutel, R.L., Lestrade, J.F., Preston, R.A., and Phillips, R.B. (1984b)

Astrophys. J., in press.
Noyes, R.W., Hartmann, L.W., Baliunas, S.L., Duncan, D.K., and Vaughan, A.H. (1984) Astrophys. J. 279, 763.
Pallavicini, R. (1984) in Frontiers of Astronomy and Astrophysics (R. Pallavicini ed.), p. 83.
Pallavicini, R., Golub, L., Rosner, R., and Vaiana, G.S. (1982) in Cool Stars, Stellar Systems and the Sun (M.S. Giampapa and L. Golub eds.), Vol. II, p. 77.
Pallavicini, R., Golub, L., Rosner, R., Vaiana, G.S., Ayres, T., and Linsky, J.L. (1981a) Astrophys. J. 248, 279.
Pallavicini, R., Peres, G., Serio, S., Vaiana, G.S., Golub, L., and Rosner, R. (1981b) Astrophys. J. 247, 692.
Pallavicini, R., Sakurai, T., and Vaiana, G.S. (1981c) Astron. Astrophys. 98, 316.
Pallavicini, R., Vaiana, G.S., Tofani, G., and Felli, M. (1979) Astrophys. J. 229, 375.
Parker, E.N. (1979) in Cosmical Magnetic Fields, Oxford Univ. Press.
Phillips, M.J., and Hartmann, L. (1978) Astrophys. J. 224, 182.
Poletto, G., Vaiana, G.S., Zombeck, M.V., Krieger, A.S., and Timothy, A.F. (1975) Solar Phys. 44, 83.
Ramsey, L.W., and Nations, H.L. (1980) Astrophys. J. Letters 239, L121.
Robinson, R.D., Worden, S.P. and Harvey, J.W. (1980) Astrophys. J. Letters 236, L155.
Rosner, R., Golub, L., Coppi, B., and Vaiana, G.S. (1978) Astrophys. J. 222, 317.
Schmahl, E.J., Kundu, M.R., Strong, K.T., Bentley, R.D., Smith, J.B.Jr., and Krall, K.R. (1982) Solar Phys. 80, 253.
Schrijver, C.J., Mewe, R., and Walter, F.M. (1984) Astron. Astrophys. in press.
Scheeley, N.R. (1967) Astrophys. J. 147, 1106.
Simon, T., Linsky, J.L., and Schiffer, F.H.III (1980) Astrophys. J. 239, 911.
Skumanich, A. (1972) Astrophys. J. 171, 565.
Skumanich, A., Smythe, and Frazier, E.M. (1975) Astrophys. J. 200, 747.
Spangler, S.R., and Moffett, T.J. (1976) Astrophys. J. 203, 497.
Spangler, S.R., Shawhan, S.D., and Rankin, J.M. (1974) Astrophys. J. Letters 190, L129.
Stencel, R.E., and Mullan, D.J. (1980) Astrophys. J. 238, 221.
Stern, R.A. (1983) Adv. Space Res. 2, 39.
Swank, J.H., White, N.E., Holt, S.S., and Becker, R.H. (1981) Astrophys. J. 246, 208.
Topka, K., and Marsh, K.A. (1982) Astrophys. J. 254, 641.
Uchida, Y., and Sakurai, T. (1983) in Activity in Red-dwarf Stars (P.B. Byrne and M. Rodonò eds.), p. 629.
Ulmschneider, P. (1979) Space Sci. Rev. 24, 71.
Vaiana, G.S. et al. (1981) Astrophys. J. 245, 163.
Vaiana, G.S., and Rosner, R. (1978) Ann. Rev. Astron. Astrophys. 16, 393.
Vaughan, A.H. (1980) Publ. Astron. Soc. Pacific 92, 392.
Vaughan, A.H. (1983) in Solar and Stellar Magnetic Fields: Origins and Coronal Effects (J.O. Stenflo ed.), p. 113.

Vaughan, A.H., Baliunas, S.L., Middelkoop, F., Hartmann, L., Mihalas, D., Noyes, R.W., and Preston G.W. (1981) Astrophys. J. 250, 276.
Velusamy, T., and Kundu, M.R. (1981) Astrophys. J. Letters 243, L103.
Vilhu, O., and Rucinski, S.M. (1983) Astron. Astrophys. 127, 5.
Vogt, S.S. (1983) in Activity in Red-dwarf Stars (P.B. Byrne and M. Rodonò eds.), p. 137.
Walter, F.M. (1981) Astrophys. J. 245, 677.
Walter, F.M. (1982) Astrophys. J. 253, 745.
Walter, F.M., Cash, W., Charles, P.A., and Bowyer, C.S. (1980) Astrophys. J. 236, 212.
Walter, F.M., Gibson, D.M., and Basri, G.S. (1983) Astrophys. J. 267 665.
Weber, E., and Davis, L. (1967) Astrophys. J. 148, 217.
Wentzel, D.G. (1981) in The Sun as a Star, (S. Jordan ed.), p. 331.
White, O.R., and Livingston, W.C. (1981) Astrophys. J. 249, 798.
Wilson, O.C. (1963) Astrophys. J. 138, 832.
Wilson, O. (1978) Astrophys. J. 226, 379.
Willson, R.C., Gulkis, S., Janssen, M., Hudson, H.S., and Chapman, G.A. (1981) Science 211, 700.
Withbroe, G.L., and Noyes, R.W. (1977) Ann. Rev. Astron. Astrophys. 15, 363.
Yoshimura, H. (1975) Astrophys. J. Suppl. 29, 467.
Yoshimura, H. (1978a) Astrophys. J. 220, 692.
Yoshimura, H. (1978b) Astrophys. J. 226, 706.
Zheleznyakov, V.V. (1970) Radio Emission of the Sun and Planets, Oxford, Pergamon Press.

THE HR DIAGRAM FOR NORMAL RADIO STARS

D. M. Gibson[1]
Department of Physics and Research and Development Division
New Mexico Institute of Mining and Technology
Socorro, NM 87801
USA

ABSTRACT. It is found that nonthermal radio emission is associated with stars in very specific locations on the HR diagram. The four classes of objects are typified by early-type mass-loss stars (O5/WR), late-type giants and supergiants (M2II), subgiant K-stars (K0 IV-III), and flare stars (dMe). The members of each class exhibit about the same maximum radio luminosities, log (L_R/L_{bol}), and flaring timescales, spectra, and polarizations. Membership in a binary system is not found to be a necessary condition for detectable nonthermal emission.

1. THE SAMPLE

From a survey of the literature as well as from unpublished or prepublication results I have identified 77 normal stellar objects as nonthermal radio emitters. I list them in Table 1: 59 are incoherent emitters, 9 are coherent, and 9 have exhibited both types of emission. With the exception of the RS CVn binaries 39 Cet and V711 Tau, all of the coherent emitters can be identified with dMe (flare) stars. Of the sample, 42 are close (interacting) binaries, and 35 are single stars or wide binaries in which the radio emitting component has not yet been identified. Most of the close binaries (33) are listed in the updated version of the Hall Catalog of RS CVn Binary Star Systems.[2] Of the remainder, all have components similar to the "active" components in RS CVn systems or to dMe stars. The single stars include 8 very-hot, mass-loss stars, 3 red giants and supergiants, 2 (post) T Tauri stars, and 21 dMe (flare) stars.

---

[1] Visiting Fellow (1983-1984) at the Joint Institute for Laboratory Astrophysics, Boulder, CO. JILA is operated by the U. of Colorado and the National Bureau of Standards.

[2] Edited by D. S. Hall, M. Zeilik, and E. R. Nelson; available from M. Zeilik, Dept. of Phys. and Astron., U. of New Mexico.

R. M. Hjellming and D. M. Gibson (eds.), Radio Stars, 213–218.

Table 1. "NORMAL" STARS EXHIBITING NONTHERMAL RADIO EMISSION

Name	HR	HD	Gliese	Sp	$\log L_R$[a]	Ref.	C[b]	Ref.	Notes*
39 Cet	373	7672		wd+G5III	16.9	37	C	37	
UV Psc		7700		G2IV-V+K3IV-V	17.4	43			
L726-8A			65A	dM5.5e	14.0	13	C	13	
UV Cet			65B	dM6e	13.3	18	C	13	
Algol	936	19356		B8V+G5III	17.6	21			*
UX Ari		21242	14.1	G5V+K0IV	$(17.9)_{10}$	11			
V711 Tau	1099	22468		G5IV+K1IV	18.0	12	C	4	*
V471 Tau				wd+K7Ve	15.3	34			
		26337		G5IV+?	16.3	30			
b Per	1324	26961		A2IV+G	$(16.5)_{8}$	23			
V410 Tau				K3V-IVe	17.7	6			*
RZ Eri		30050		A5Vm+G8IV	$(16.7)_{8}$	32			
V1005 Ori			182	dM0.5e	$(15.4)_{15}$	33			*
12 Cam	1623	32357		K0III+?	16.8	8			
V371 Ori			207.1	dM3e			C	41	
σ Ori E	1932	37497		B2Vp	17.9	5			
		37847		F+G8III	16.9	39			
χ^1 Ori AB	2047	39587	222AB	G0V+dM	14.1	28			
α Ori	2061	39801		M2Iab	$(18.7)_{15}$	25			
π Aur	2091	40239		M3II	$(18.0)_{10}$	36			
		51268		K2IIIp+?	18.1	39			
AR Mon		57364		G8III+K2-3III	17.5	30			
YY Gem			278C	dM1.5e+dM1.5e	15.6	15			
σ Gem	2973	62044		K1III+?	16.9	22			
YZ CMi			285	dM4.5e	15.0	16	C	16	
54 Cam	3119	65626		G0V+G2V	16.3	43			
RX Pup		69190		(Symbiotic)	—	35			*
TY Pyx		77137		G2IV+K0IV	16.0	32			
		81410		F5V+K0III	18.1	39			
AD Leo			388	dM2.5e	$(15.4)_{1.4}$	27	C	27	
DM UMa				G8V-IV+?	16.8	30			
CN Leo			406	dM8e			C	31	*
o Hya	4494	101379		A0V+K3III	17.8	7			
Ross 128			447	dM5			C	31	
DK Dra	4665	106677		K0III+K0III	16.5	30			
Wolf 424			473	dM5.5e			C	44	
RS CVn		114519	501.1	F4IV-V+K0IV	17.2	19			
BH CVn	5110	118216		F2IV+KIV	$(18.1)_{10}$	11			
Proxima Cen			551	dM5e	13.8	40			
RV Lib		128171		G2-5+K5	17.3	30			
BD+16°2708			569	dM0e			C	31	
		137164		K1IV+K2IV	16.9	29			
σ^2 CrB	6064	146361	615.2A	F8V+G0V	$(16.5)_{15}$	22			
DoAr 21				dGe	18.1	10			*
WW Dra		150708		G2IV+K0IV	17.1	26			
Wolf 630 AB		152751	644AB	dM4e+dM4e	15.0	18	C	31	
		155638		G8IV+?	16.1	30			
29 Dra		160538		K0-2III+?	18.0	8			
9 Sgr	6736	164794		O4V	18.8	2			
AM Her				wd+dM4	15.6	9	C	9	
		167971		O8If	19.8	1			
		168112		O5III	18.7	1			
V1216 Sgr			729	dM4.5e			C	31	
V1285 Aql			735	dM2e		47	C	31	
R Aqℓ		177940		gM5e-8e	$(18.8)_{10}$	39			
MR 93				WC7	18.8	1			
		185510		K0IV-III+?	16.4	8			
VW Cep				G6V+K1V	15.0	24			
		193793		WC7	19.6	14			
Cyg OB2 No. 9				O5f	19.5	2			
Cyg OB2 No. 8A				O6Ib	18.7	1			
WR 147				WN7	20.1	1			
AT Mic		196982	799AB	dM4.5e+dM4.5e	15.2	42			
AU Mic		197481	803	dM1.6e	15.3	18			

Name	HR	HD	Gliese	Sp	log L_R[a]	Ref.	C[b]	Ref.	Notes*
FF Aqr				sdOB+G8III	17.7	8			
RT Lac		209318		G9IV+K1IV	17.8	20			
HK Lac		209813		FIV+K0III	$(17.8)_{10}$	11			
AR Lac		210334		G2IV+K0IV	$(17.9)_{10}$	11			
V350 Lac	8575	213389		K2IV-IIIp+?	$(17.1)_{10}$	11			
DM-21°6267AB			867AB	dM2e+dM4e			C	31	
EV Lac			873	dM4.5e			C	29	
IM Peg	8703	216489		K1IV-IIIp+?	18.2	43			
SZ Psc		219113		F8V+K1IV	$(18.1)_{10}$	32			
EQ Peg A			896A	dM3.5e	$(14.1)_{15}$	18	C	31	
EQ Peg B			896B	dM4.5e	13.3	45			
λ And	8961	222107		G8IV-III+?	16.6	3			
II Peg		224085		K2-3V-IV+?	$(17.4)_{10}$	11			

[a] L_R is the maximum observed 5 GHz incoherent radio luminosity of the star (ergs s^{-1} Hz^{-1}). L_R's measured at other frequencies are indicated in parentheses with the frequency (in GHz) indicated as a subscript.
[b] C indicates the star has been detected as a coherent emitter.
*Notes:
Algol (=β Per); 5 GHz max interpolated
V711 Tau; max interpolated
V410 Tau; a T Tau or post T Tau star
V1005 Ori; uncertain whether event was incoherent
RX Pup; the distance is quite uncertain; $S_{8.7}^{max}$ = 50 mJy; if d = 1 kpc, log L_R = 19.7
CN Leo; the ID is somewhat questionable because the field is heavily confused
DoAr 21 (= 10 Oph); a T Tau or post T Tau star in the Rho Oph cloud

2. DISCUSSION

It is important to point out that the sample is incomplete in the sense that it is not luminosity-limited even within a particular class of stars. Thus, any conclusions we draw here must be viewed with some skepticism. However, it is also worthwhile noting that the trends which we note here began to emerge when the number of detections was substantially less.

I have plotted the location(s) of those stars (or systems) which are detected radio emitters on a variety of HR diagrams. Plotting both components of binaries or arbitrarily plotting either component together with the single stars and the known components of single-line spectroscopic binaries leads to HR diagrams in which there is a lot of scatter. In contrast, the tightest HR diagram is that for which I choose the component of double-line systems to be most like those in single or single-line of systems (see Fig. 1). A number of observational and physical conclusions follow from this grouping:

1) There are four regions in the HR diagram where nonthermal radio stars are found: A - the WR/O5 region; B - the M2II region; C - the K0IV-III region; and D - the dMe region. The magnetic B star, σOri E (B2Vp), is at present unique. The large "gap" in spectral types B, A, and F appears to be real. That a star's location in one of these four regions is a necessary -- though not sufficient -- condition for its detection at today's sensitivities can be tested by a thorough examination of existing data including upper-limits and/or new observations.

2) While the range in peak incoherent 5 GHz luminosities L_R (ergs s^{-1} Hz^{-1}; see Table 1; also indicated by the size of the "bubble" in Fig. 1) is ~10^7 for all stars in the sample it shows a surprisingly small range for stars within each class ($\leqslant 10^2$). In addition, the L_R's seem to "pile up" near the maximum L_R among stars within each class. A maximum L_R for a particular class may result if their coronae can only support a maximum energy density u_{rel} of radiating particles, one which is just below the level at which the corona would be disrupted.

3) The fraction of stellar energy that escapes as radio emission can be estimated by comparing the integrated maximum radio luminosity L_R^* to the bolometric luminosity. We form L_R^* by multiplying L_R by an "effective" bandwidth of 15 GHz, assuming the radio emission is broad-band and the spectrum is relatively flat. Typical ratios of L_R^*/L_{bol} for each of the stellar classes are:

Class :	A	B	C	D
L_R^*/L_{bol} :	10^{-10}	10^{-9}	10^{-7}	10^{-7}

That RS CVn and dMe stars comprise the bulk of the detected nonthermal stars would seem to be a reflection of their relatively high L_R^*/L_{bol} ratios. The same stars also have relatively high ratios of L_x/L_{bol}, ~$10^{-3.5}$ and ~10^{-3} respectively. The ratio L_R^*/L_x ~10^{-4} suggests that the same or a closely-linked process is responsible for heating the coronal gas and accelerating the radio-emitting electrons.

4) There are no apparent differences in L_R between binaries with two cool components, binaries with one hot and one cool component, and single stars for classes C and D. (All the stars in classes A and B are apparently single.) This sugggests that flares need not be triggered and sustained by interacting magnetospheres (cf. Simon et al. 1980; Uchida and Sakurai (1983) but rather are intrinsic to the active component(s). This further strengthens my association of the radio emission with the late-type active component in binaries containing both hot and cool stars.

5) The late-type stars (Classes B, C, and D) are located in parts of the HR diagram where there is good reason to suspect that the surfaces of the stars are being "braked" - magnetically, by strong winds, or by angular momentum transfer due to recent post-main sequence expansion -- with respect to their interiors. It remains to be evaluated whether radial differential rotation is a necessary (though probably not sufficient) condition for the presence of nonthermal radio emission in late-type stars.

3. REFERENCES

1 Abbott, D.C., Bieging, J. and Churchwell, E. 1985, this volume.
2 Abbott, D.C., Beiging, J. and Churchwell, E. 1984, Ap. J., **280**, 67.

3 Bath, G.T. and Wallerstein, G. 1976, Pub.A.S.P., **88** 754.
4 Brown, R.L. and Crane, P.C. 1978, A. J., **83**, 1504.
5 Churchwell, E., Abbott, D.C. and Beiging, J. 1984, private communication.
6 Cohen, M. and Beiging, J. 1984, private communication.
7 Collier, A.C. et al. 1982, M.N.R.A.S., **200**, 869.
8 Drake, S.A., Simon, T. and Linsky, J.L. 1984, in preparation.
9 Dulk, G.A., Bastian, T.S. and Chanmugan, G. 1983, Ap. J., **273**, 249.
10 Feigelson, E. 1984, private communication.
11 Feldman, P.A. 1983, in Activity in Red Dwarf Stars, eds, P.B. Byrne and M. Rodono (Reidel:Dordrecht), p. 429.
12 Feldman, P.A. et al. 1978, A. J., **83**, 1471.
13 Fisher, P.L. and Gibson, D.M. 1982, in Second Cambridge Workshop on Cool Stars, Stellar Systems, and the Sun, SAO Sp. Rept. 393), eds.
14 Florkowski, D. and Johnston, K. 1984, private communication.
15 Gary, D.E. and Linsky, J.L. 1984, private communication.
16 Gibson, D.M. 1984, Proc. Southwest Reg. Conf. Astron. Astrophys., **IX**, 35.
17 Gibson, D.M. 1983, in Activity in Red Dwarf Stars, eds. P.B. Byrne and M. Rodono (Dordrecht:Reidel), p. 273.
18 Gibson, D.M. and Cox, J.J. 1984, in preparation.
19 Gibson, D.M. and Newell, R.T. 1979, IAU Circ. 3337.
20 Gibson, D.M., Owen, F.N. and Hjellming, R.M. 1978, Pub. A.S.P., **90**, 751.
21 Gibson, D.M., Viner, M.R. and Peterson, S.D. 1975, Ap. J., **200**, L143.
22 Hjellming, R.M. and Gibson, D.M. 1980, in IAU Symp. 86: Radio Physics of the Sun, eds. M.R. Kundu and T.E. Gergely (Reidel:Dordrecht), p. 209.
23 Hjellming, R.M. and Wade, C.M. 1973, Nature, **242**, 716.
24 Hughes, V.A. and McLean, B.J. 1984, Ap. J., **278**, 716.
25 Kellermann, K.I. and Pauliny-Toth, I.I.K. 1966, Ap. J., **145**, 953.
26 Lang, K.R. and Willson, R. 1983, in Proc. 7th Eur. Reg. Astr. Meeting.
27 Lang, K.R. et al. 1983, Ap. J., **272**, L15.
28 Linsky, J.L. and Gary, D.E. 1983, Ap. J., **274**, 776.
29 Lovell, B. and Chugainov, P.F. 1964, Nature, **203**, 1213.
30 Mutel, R.L. and Lestrade, J.F. 1984, Ap.J., in press.
31 Nelson, G.J. et al. 1979, M.N.R.A.S., **187**, 405.
32 Owen, F.N. and Gibson, D.M. 1978, A. J., **83**, 1488.
33 Rodono, M. et al. 1984, in Proc. Fourth Eur. IUE Conference, ESA SP-218, p. 247.
34 Sanders, W.T. and Sramek, R.A. 1984, private communication.
35 Seaquist, E.R. 1977, Ap. J., **211**, 547.
36 Seaquist, E.R. 1967, Ap. J., **148**, L23.
37 Simon, T., Fekel, F.C. and Gibson, D.M. 1984, Ap. J., submitted.
38 Simon, T., Linsky, J.L. and Schiffer, F.H. 1980, Ap. J., **239**, 911.
39 Slee, O.B., Haynes, R.F. and Wright, R.E. 1984, M.N.R.A.S., in press.

40 Slee, O.B. and Page, A.A. 1979, in IAU Coll. 46: Changing Trends in Variable Star Research, eds. F.M. Bateson, J.Smak and I.H. Urich (Univ. Waikato), p. 150.
41 Slee, O.B., Solomon, L.H. and Patson, G.E 1963, Nature, **199**, 991.
42 Slee, O.B. et al. 1981, Nature, **292**, 220.
43 Spangler, S.R., Owen, F.N. and Hulse, R.A. 1977, A. J., **82**, 989.
44 Spangler, S.R., Shawhan, S.D. and Rankin, J.M. 1974, Ap. J., **190**, L129.
45 Topka, K. and Marsh, K.A. 1982, Ap. J., **254**, 641.
46 Uchida, Y. and Sakurai, T. 1983, in Activity in Red Dwarf Stars, eds. P.B. Byrne and M. Rodono (Dordrecht:Reidel), p. 629.
47 Woodsworth, A. and Hughes, V.A. 1973, Nature Phys. Sci., **246**, 111.

OBSERVATIONS OF NONTHERMAL EMISSION FROM EARLY-TYPE STARS

D. C. Abbott,[1] J. H. Bieging[2] and E. Churchwell[3]
[1]Joint Institute for Laboratory Astrophysics, National Bureau of Standards and University of Colorado, Boulder, Colorado 80309
[2]Radio Astronomy Laboratory, University of California, Berkeley, California 94720
[3]Washburn Observatory, University of Wisconsin, Madison, Wisconsin 53706

ABSTRACT. Observations of a distance-limited sample of 44 Wolf-Rayet (WR) and 25 OB stars were made at 6 cm with the VLA. The stronger sources were reobserved at 2 cm and 6 cm. Based on the criteria of source strength, variability, spectral index, and spatial extent, we conclude that the emission from 5 OB stars and 2 WR stars is produced by nonthermal processes. The ratio of probable nonthermal emitters to probable thermal emitters in our sample is 0.6 in the OB stars and 0.2 in the WR stars.

1. INTRODUCTION

The discovery of nonthermal radio emission from two luminous, O-type stars by Abbott, Bieging and Churchwell (1984a), demonstrated the existence of a new and unexpected physical process in at least some early-type stars. Further observations, some of which are reported here, show that nonthermal radio emission is quite widespread in these stars. The following two types of nonthermal emitters are already identified:

1) "Stellar wind" nonthermal emitters. These stars are all losing mass at high rates, which makes their winds opaque out to hundreds of stellar radii. The nonthermal emission must therefore be generated *in situ* by processes within the wind itself.

2) "Surface" nonthermal emitters. By contrast, these stars have very weak, or no measurable mass loss, so their winds are transparent at radio wavelengths. The source of the emission is likely close to the stellar surface, which implies high brightness temperatures. The probable examples of "surface" emitters are σ Ori A (O9.5 V), σ Ori E (B2 V), and HD 37017 (B2 V) from Drake *et al.* (1985), θ^1 Ori A (B1) Garay (1984), and HD 26676 (B8 V) Strom and Harris (1977). All but one were serendipitous

R. M. Hjellming and D. M. Gibson (eds.), Radio Stars, 219–224.

discoveries, so it is not clear how prevalent this behavior is, or even if these objects are related. Two stars are members of the helium-variable spectral class with optically detected, kilogauss magnetic fields, which suggests a synchotron origin.

Since many of the spectral classes of early-type stars remain unstudied at radio wavelengths, more examples of nonthermal emitters will probably be forthcoming.

This paper presents further observations with the NRAO[4] Very Large Array (VLA) of the "stellar wind" nonthermal emitters. We undertook a survey of continuum radio emission from a distance-limited sample of Wolf-Rayet (WR) and the most luminous OB stars. All have very dense winds. We present here summary results of the nonthermal emitters in our sample.

2. OBSERVATIONS

2.1. The Survey

The 6 cm survey covered 44 WR stars within 3 kpc and above declination $\delta \geq -47°$, and 25 OB stars within 2.5 kpc, with bolometric luminosity $L_* \geq 10^6 L_\odot$, and above declination limit $\delta \geq -47°$. Both stellar samples are complete to the indicated distance. The stronger sources were re-observed at two wavelengths. All sources were observed to a limiting sensitivity of 0.1 mJy, except where confusing sources imposed a higher limit. Figure 1 shows example intensity maps of new nonthermal sources.

A major difficulty is to distinguish thermal wind emission from nonthermal mechanisms. We have characterized the radio emission from each of our stars following the empirical criteria of Abbott (1984). The results are summarized in Table 1. Nonthermal emission dominates in 24% of the OB stars and 10% of the WR. The phenomenon may be even more widespread given the large number of undetermined stars. The lower percentage for the WR stars is suggestive that the nonthermal mechanism weakens as a star evolves. However, it may also be a contrast effect, because these stars are all very strong thermal wind emitters.

2.2. The Stellar Wind Nonthermal Radio Sources

Table 2 presents the observations of the seven sources considered to be "definite" nonthermal radio emitters on the basis of their radio emission strength, variability, spectral index, and/or spatial extent. For the O-type stars, we have not found any identifiable characteristic in the optical or UV spectrum which distinguishes the nonthermal

[4]The National Radio Astronomy Observatory is operated by Associated Universities, Inc., under contract with the National Science Foundation.

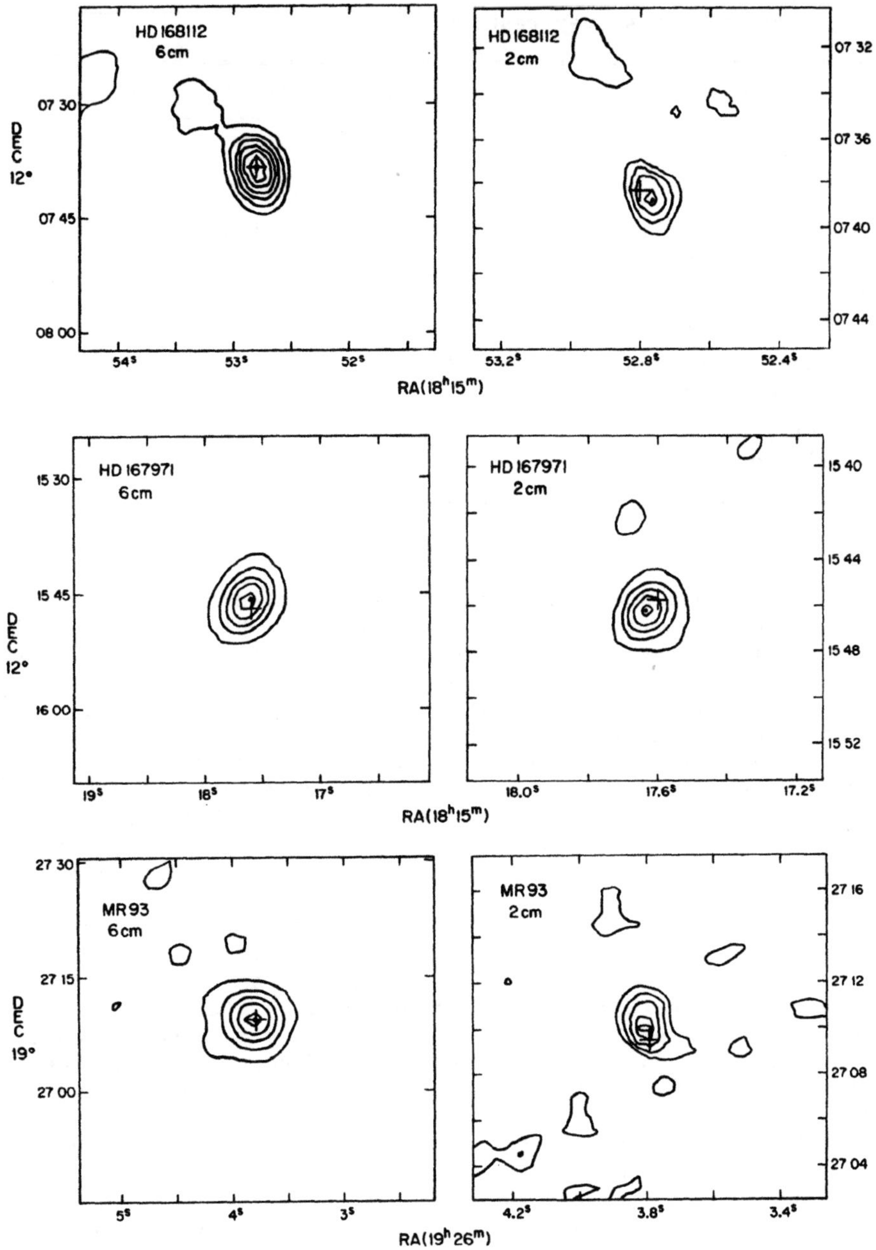

Fig. 1. VLA intensity maps at 2 cm and 6 cm of three of the new sources of nonthermal radio emission. Crosses indicate the optical position of the star. The contours represent 10%, 30%, 50%, 70%, 90%, and 98% of the peak flux densities given in Table 2.

Table 1. Summary of survey results.

Category	No. OB Stars	No. WR Stars
1. Definite thermal wind emission	6 (24%)	7 (16%)
2. Probable thermal wind emission	4 (16%)	16 (36%)
3. Unknown, undetected, or not observed	9 (37%)	17 (39%)
4. Probable nonthermal wind emission	1 (3%)	2 (5%)
5. Definite nonthermal emission	5 (20%)	2 (5%)

emitters from other stars of similar spectral type. By contrast, the two WR nonthermal emitters have optical spectra which are obviously peculiar (Torres, 1984, priv. comm.). All the stars in Table 2 have high rates of mass loss, as determined from their UV or optical spectra, so the source of the emission must be located far from the stellar photosphere ($R/R_* \gtrsim 100$).

One object, WR 147 (WN7), was included as a nonthermal source in an earlier contribution describing our survey (Abbott, Bieging and Churchwell 1984b). Subsequent analysis has shown that the 6 cm emission from WR 147 is resolved, and we now consider it to be definitely thermal, although it deviates from the standard model of thermal wind emission.

No linear or circular polarization was detectable in any star. The Stoke's parameters Q, U, and V were always less than the 2 sigma level, which implies upper limits to the percentage polarization of <6%. The radio wavelength luminosities are very similar, typically 10^{19} ergs s^{-1} Hz^{-1} at 5 GHz, although this could be a selection effect, because sources of low luminosity would not usually be detectable due to lack of sensitivity or the presence of thermal wind emission.

The level of the nonthermal fluxes of these sources are usually stable on time scales of months and years. The spectral index exhibits somewhat larger amplitude variability. However, it is clear from the star Cyg OB2 No. 9 that major fluctuations in the flux levels can occur.

Table 2. "Stellar wind" nonthermal radio emitters.

Star	Spectral Type	Date of Observation	Flux Density m(Jy) 2 cm	6 cm	20 cm	Spectral[a] Index	Representative 6 cm Luminosity
9 Sgr	O4 V	79 Jul 13	---	1.0 ± 0.4	---	---	
		80 May 22	---	1.8 × 0.3	---	---	
		82 Feb 9	≲2.4	2.5 ± 0.3	---	≲0.0	
		82 May 26	---	2.4 ± 0.3	3.6 ± 0.3	−0.3 ± 0.1	
		83 Aug 22	≲0.8	1.5 ± 0.2	3.9 ± 0.4	−0.8 ± 0.2	6×10^{18} ergs s^{-1} Hz^{-1}
Cyg OB2 No. 9	O5f	80 May 22	---	7.1 ± 0.4	---	---	
		81 Oct 16	≲4.9	4.3 ± 0.4	---	≲0.1	
		82 Aug 26	---	4.6 ± 0.2	4.5 ±0.2	0.0 ± 0.1	
		83 May 9	1.2 ± 0.3	0.7 ± 0.2	≲1.0	0.6 ± 0.3	
		83 Aug 22	---	1.6 ± 0.2	---	---	
		84 Mar 4	---	5.6 ± 0.1	---	---	
		84 Apr 4	4.2 ± 0.2	6.0 ± 0.2	---	0.3 ± 0.1	2×10^{19}
HD 168112	O5 III	84 Mar 9	---	1.3 ± 0.1	---	---	
		84 Apr 4	1.2 ± 0.1	1.9 ± 0.1	---	−0.4 ± 0.2	6×10^{18}
Cyg OB2 No. 8A	O6Ib	80 May 22	---	1.0 ± 0.2	---	---	
		84 Mar 4	---	0.8 ± 0.1	---	---	
		84 Mar 9	0.5 ± 0.2	0.7 ± 0.1	---	−0.4 ± 0.4	3×10^{18}
HD 167971	O8I	84 Mar 4	---	15.4 ± 0.2	---	---	
		84 Apr 4	7.0 ± 0.2	13.8 ± 0.2	---	−0.6 ± 0.1	7×10^{19}
MR 93	WC7	82 Aug 20	---	1.3 ± 0.1	---		
		84 Apr 4	0.9 ± 0.1	1.5 ± 0.1	---	−0.5 ± 0.2	7×10^{18}
HD 193793	WC7+Abs	(1),(2),(3)	Variable	Variable	---	Variable	3×10^{19}

[a]Defined by $S_\nu \propto \nu^\alpha$.

1. Florkowski (1982).
2. White (1984, private communication).
3. Florkowski and Johnson (1984, private communication).

3. CONCLUSIONS

To summarize, we have identified seven O-type and WR-type stars which exhibit nonthermal emission at radio wavelengths. We believe these stars represent a class of object, which we call "stellar wind" nonthermal emitters, with the following characteristics:

- Strong stellar winds which are opaque at radio wavelengths.
- A luminosity at 5 GHz of $\approx 10^{19}$ ergs s^{-1} Hz^{-1}.
- A spectrum in the range $0.0 \geq \alpha \geq -0.7$ ($S_\nu \propto \nu^\alpha$).
- No measurable circular or linear polarization.
- Lack of pronounced variability, although order of magnitude changes can happen (e.g. Cyg OB2 No. 9).

We acknowledge the support of National Science Foundation Grants AST82-18375 to the University of Colorado, AST81-14717 to the University of California, and AST79-05578 to the University of Wisconsin.

REFERENCES

Abbott, D. C. 1985, these proceedings.
Abbott, D. C., Bieging, J. H., and Churchwell, E. 1984a, Ap. J., **280**, 671.
_____. 1984b, to appear in Proceedings, Workshop/Colloquium on "The Origin of Non-Radiative Heating/Momentum in Hot Stars," ed. A. B. Underhill.
Drake, S. A., Abbott, D. C., Bieging, J. H., Churchwell, E., and Linsky, J. L. 1985, these proceedings.
Florkowski, D. R. 1982, in Wolf-Rayet Stars: Observations, Physics, and Evolution, eds. C. de Loore and A. Willis (Reidel: Dordrecht), p. 63.
Garay, G. 1985, these proceedings.
Strom, R. G. and Harris, D. E. 1977, Nature, **269**, 581.

RADIO EMISSION FROM AM HERCULIS

T. S. Bastian and G. A. Dulk
Department of Astrophysical, Planetary, and Atmospheric Sciences
University of Colorado, Boulder, 80309

G. Chanmugam
Department of Physics and Astronomy
Louisiana State University, Baton Rouge, 70803

ABSTRACT. Observations of the quiescent microwave emission of the magnetic cataclysmic variable AM Herculis are presented. The emission, which declined from a mean value of 0.58 mJy at 4.9 GHz to ≈0.3 mJy, in rough coincidence with the entry of AM Herculis into an optical low-state (mid-1983), is explained in terms of optically thick gyrosynchrotron emission. It is noted that the observation of a coherent outburst at 4.9 GHz, interpreted as the result of a cyclotron maser on the red dwarf secondary, indicates that the secondary is magnetized. Possible implications are briefly explored. Comparisons between this system and other stellar continuum radio sources are made.

1. INTRODUCTION

AM Herculis is the prototype of a subclass of cataclysmic variables which are characterized by the circularly polarized optical radiation they emit (Tapia, 1977). They number ≈10 and are believed to be close binaries composed of a strongly magnetized (few $\times 10^7$ gauss) white dwarf primary accreting mass from a late-type red dwarf secondary. The magnetic field prevents the formation of an accretion disk. Instead, matter is funneled directly onto the magnetic pole(s) via an accretion column. Some distance above the pole, the accreting plasma forms a strong, standing shock. Cyclotron emission at optical and UV wavelengths is produced in the foreshock region, and hard X-rays (≈30 keV) are produced by bremsstrahlung in the hot post-shock region.

Emission at radio wavelengths may have no direct bearing on the dynamics and energetics of the accretion process, but serves as a useful probe of the white dwarf magnetosphere and its interaction with the red dwarf secondary. Our purpose in this paper is to present and interpret the observed radio emission from AM Her, and to compare it with other stellar radio sources, e.g., RS CVn's and dMe flare stars.

R. M. Hjellming and D. M. Gibson (eds.), Radio Stars, 225–228.

2. OBSERVATIONS

All observations of the microwave emission from AM Her have been made with the Very Large Array. Table I summarizes the 4.9 GHz observations from late-1981 to March, 1984. The emission declined moderately from late-1981 to mid-1983 when it dropped below the limits of detectability for a time, before reappearing at a reduced level in early-1984. The abrupt decline in mid-1983 was in rough coincidence with the entry of AM Her into an optical low-state and a change in state of the soft X-ray emission (J. Heise, private communication). AM Her was detected at 14.9 GHz on 19 March 1983 and upper limits on the emission at 1.45 GHz and 14.9 GHz were established on 8 July 1982. We also note that the observations of 8 July 1982 included the detection of an outburst from the red dwarf secondary. This emission was characterized by very high brightness, $T_b > 10^{10}$K, strong RH polarization, large variations over short timescales (10 s), and a peak flux density ≈20 times that of the (unpolarized) quiescent emission.

TABLE I

Date	Freq.(GHz)	Flux Density(mJy)	Array
10-14-81	4.9	0.67 ± 0.05	C/D
7-8-82	4.9	0.55 ± 0.05	A
	1.45	<0.24	
	14.9	<1.11 (3σ)	
3-19-83	4.9	0.52 ± 0.06	C
	14.9	0.52 ± 0.10	
9-10-83	4.9	<0.30	A
1-23-84	4.9	<0.29 (3σ)	B
3- 7-84	4.9	0.32 ± 0.04	C

3. INTERPRETATION

The brightness temperature of the observed quiescent microwave emission (≈0.6 mJy at 4.9 GHz before entry into the low-state) is too high to be explained by thermal bremsstrahlung. Therefore we have modeled the emission in terms of optically thick gyrosynchrotron radiation. We assume the emission results from energetic (≈500 keV) electrons, which are injected into the magnetosphere at a steady rate with an initial power law distribution:

$$N(\varepsilon) = A\varepsilon^{-\alpha}, \; A = (\alpha-1)\, N_o\, \varepsilon_o^{\alpha-1}$$

where N_o is the number of electrons with $\varepsilon > \varepsilon_o$, ε_o taken to be 10 keV in the present context. Balancing the steady injection of a power law electron spectrum (α taken to be constant in time), against synchrotron losses yields a time-independent differential number density of (Melrose, 1980)

$$N(\varepsilon, r_9) = 1.7\times10^{-12}\; \varepsilon_o^{\alpha-1}\; N_o\; r_9^{\,6}\; \varepsilon^{-\alpha}/(\varepsilon+2m_ec^2)$$

where r_9 is the radius from the white dwarf in units of 10^9 cm and a dipolar magnetosphere with $B = 2\times10^7\ r_9^{-3}$ gauss has been assumed. The number density depends sensitively on radius, with the inner magnetosphere being depleted of energetic electrons relative to the outer magnetosphere. This perhaps explains the peak in the microwave spectrum somewhere between 4.9 and 14.9 GHz; the latter emission must necessarily arise from deeper in the magnetosphere.

Dulk, Bastian, and Chanmugam (1983) show that the outburst observed at 4.9 GHz was probably due to the action of a cyclotron maser operating at the second harmonic of the gyrofrequency on or near the red dwarf secondary. Field strengths of $\approx$1000 gauss are implied, indicating that the secondary is magnetized. Comparable field strengths are known to occur on the Sun, whose magnetic moment is $\approx 10^{-2}$ that of the white dwarf primary. Therefore the field of the secondary need not be a large perturbation on the primary's magnetosphere. On the other hand, late-type red dwarfs are known to be approaching a fully convective state. This, coupled with the strictly enforced rapid rotation of the system (Period = 3.1 hr) may quite possibly yield a much larger magnetic moment for the secondary, and conceivably alter the magnetospheric configuration in a drastic manner.

A preliminary investigation into this possibility in terms of simple aligned or anti-aligned dipole pairs has shown that the observed optically thick gyrosynchrotron emission is not inconsistent with the presence of an anti-aligned secondary magnetic moment 0.2 times that of the primary. The effect of introducing a secondary dipole into the white dwarf's magnetosphere is to enhance the 4.9 GHz source size relative to the 14.9 GHz source, again offering a possible explaination for the turn-over in the microwave spectrum. It must be stressed, however, that this preliminary investigation did not include the effect of synchrotron losses, which have been shown to play an important role in determining the spatial distribution of energetic electrons.

An interesting consequence of this second model (sketched above, but explored in more detail by Bastian, Dulk, and Chanmugam, 1985), is the possibility of particle acceleration by magnetic reconnection. This may take place in a global sense, as described by Bahcall, Rosenbluth, and Karlsrud (1973), or in a more localized fashion as emerging flux on the secondary reconnects into the ambient field. Another interesting possibility is the capacity of the secondary to modulate the transfer of mass onto the white dwarf primary by means of a magnetic cycle similar to that observed on the Sun and other stars. The rough coincidence between the entry of AM Her into an optical low-state and a radio low-state is intriguing for this reason.

4. COMPARISON WITH OTHER STELLAR RADIO SOURCES

A comparison can be made between AM Her and other stellar gyrosynchrotron sources. The quiescent radio luminosity of AM Her is 3.4 Lu (1 Lu = 10^{15} erg s^{-1} Hz^{-1}), whereas a radio luminosity of 65 Lu was attained during the outburst. The binary system L726-8, composed of L726-8 A (dM5.5e) and UV Cet (dM6e), has been detected at 4.9 GHz both in quiescence and in an outburst similar in many respects to that observed from AM Her (Gary, Linsky, and Dulk, 1982). The radio luminosities are 0.012 Lu and 0.15 Lu, respectively. In contrast, two nearby RS CVn's, V711 Tau and AR Lac, observed at 2.965 and 8.085 GHz, have quiescent radio luminosities of 20-40 Lu (Owen and Gibson, 1978). Brown and Crane (1978) observed a strongly polarized and rapidly varying outburst on V711 Tau which reached a peak radio luminosity of 320 Lu.

AM Her, both in quiescence and outburst, lies between the dMe binary and the RS CVn's in terms of radio luminosity. The question to be asked is whether the enhanced radio luminosity of AM Her relative to L726-8 is a result of the M component being forced to rotate rapidly, or whether AM Her is under-luminous relative to the RS CVn's because it is a close binary.

5. CONCLUSIONS

The radio emission of AM Her is consistent with optically thick gyrosynchrotron emission from a power law injection spectrum of energetic ($\approx$500 keV) electrons. The observed outburst at 4.9 GHz of coherent radiation from the secondary implies that it is magnetized. While it need not possess a large magnetic moment, the deep convective envelope and the rapid rotation of the secondary lead us to believe that it may indeed possess a large magnetic moment, with consequences that have by no means been sufficiently explored.

Comparisons of the AM Her magnetic cataclysmic variable with quiescent and flaring radio emission of dMe flare stars and the RS CVn's places Am Her between the two classes. AM Her has attributes of both classes of radio source.

ACKNOWLEDGEMENTS

This work was supported in part by NASA grants NSG-7287 and NAGW-91 to the University of Colorado.

The National Radio Astronomy Observatory Very Large Array is operated by Associated Universities, Inc., under contract with the National Science Foundation.

REFERENCES

Bahcall, J.N., Rosenbluth, M.N., and Karlsrud, R.M. (1973) Nat. Phys. Sci., **243**, 27.

Bastian, T.S., Dulk, G.A., and Chanmugam, G. (1985), to be submitted.

Brown, R.L., and Crane, P.C. (1978), Ap. J., **83**, 1504.

Dulk, G.A., Bastian, T.S., and Chanmugam, G. (1983), Ap. J., **273**, 569.

Gary, D.E., Linsky, J.L., and Dulk, G.A. (1982), Ap. J. (Lett.), **273**, L79.

Melrose, D.B. (1980), Plasma Astrophysics, v.2, Gordon and Breach, London.

Owen, F.N., and Gibson, D.M. (1978), A. J., **83**, 1488.

Tapia, S. (1977), Ap. J. (Lett.), **212**, L125.

MICROWAVE EMISSION FROM LATE TYPE DWARF STARS UV Ceti and YZ CMi

M. R. Kundu and R. K. Shevgaonkar[1]
Astronomy Program, University of Maryland
College Park, MD 20742

ABSTRACT. We present simultaneous VLA observations of two late type dwarf stars UV Ceti and YZ CMi at 6 and 20 cm. Multiwavelength observations put sufficient constraints on existing interpretations of quiescent radio emission from these stars. We find that the microwave emission is due to gyro-synchrotron radiation of nonthermal electrons having a power law energy distribution. This emission originates from a source whose size is 2-3 times larger than the star. From the lifetime of several hours of the nonthermal particles against radiation and collisional losses we estimate a magnetic field of ~ 140 G and a density of $\lesssim 2 \times 10^8$ cm^{-3} in the microwave source.

1. OBSERVATIONS

In this paper, we report microwave observations of two late type dwarf stars UV Ceti and YZ CMi, using the Very Large Array (VLA). The observations were made on August 12 and 13, 1983 with a resolution of 1".6 x 1" and 0".65 x 0".45 at 20 and 6 cm for UV Ceti and 1".2 x 1".2 and 0".5 x 0".5 at 20 and 6 cm for YZ CMi respectively. With a resolution of ~ 1" arc UV Cet and its binary companion are clearly resolved. The flux of UV Cet remains constant at a value of 1.05 ± 0.15 mJy at 20 cm and 1.27 ± 0.15 mJy at 6 cm respectively. The polarization of UV Cet at 6 cm was less than ~ 20% on both days. On the other hand, the binary companion of UV Cet, L726-8A showed large variations in total intensity over the two-day period. The fluxes of 3.01 ± 0.15 and 0.6 ± 0.1 mJy at 20 and 6 cm on the first day decrease to 2.17 ± 0.15 and 0.3 ± 0.1 mJy respectively on the second day. Although the flux of the source changes by a factor of ~ 1.5 over two days, the degree of circular polarization at 20 cm remains fairly constant at a value of ~ 65 % ± 10%.

We have made 2-hour synthesis maps at 6 and 20 cm for the source YZ CMi. The flux density of this source changes by a factor of ~ 5 over

[1]On leave of absence from Indian Institute of Astrophysics, Bangalore, India

R. M. Hjellming and D. M. Gibson (eds.), Radio Stars, 229–232.

two days. On the first day, the fluxes at 20 and 6 cm are 3.15 ± 0.15 and 0.88 ± 0.1 mJy respectively and the degree of polarization at 20 cm is as high as 85%; on the second day the flux at 20 cm is reduced to 0.6 mJy and the degree of polarization is $\lesssim$ 20%. The corresponding 6 cm flux is below the 3σ level.

2. DISCUSSION

Assuming that the flux of 1.55 mJy of UV Cet at 6 cm is the steady component, it has been argued in the past (Gary and Linsky 1981; Linsky and Gary 1983) that the thermal bremsstrahlung cannot account for the observed high brightness temperature. The use of gyroresonant thermal emission has shown that for coronal temperatures of $\sim 10^7$ K, the microwave source at 6 cm must be 6-7 times bigger than the stellar photosphere. The temperature can not be increased to too high values, since the soft X-ray observations of some stars (Haisch et al 1980; Swank and Johnson 1982) indicate temperatures in the range of a few times 10^6 K. To overcome this difficulty it has been proposed that the coronae of these stars have a multi-temperature distribution (Linsky and Gary 1983). The plasma observed in soft X-rays has a temperature a few times 10^6 K and a high electron density $\sim 10^8$ cm^{-3}, while the high temperature microwave component has a temperature of $\sim 10^8$ K and density of $\sim 10^5$ cm^{-3}. With a high temperature of $\sim 10^8$ K, the 6 cm source size is only $\sim$ 2 times the star radius and the observations fit with the theoretical model quite well. However, the simultaneous 6 and 20 cm observations are not quite consistent with the multi-temperature model, and it is important to investigate nonthermal radiation as a possible generating mechanism along with the gyroresonant thermal emission.

If the source is circular in shape, one gets a relation between the source radii at two wavelengths,

$$r_{20cm} \approx 3.3 \left(S_{20cm} / S_{6cm}\right)^{1/2} r_{6cm} \qquad (1)$$

where, S_{20cm} and S_{6cm} are the fluxes of the source at 20 and 6 cm respectively.

For UV Cet the ratio of flux densities at the two wavelengths gives $S_{20cm}/S_{6cm} \sim 0.83$ and so $r_{20cm} \approx 3\ r_{6cm}$. If we assume that the emission at both wavelengths is due to gyrosynchrotron radiation from the high temperature thermal electrons and if the 6th harmonic becomes optically thick (as argued by Gary and Linsky 1981), the ratio of magnetic fields at the two levels is $\sim$ 3. This implies that if we assume that the corona of the star is isothermal and the emission is gyroresonant, the magnetic field above the phtosphere of the star varies as r^{-1}. This variation of magnetic field with height does not represent a potential configuration. Thus, although the thermal model is consistent with the observations at 6 cm, it encounters certain difficulties in explaining the observations at more than one wavelength.

If the emission is due to nonthermal electrons having a power law energy distribution, the effective source temperature is a function of wavelength and even for optically thick emission $T_{b20cm} \neq T_{b6cm}$.

Therefore eq. (1) becomes

$$r_{20cm} = 3.3 \left\{\frac{S_{20cm}}{S_{6cm}} \frac{T_{b6cm}}{T_{b20cm}}\right\}^{1/2} r_{6cm} \qquad (2)$$

Since the observations were carried out over a period of a few hours, the lifetime of the nonthermal particles should be at least of the order of the observing period. Assuming that the electrons lose their energy due to radiation, a lifetime of ~ 6 h requires a magnetic field $\lesssim$ 140 G in the microwave source (Bekefi 1966). For this magnetic field and $\delta \simeq 3$ it can be shown (see e.g. Shevgaonkar and Kundu 1984; Dulk and Marsh 1982) that the 6 cm emission is optically thin with $T_b \sim$ a few times 10^7 K whereas the 20 cm emission could be optically thick with $T_b \sim$ a few times 10^8 K.

Thus, $$T_{b20cm}/T_{b6cm} \approx 10 \qquad (3)$$

and $$r_{20cm} \approx (S_{20cm}/S_{6cm})^{1/2} r_{6cm} \qquad (4)$$

For a source like UV Cet for which the ratio of the two flux densities is ~ 1, the source sizes at the two wavelengths are equal, which implies that both 6 and 20 cm emissions originate from the same physical volume of radius ~ 2.7 R_*, where R_* is the radius of the star. Assuming that the whole corona participates in emitting microwave radiation, the magnetic field at 2.7 R_* is ~ 140 G; therefore with a simple radially outward magnetic field with r^{-2} variation, the photospheric magnetic field comes out to be 1000 G, which is a reasonable estimate for the average magnetic field on the star's surface.

Our study indicates that the microwave flux density of a few mJy which we assume as steady flux may not be a steady component of emission in the strictest sense. It is conceivable that there are low level flare activity at regular or quasi-regular intervals which keep the stellar corona filled with nonthermal particles. An additional justification for this idea is that our measured flux of 1.27 ± 0.15 mJy of UV Cet at 6 cm is at the lower limit of the error bar given for the flux by Linsky and Gary (1983). If we assume that our flux value is truly smaller than the flux of 1.55 ± 0.27 mJy, then it is difficult to assume that this flux is the steady component of the star. Thus it is possible that the flux measured by us is not the quiescent or steady component and that we have been observing a time integrated flux of short lived mini flares. The production of nonthermal particles could occur in a manner analogous to that responsible for radio noise storms from non-flaring active regions. If the flaring events occur at the rate of once every few hours, the nonthermal particles will survive for a few hours against radiation losses if the magnetic field is $\lesssim$ 140 G. However, although the electrons do not lose their energy due to radiation they will thermalize quickly if the ambient coronal electron density N_e is high. The thermalization time due to energy exchange of the nonthermal electrons is

$$\tau_\varepsilon \approx 1.2 \times 10^{-18} \frac{v^3}{N_e} \qquad (5)$$

where, v is the velocity of the nonthermal electrons. For $v \approx 0.4$ c and $\tau_\varepsilon \sim$ a few hours, the ambient electron density should be $\lesssim 2 \times 10^8$ cm^{-3}, which is similar to the estimate given by Linsky and Gary (1983). The arguments presented above could also be applicable to the sources L726-8A and YZ CMi.

3. CONCLUSIONS

High resolution observations of two late type dwarf stars UV Ceti and YZ CMi simultaneously at 6 and 20 cm are presented here. These multiwavelength observatios put enough constraints on existing interpretations to conclude that the quiescent microwave emission from these stars is of nonthermal origin, namely gyrosynchrotron radiation. This quiescent emission is probably an integrated effect of many short lived mini flares or it could be analogous to the noise storm radiation observed from nonflaring active regions on the Sun. Assuming that the whole stellar surface participates in producing the microwave emission, the radius of the radio star is found to be ~ 2-3 times its bolometric radius.

Since the microwave emission is due to nonthermal particles, its observed duration of several hours requires that these particles should survive for few hours against radiation as well as collisional losses. The lifetime of the particles against radiation losses provides an estimate of the magnetic field of ~ 140 G in the emitting source and the lifetime against collisional losses gives an upper limit of $\sim 2 \times 10^8$ cm^{-3} for the density of the ambient plasma. Assuming r^{-2} variation a magnetic field of ~ 140 G in the microwave source gives an average magnetic field of ~ 1000 G at the photosphere of the stars.

This research was supported by NSF grant ATM 81-03089.

REFERENCES

Bekefi, G., 1966, "Radiation Processes in Plasmas", J. Wiley and Sons, New York.
Dulk, G. A., and Marsh, K. A., 1982, Ap. J., **259**, 350.
Gary, D. E. and Linsky, J. L., 1981, Ap. J., **250**, 284.
Haisch, B. M., Linsky, J. L., Harnden, F. R., Jr., Rosner, R., Seward, F. D., and Vaiana, G. S., 1980, Ap. J. (Letters), **242**, L99.
Linsky, J. L., and Gary, D. E., 1983, Ap. J., **274**, 776.
Shevgaonkar, R. K., and Kundu, M. R., 1984, to appear in Ap. J.
Swank, J. H., and Johnson, H. M., 1982, Ap. J. (Letters), **259**, L67.

THERMAL EMISSION AND POSSIBLE ROTATIONAL MODULATION IN AU MIC

J. J. Cox and D. M. Gibson
New Mexico Tech
Physics Department
Socorro, NM 87801
USA

ABSTRACT. We have made VLA observations of the dM2.5e star AU Mic at 2, 6, and 20 cm in an attempt to detect rotational modulation of its quiescent emission. There appears to be weak evidence for such a modulation. We have found that the quiescent emission at 6 and 20 cm is nonthermal while the emission at 2 cm is thermal. The small amplitude of variations at 2 cm suggests a large coronal filling factor. Such an assumption leads to a brightness temperature at 2 cm approximately equal to the X-ray temperature.

1. INTRODUCTION

The flare star AU Mic (dM2.5e) has been detected previously as a strong X-ray (Caillault, 1982) and radio (Gibson, 1982) source. It has a known optical rotational period of 4.865 days, a radius of 0.59 $R_{\odot}$, and is located at a distance of 8.8 pc (Bopp and Fekel, 1979). Since AU Mic has had a large photometric wave due to spot(s) and is a strong quiescent radio source, even at 2 cm, we believed it to be a good candidate for modulation in its radio continuum.

2. OBSERVATIONS

The observations of AU Mic with the VLA took place December 9, 12, 15, 19, and 23, 1983; each lasted three hours (19 to 22 LST). The observations were made sequentially at 2, 6, and 20 cm (14940, 4860, and 1490 MHz, respectively). The data are presented in three ways:

1) a plot of the flux densities S_{λ} versus UT Date (Fig. 1),
2) a plot of the S_{λ}'s as a function of rotational phase, where we have adopted 0^{h} UT on 9 Dec. as $\emptyset = 0.0$ (Fig. 2), and
3) a plot of the dynamic spectra (Fig. 3).

R. M. Hjellming and D. M. Gibson (eds.), Radio Stars, 233–236.

3. DISCUSSION

3.1 Rotational modulation

We interpret the enhanced 6 and 20 cm flux densities seen in AU Mic on Dec. 9 (and possibly Dec. 12) as weak flaring. While the first few days of observations give the impression that we are seeing the decay of a single flare we note that the timescales of these variations are much longer than typical flaring timescales for other dMe stars. Thus we conclude that the detected emission is probably due to several independent mini-flares.

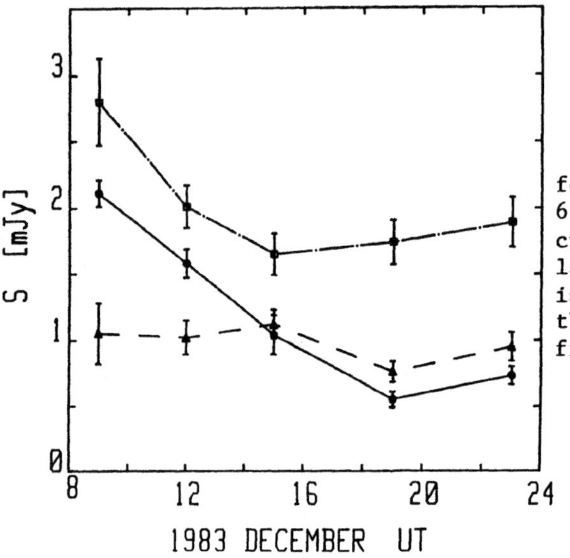

Figure 1: Light curves for AU Mic at 2 cm (triangles), 6 cm (filled circles), and 20 cm (squares). The emission is largely quiescent though there is evidence for weak flaring at the longer wavelengths on the first two days.

If we remove the Dec. 9 data from further consideration we can ask whether the emission is modulated in a way similar to the Slowly-Varying Component (S-Component) seen in the Sun. An examination of the 6 cm data in Fig. 2 shows strong evidence for a "sinusoidal" variability at a significance level of 99%. The 20 cm data show a possible (2σ) wave. We would be inclined to dismiss it entirely except that the 20 cm "wave" is in phase with the one at 6 cm and there are good reasons to suspect that the amplitude of the modulation would be less at 20 cm than at 6 cm (Gibson, 1984).

At 2 cm the amplitude of the variations is not as pronounced as those at 6 or 20 cm if present at all. This small amplitude variation suggests the filling factor of the quiescent corona is near unity. Under this assumption the brightness temperature, T_B, can be determined from

$$T_B = 2.12 \times 10^7 S_\lambda \lambda^2 d_*^2 / R_*^2$$

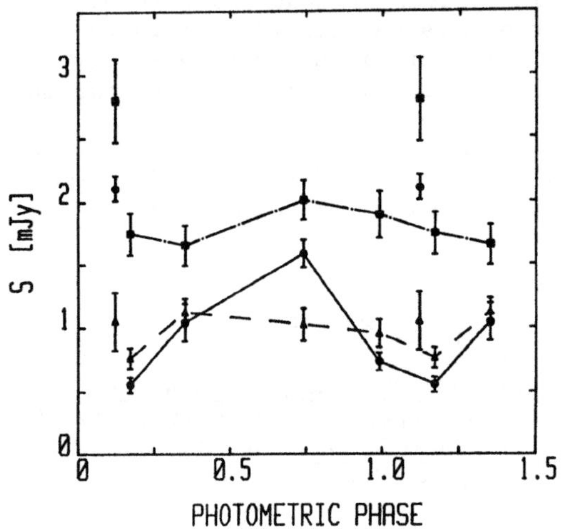

Figure 2: The 6 and 20 cm data (circles and squares respectively) show a "sinusoidal" variability, i.e. the possible existence of rotational modulation. The 2 cm data (the triangles) show a lack of amplitude variation.

where (cm) is the observing wavelength, R (R_o) is the source radius, and d (pc) is the distance to the star (Gibson, 1983). The brightness temperature for AU Mic was calculated to be $2x10^7$ K, which is about that found at X-ray wavelengths (Caillault, 1982). We conclude that AU Mic is a thermal coronal emitter at 2 cm, making it the first star for which the association has been made reliably.

3.2 Spectral information

The dynamic spectra (Fig. 3) indicate the probability of two spectral components. Between 6 and 20 cm the spectral index is always negative implying the long wavelength emission is nonthermal and optically thin.

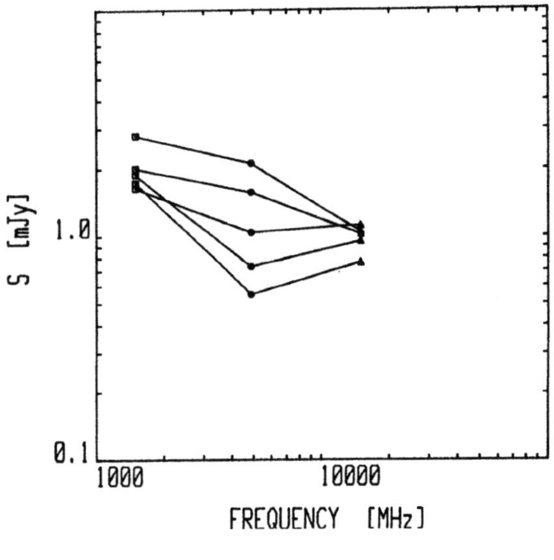

Figure 3: The dynamic spectra for AU Mic shows evidence nonthermal emission at 6 and 20 cm (circles and squares respectively), and thermal emission at 2 cm (the triangles). The dates of the observations (top-to-bottom at 6 cm) are 1983 Dec. 9, 12, 15, 23, and 19.

The mean spectral index $\alpha = -0.56$ is typical of that found in other stellar sources in which the probable emisson mechanism is gyro-synchrotron radiation. However, since there is no detectable circular polarization at any wavelength the magnetic structure of the star must be reasonably symmetric for this to be the case.

On Dec. 15, 19, and 23, $\alpha > 0$ between 2 and 6 cm consistent with our interpretation that the 2 cm emission is thermal. They are the first detections of a "U-shaped" radio spectra in another star.

REFERENCES

Bopp, B. W. and Fekel, F. 1977, Astron. J., **82**, 490.

Caillault, J.-P. 1982, Astron. J., **87**, 558.

Gibson, D. M. 1984, in Cool Stars, Stellar Systems, and the Sun, ed. S. L. Baliunas and L. Hartmann, (Springer-Verlag:Berlin), p. 197.

Gibson, D. M. 1983, in Activity in Red Dwarf Stars, ed. P. B. Byrne and M. Rodono, (Reidel:Dordrecht), p. 200.

FLARE ACTIVITY AND THE QUIESCENT X-RAY EMISSION IN dMe STARS

J. G. Doyle and C. J. Butler

Armagh Observatory
Armagh BT61 9DG
N. Ireland

ABSTRACT. Einstein observations of the X-ray flux of quiescent dwarf Me stars are correlated with the time-averaged energy emitted by flares in the Johnson U band. It is shown that the energy emitted by the coronae of these stars in X-rays is about an order of magnitude greater than the U band flare energy. From an estimate of the ratio of the total radiation emitted to the U band flux, it seems possible that if a similar amount of mechanical energy was dissipated in the stellar atmosphere, then the observed flare events can provide sufficient energy to heat the coronae of these stars.

1. INTRODUCTION

From a study of main-sequence stars (Vaughan et al, 1981) it had been found that the excess emission above the continuum in CaII H and K decayed exponentially as the rotation period increased. More recently (Noyes et al. 1984), a further dependence on spectral type was established, they found that the CaII H and K excess emission was well correlated with the parameter P_{obs}/τ_c, where P_{obs} is the observed rotation period and τ_c (B-V) a theoretically derived convective turn-over time.

However, the correlation of the X-ray emission with rotation is less clear. For example (Ayres and Linsky, 1980) the X-ray to bolometric luminosity (L_x/L_{bol}) scaled as $(V \sin i)^3$, whereas others (Pallavicini et al 1981, and Walter 1982) found $L_x \propto (V \sin i)^2$, or the data to be well represented by an exponential dependence of L_x/L_{bol} upon P_{obs}, or a power-law distribution with a break in the slope near 12 days. For the active dKe and dMe stars, it has been shown (Linsky et. 1982) that the transition zone lines are enhanced relative to the chromospheric lines and that there is a power-law dependence of L_x/L_{bol} versus rotation period (Byrne et al, 1984a), with a possible change in the slope at 10 days similar to that found in Walter for G and K stars. The scatter for the dKe and dMe stars was however much greater. These dwarf emission line stars have been long known to undergo flare activity, whose flare energy often far exceeds that of the largest known solar

R.M. Hjellming and D.M. Gibson (eds.), Radio Stars, 237–242.

flare (Byrne et al. 1984b). Here we investigate whether a correlation exists between the flare rate and the quiescent X-ray flux, and to discuss its importance in the context of coronal heating.

2. OBSERVATIONAL DATA

An extensive literature search was undertaken to compile a list of stars whose flare activity rate in the U band and 'quiescent' X-ray flux were known. The time-averaged flare energy L_u^* is defined as the summation of the total flare energy divided by the total monitoring time, L_x is the X-ray flux observed by the IPC on board Einstein in the energy range 0.2-4 keV. Further details on the observational data is given elsewhere (Doyle and Butler, 1984).

3. RESULTS AND DISCUSSION

In Figure 1 we plot the data points, open circles represent data points which are based on less than 15 hours of observation and thus are probably uncertain. Fitting a straight line through the reliable data we derive

$$\log L_x = 1.0 \log L_u^* + 1.12 \quad (1)$$

In deriving this relation we have excluded the uncertain data points for Gl 234A and Gl 735, see Doyle and Butler (1984) for further details.

The relationship between the flare activity rate and the 'quiescent' X-ray flux may be understood in terms of a solar analogy. In a series of plots, a correlation between the increase in sunsopt numbers, Ca plage intensity, X-ray flux (excluding the contribution due to flares), and the flare rate over a six month period was shown (Vaiana and Rosner, 1978). From their curves, a clear correlation existed between the increase in sunspot number, hence active region numbers, and the total 'quiescent' X-ray flux. A correlation was also evident between the increase in the number of active regions and the flare activity rate.

Applying this to Figure 1, would imply that the more active a particular star (i.e. the larger L_u^*), the larger fraction of the star's surface is covered with active regions, hence the larger is the X-ray flux. Some of the stars in our list have known rotation periods, which have been derived from the observed modulations in the photometric light curve. These modulations have been interpreted in terms of star-spots. It is interesting to note that the stars with known rotational modulation in our list are those with the largest time-averaged flare energy. Although the stars with the smaller L_u^* values have been studied for modulations in their light curve, none as yet have shown variations. From the data presented here, this may be understood in terms of the number of active regions; the stars with a smaller L_u^* having a smaller number of star-spots, and hence less likely to have modulations in their optical flux.

The fact that the slope of the line drawn in Fig.1 is unity

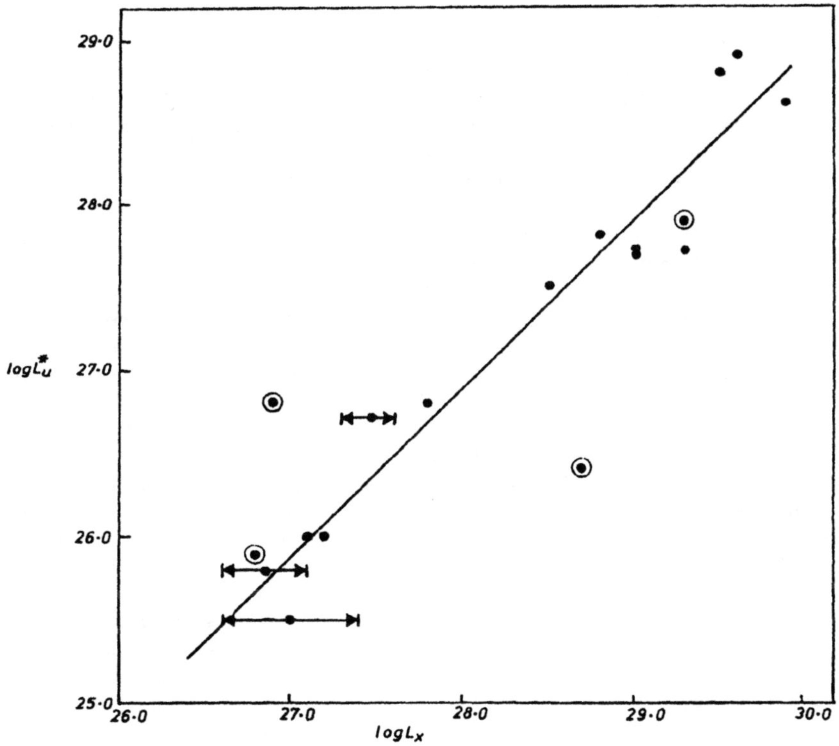

Figure 1. A plot of the time-averaged flare energy in the U band, L_u^*, versus the "quiescent" x-ray flux L_x. Both L_x and L_u^* are in units of erg s^{-1}.

implies that L_X is directly proportional to L_u^*, i.e. the quiescent coronal emission is proportional to the time-averaged energy of the flares observed in the U band. Rewriting equation (1) as

$$L_x = 13.2\ L_u^* \tag{2}$$

we see that the quiescent X-ray flux from the coronae of the dMe stars is only about an order of magnitude greater than the time-averaged flare output in U. This raises the question as to whether all of the energy contained in the corona, and radiated away in X-rays, could originate in stellar flares. In order to answer this important question we need to assess the ratio of the total energy released in a stellar flare to that emitted in the U band. Unfortunately this ratio cannot be determined directly for any flares yet observed on dMe stars. Even though stellar flares have now been observed in X-rays (Einstein satellite), ultraviolet (IUE), radio (VLA) and optically.

Making several assumptions we estimate that the total optical, ultraviolet and X-ray energy emitted from a flare to be $\sim 14L_u$. Details of how this estimate was made and its uncertainty can be found elsewhere (Butler and Doyle, 1984). Thus if only twice the amount of energy emitted by flares in electromagnetic radiation were to be 'absorbed' by the corona of the dMe stars there is sufficient energy available in flares alone to heat the coronae of these stars. In fact the amount of mechanical energy deposited in the stellar atmosphere by the flare event may far exceed twice the radiated energy. The only flare for which the deposition of mechanical energy has been estimated is the solar flare observed by Skylab on 5 September 1973. For this event (Cranfield et al. 1980 and Webb et al. 1980), the ratio of mechanical energy to radiated energy was 50.

From Figure 1 we see that the individual points do not deviate more than a factor of two from the straight line, even though both L_u^* and L_X range over three orders of magnitude in energy. The small scatter of the data points from the mean relation implies that any short term (of the order of days) or long term variability (of the order of years) in the X-ray flux be less than a factor of two. This is consistent with a maximum degree of variability in X-ray flux of less than a factor of 2 for these stars (Vaiana, 1983).

The low degree of variability in X-rays detected and implied by the scatter in Figure 1 for dMe stars contrasts with the large rotational modulation of X-ray flux from the solar corona (Vaiana and Rosner, 1978). This suggests that perhaps the coronae of the dMe stars are more uniform than the solar corona which, from X-ray imaging and rotational modulation of X-rays, is seen to be highly inhomogeneous. A further consequence of the small scatter about the mean line in Figure 1 is that long period activity cycles in X-ray flux ($P \sim$ years) analogous to the solar 11 year cycle are either not present on dMe stars or have smaller amplitude than on the Sun.

4. CONCLUSION

Over recent years the suggestion that rotation and its interaction with convection is responsible for the generation of magnetic fields in late type stars has gained wide acceptance. The re-connection of these magnetic fields is presumed to give rise to the various manifestations of stellar activity including the dramatic release of energy in flares.

The correlation of the mean U band flare energy with the coronal X-ray emission for dMe stars implies that, if about twice the total energy radiated by these stars in flares were to be dissipated in their upper atmospheres, enough energy would be available from flares to heat their coronae and produce the observed X-ray flux. Thus while less dramatic re-connection of magnetic fields may take place in the atmospheres of these stars it appears that the actual flares may be sufficient to heat the corona.

In fact, it has been recently suggested (Brueckner, 1983) that the energy requirements of the solar corona is perhaps provided by high velocity jets of material from the chromosphere. These jets have been observed with a high spatial and temporal resolution instrument. The typical velocity of these jets is approximately 400 km s^{-1} and are confined to a small area, $\sim$ 2". These jets are thought to be exploding small loops, perhaps involving the reconnection of two loops as proposed for flare theory (Priest, 1981). Since the velocity of these jets are much greater than the Alfven velocity, shock waves will be generated in the corona, resulting in it being heated.

We thus may have the situation where in the dKe-dMe that the flares produce high velocity directed mass motions which results again in the heating of the corona through shock waves. High velocity mass upflows have been observed during solar flares (Doyle et al, 1984).

The small scatter about the mean line is unlikely to be fortuitous and suggests that rotational modulation and long term cyclic variability in the X-ray flux from dMe stars is less obvious than would be expected implying the presence of coronae that are relatively homogeneous and stable.

5. REFERENCES

Ayers, T.R. and Linsky, J.L., Astrophys. J., 241, 279-299 (1980)

Brueckner, G.E. and Bartoe, J.D.F., Astrophys. J., 272, 329-348 (1983).

Butler, C.J. and Doyle, J.G., Irish Astron. J. (in press) (1984).

Byrne, P.B., Doyle, J.G., Butler, C.J. and Andrews, A.D., Mon. Not. R. Astr. Soc. (in press) (1984).

Byrne, P.B., Doyle, J.G. and Butler, C.J., Mon. Not. R. Astr. Soc., 206, 907-918 (1984).

Cranfield, R.C., Cheng, C.-C., Dere, K.P., Dulk, G.A., McLean, D.J.,

Robinson, R.D. Jr., Schmahl, E.J. and Schoolman, S.A., "Solar Flares", Colorado Univ. Press, ed. P.A. Sturrock, 451-469 (1980).

Doyle, J.G. and Butler, C.J., Nature (submitted) (1984).

Doyle, J.G., Byrne, P.B., Dennis, B.R., Emslie, A.G., Poland, A.I. and Simnett, G.M., Solar Phys. (submitted) (1984).

Linsky, J.L., Bormann, P.L., Carpenter, K.G., Wing, R.F., Giampapa, M.S., Worden, S.P. and Hege, E.K., Astrophys. J., 260, 670-694 (1982).

Noyes, R.W., Hartmann, L.W., Baliunas, S.L., Duncan, D.K. and Vaughan, A.H., Astrophys. J., 279, 763-777 (1984).

Pallavicini, R., Golub, L., Rosner, R. and Vaiana, G.S., Astrophys. J., 248, 279-290 (1981).

Priest, E.R., "Solar Active Regions", Skylab Workshop III, Colorado Univ. Press, ed. F.G. Orrall, (1981).

Vaiana, G.S., IAU Sym. No.102, "Solar and Stellar Magnetic Fields: Origins and Coronal Effects", ed. J.O. Stenflo, 165-186 (1983).

Vaiana, G.S. and Rosner, R., Ann. Rev. Astr. Ap. 16, 393-428 (1978).

Vaughan, A.H., Baliunas, S.L., Middelkoop, F., Hartmann, L.E., Mihalas, D., Noyes, R.W. and Preston, G.W., Astrophys. J., 250, 276-283 (1981).

Walter, F.M., Astrophys. J., 253, 745-751 (1982).

Webb, D.F., Cheng, C.-C., Dulk, G.A., Edberg, S.J., Martin, S.F., McKenna-Lawlor, S. and McLean, D.J., "Solar Flares", Colorado Univ. Press, ed. P.A. Sturrock, 471-499 (1980).

CORONAL-HOLE DETECTABILITY ON SOLAR-TYPE STARS

Richard C. Altrock
Air Force Geophysics Laboratory
National Solar Observatory
Sunspot, New Mexico 88349 USA

ABSTRACT. It is shown that light from the solar corona, which is integrated over the visible disk (coronal flux or irradiance), can be used to infer the disk passage of large coronal holes. Observations above the limb of Fe XIV 5303Å are used to produce a synoptic intensity map of the solar disk as it would appear in coronal light. The intensity at each point on the map is summed to produce a daily value of coronal irradiance. The time variation of this quantity shows a decrease of 28%, followed by recovery, as a large coronal hole transits the disk from 21 March through 7 April 1984. The occurrence of a coincident geomagnetic disturbance implies that the associated high-speed solar-wind stream strikes the earth. Other solar data sets, specifically sunspot number and 10.7 cm radio flux, do not have unambiguous coronal hole signatures during this period. This technique suggests that coronal holes might be observed on stars, if a suitable method for isolating coronal radiation is used; e.g., radio or EUV.

1. INTRODUCTION

Coronal holes, regions of slightly lower-than-average temperature and density in the solar corona, were first firmly identified in soft x-ray pictures of the sun taken by Skylab in 1973 and 1974. Research at that time showed that these areas, which appear very dark in the x-ray pictures, are the source of high-speed solar-wind streams, due to unipolar magnetic fields that open out into interplanetary space. Subsequently, it was found that these interesting regions could be observed from the ground, both in He I 10830Å and coronal radiation, most notably in white light, Fe XIV 5303Å, and radio. Since these regions give us information about the magnetic-field configuration and the presence of wind streams in the corona and heliosphere, their detection on other stars could hope to increase our knowledge of the physics of these stars, as well as to give us indirect evidence of stellar winds that have perhaps not yet been detected. A necessary condition for the detection of stellar coronal holes is the presence of a signature in some wavelength range of stellar flux. Thus it is of some interest to determine if coronal holes

R.M. Hjellming and D.M. Gibson (eds.), Radio Stars, 243–246.

are detectable in one or more wavelengths of solar flux. Of the wavelengths in which solar coronal holes have been detected, at least one appears to be unsuitable for detection in stellar flux, due to contamination by chromospheric radiation. He I 10830Å, which shows coronal holes with very low contrast on the solar disk, has properties that make its integrated light dominated by active regions, and it therefore does not exhibit a coronal-hole signature (Richard Donnelly, private communication).

This paper explores the question of whether coronal holes are visible in the flux of the coronal emission line at 5303Å. Necessary conditions for detection are high contrast and large size. On the sun, these conditions appear in 5303Å synoptic maps only at times of low solar activity. This is due to the fact that this line is optically thin and can only be observed over a long line-of-sight above the solar limb. Thus the conditions for high contrast are only met when there are no active regions along the line of sight, which generally only occurs away from solar maximum. In addition, observations over two solar cycles now show that the largest coronal holes occur during the declining phase of solar activity. Therefore, the best chance to detect coronal holes in 5303Å flux is during this declining phase, and this paper deals with that time period.

2. OBSERVATIONS

Daily observations of the solar corona are made at the National Solar Observatory facility at Sacramento Peak (NSO/SP) with the Photoelectric Coronal Photometer (Fisher, 1973; Smartt, 1982). The 40-cm-aperture coronagraph is used to form an occulted image of the corona, which passes through a narrow-band filter that spectrally discriminates at 100 kHz between the corona in 5303Å and an off-band wavelength. This technique allows the sky background contribution to be electronically subtracted. The entrance aperture of 1.1' is scanned around the limb at various radius vectors, and the output is sensed by a photomultiplier, digitized and recorded. Data points are recorded every three degrees in latitude. These scans have been published by Altrock, DeMastus and Gilliam (1977-1983) and Altrock and Gilliam (1983-1984). Figure 1 shows the sun as it would have appeared in 5303Å on March 30, 1984. Fifteen continuous days (including some one-days gaps) of east-limb data (with one imbedded day of data from the previous west-limb passage) are used to produce this map, which is transformed from the usual rectangular longitude-latitude format into a spherical format. The calibration bar on the right shows the intensities in units of millionths of the brightness of the solar disk center at 5303Å. The large coronal hole in the north center is the subject of this study.

3. ANALYSIS

Maps similar to that in Figure 1 were produced for each day in the interval 18 March (day-of-the-year [doy] 78) through 11 April (doy 102)

1984. Flux was computed by dividing a map into equal areas of angular extent as observed from the earth and computing the average intensity over the map. Figure 2 shows the results. The 5303Å flux shows a 28% decrease as the coronal hole transits the disk. Two other possible coronal-hole indicators, the sunspot number and the 10.7 cm radio flux obtained from Solar Geophysical Data, do not show any clear signature of coronal-hole passage. The radio flux would be expected to show a decrease similar to the optical flux, if any signature were to be seen (see below). The sunspot number might be expected to correlate with coronal flux, since the brightest coronal regions are found over active regions containing the sunspots. However, no such correlation is seen during this period. Increased geomagnetic activity (Fredricksburg A, obtained from Solar Geophysical Data) during the time that the coronal hole was on the disk (roughly doy 82 to 98) indicates that an associated solar wind stream intersected the earth.

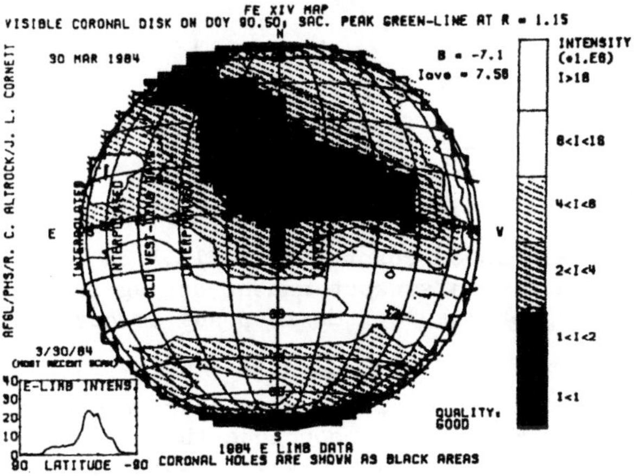

Figure 1. A synoptic map of the intensity of Fe XIV 5303Å at 0.15 solar radii above the limb. See text for details.

4. CONCLUSIONS

The disk passage of a large coronal hole during March and April 1984 produced a 28% decrease in a pseudo-coronal-flux parameter obtained from synoptic maps of Fe XIV 5303Å. This analysis, made during a period of optimum conditions for detecting such a coronal-hole signature, indicates at least that detecting such a disk passage on a solar-type star is possible. The interesting physics to be obtained from such an observation makes it highly desirable to search for such signatures on nearby near-solar-type stars. However, the technique presented here cannot be easily used on stars, because the optically-thin nature of the corona in visible light requires the use of a coronagraph to block out the light of the stellar disk. Such observations are only just now becoming possible. Space-based observations in soft x-rays or extreme ultraviolet might be a possiblity, although the flux technique has not been tested on the solar observations.

A more promising technique would seem to be observations of radio flux, if the right wavelength is chosen. In the decimeter range the solar corona begins to become optically thick. Papagiannis and Baker (1982) indicate that the best wavelength to observe solar coronal holes with is near 40 to 50 cm. A brightness-temperature decrease of

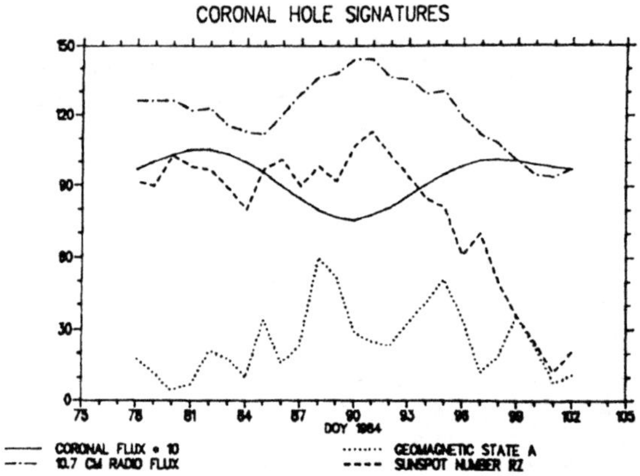

Figure 2. Fe XIV 5303Å coronal flux, 10.7 cm radio flux, geomagnetic state, and sunspot number during the analysis period. The map of Figure 1 corresponds to day of the year (doy) 90.

approximately 40% is found in coronal holes at that wavelength relative to the background corona. We may thus conclude that such a wavelength range would be appropriate for a first survey of solar-type stars in an attempt to observe coronal-hole signatures. The wavelength of maximum coronal-hole contrast should also be useful for determining rotation rates that are uncontaminated by differential rotation, since solar coronal holes have been observed to rotate rigidly with latitude.

REFERENCES

Altrock, R.C., H.L. DeMastus and L.B. Gilliam, Coronal Line Emission (Sacramento Peak), Solar-Geophysical Data, Prompt Reports (ed. H.E. Coffey), NOAA, Boulder, CO, Nos. 391-466, 1977-1983.

Altrock, R.C., and L.B. Gilliam, ibid., Nos. 467 et seq., 1983-1984.

Fisher, R.R., A photoelectric photometer for the Fe XIV solar corona, Rep. AFCRL-TR-73-0696, Air Force Geophysics Lab., Hanscom AFB, MA, 1973.

Papagiannis, M.D., and K.B. Baker, Determination and Analysis of coronal hole radio spectra, Solar Phys., **79**, 365-374, 1982.

Smartt, R.N., Solar corona photoelectric photometer using mica etalons, Proc. SPIE, **331**, 442-447, 1982.

VLA OBSERVATIONS OF A AND B STARS WITH KILOGAUSS MAGNETIC FIELDS

S. A. Drake,[1] D. C. Abbott,[1] J. H. Bieging,[2] E. Churchwell[3] and J. L. Linsky[1,4]
[1]Joint Institute for Laboratory Astrophysics, University of Colorado and National Bureau of Standards, Boulder, Colorado 80309
[2]Radio Astronomy Laboratory, University of California, Berkeley, California 94720
[3]Astronomy Department, University of Wisconsin, Madison, Wisconsin 53706

ABSTRACT. The serendipitous discovery that the star σ Ori E [B2 Vp (He Strong)] is a 3.5 mJy radio continuum source at 6 cm has stimulated a radio survey of other early-type stars with strong magnetic fields. No Ap stars have been detected of 8 observed, with typical 3σ upper limits of 0.5 mJy at 2 cm. Of 6 Bp stars examined, only HR 1890, also a helium-strong star, was detected. We discuss possible emission mechanisms for the observed radio emission, and conclude that non-thermal emission seems the most plausible, on the basis of the present data.

1. INTRODUCTION

During a VLA[5] 6 cm observation of a candidate stellar wind radio source σ Ori AB (O9.5V+B0.5V) in March 1984 we serendipitously detected at a level of 3.6 mJy the star σ Ori E (≡HR 1932; B2Vp), a visual companion 42" from the primary with which it shares a common proper motion (see Fig. 1). This interesting B star is both a prototype helium-rich star and a classical "magnetic star" of early spectral type with optically measured photospheric fields of magnitude 2-3 kilogauss (Landstreet and Borra 1978). It does not show evidence for well-developed P Cygni profiles in its ultraviolet lines, implying

[4]Staff Member, Quantum Physics Division, National Bureau of Standards.

[5]The Very Large Array is a facility of the National Radio Astronomy Observatory. N.R.A.O. is operated by the Associated Universities, Inc., under contract with the National Science Foundation.

R. M. Hjellming and D. M. Gibson (eds.), Radio Stars, 247–252.

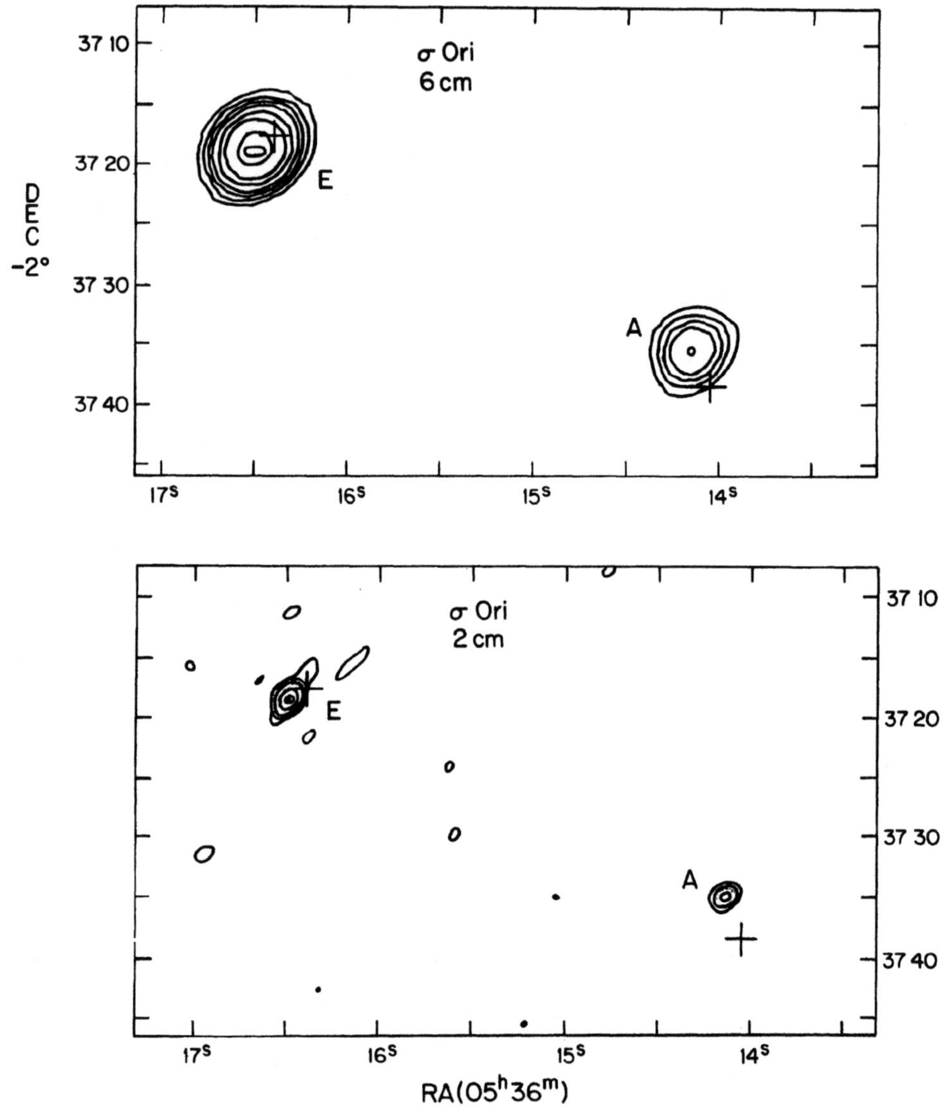

Fig. 1. VLA maps of the region of the visual double σ Orionis taken on 4 April 1984. In each map, the contour levels are the following integral multiples of 0.2 mJy: 2, 3, 4, 5, 7, 9, 13 and 15. The crosses mark the optical positions of the indicated components of the σ Orionis system.

that it does not have a stellar wind massive enough to produce the observed radio flux at its large distance (~400 pc) away from us. Groote and Hunger (1982) estimate that the mass loss rate of σ Ori E is $\lesssim 10^{-9}$ $M_{\odot}$ yr^{-1}, a factor of 10^2 too small to produce the radio flux, even if the wind velocity is as low as 100 km s^{-1}.

Subsequent re-observations of this system at 2 and 6 cm have shown that the radio source has a slightly negative spectral index, $\alpha = -0.1$, and is not significantly variable. These properties are compatible with either optically thin thermal emission or non-thermal emission. Given that σ Ori E's dominant characteristic is its abnormally strong magnetic field, we strongly suspect that the radio emission is non-thermal in origin.

We have therefore observed at 2 cm a sample of 13 additional B and A stars all of which have photospheric magnetic fields of the order of kilogauss in order to investigate whether radio continuum emission is a general property of such stars. The results of this survey are summarized in Table 1 where we list for each star its spectral type, distance, typical photospheric magnetic field strength (taken from Didelon 1983), observed radio flux and inferred radio luminosity. We have detected one additional radio source, HR 1890 ≡ HD 37017 (B1.5 Vp), which is, like σ Ori E, a helium-strong magnetic variable in the Orion complex. The inferred 2 cm monochromatic luminosities L_ν of the two stars are both about 5×10^{17} ergs s^{-1} Hz^{-1}, similar to the levels observed in active RS CVn binaries in which the radio emission is believed to be nonthermal in nature. We have obtained 3σ upper limits of about 0.5 mJy for the 12 stars that have not been detected at 2 cm.

2. OBSERVATIONS

2.1. The Helium-Strong Bp Stars

The radio observations of the two detected stars, σ Ori E and HR 1890, are listed in Table 1. In Figure 1, we show the 2 cm and 6 cm maps of the σ Ori field: both σ Ori A and E are radio sources, apparently. There is some discrepancy between the radio and optical positions for these sources, particularly σ Ori A. We do not believe that, in the case of σ Ori E, the positional disagreement is significant. A more extensive discussion of these radio sources is given by Drake et al. (1984). Two other helium-strong Bp stars (HD 37776 and HD 52860) were observed but not detected, indicating that there must be a range of at least a factor of 4 in the 2 cm radio luminosities of these stars, from $\lesssim 1.3 \times 10^{17}$ to 5.6×10^{17} ergs s^{-1} Hz^{-1}.

2.2. The Helium-Weak Bp Stars

We failed to detect the two He-weak stars that were observed despite their being closer by a factor of two to three than the He-strong stars that were detected. If typical of the class, this means that their radio luminosities must be at least an order of magnitude less

Table 1. Observational Results

Star	HD	Spectral Type	D (pc)	S_ν (mJy)	L_ν (ergs s^{-1} Hz^{-1})	λ (cm)	Date
			(a) B Stars: Detections				
HR 1890	37017	B1.5V (He strong) (B_{phot} = 2.3 kG)	500	1.81 ±0.13	5.43×10^{17}	2	1984 Jul 01
σ Ori E	37479	B2Vp (He strong) (B_{phot} = 2.8 kG)	400	3.10 ±0.11	5.95×10^{17}	2	1984 Apr 04
				2.74 ±0.19	5.26×10^{17}	2	1984 Jul 02
				3.64 ±0.06	6.99×10^{17}	6	1984 Mar 04
				3.35 ±0.06	6.43×10^{17}	6	1984 Apr 04
				3.24 ±0.13	6.22×10^{17}	6	1984 Jul 02
			(b) B Stars: Non-detections (3σ upper limits)				B_{phot} (kG)
V901 Ori	37776	B2IV/V (He strong)	500	≲0.42	$\lesssim 1.26 \times 10^{17}$	2	2.0
HR 2645	52860	B3IIIn (He strong)	800	≲0.56	$\lesssim 4.30 \times 10^{17}$	2	2.2
α Cen	125823	B7IIIpv (He weak)	160	≲0.55	$\lesssim 1.69 \times 10^{16}$	2	1.1
36 Lyn	79158	B8IIIpMn (He weak)	190	≲0.62	$\lesssim 2.69 \times 10^{16}$	2	1.3
ω Her	148112	B9p Cr	24	≲0.47	$\lesssim 3.25 \times 10^{14}$	2	1.6
			(c) Ap Stars: Non-detections				
HR 4072	89822	A0pSiSr:Hg:	23	≲0.48	$\lesssim 3.05 \times 10^{14}$	2	1.5
ε UMa	112185	A0pCr	30	≲0.42	$\lesssim 4.54 \times 10^{14}$	2	0.8
73 Dra	196502	A0pSrCrEu	55	≲0.80	$\lesssim 2.90 \times 10^{15}$	2	1.2
α^2 CVn	112413	A0pSiEuHg	37	≲0.46	$\lesssim 7.56 \times 10^{14}$	2	3.5
52 Her	152107	A2VpSrCrEu	40	≲0.40	$\lesssim 7.68 \times 10^{14}$	2	2.5
53 Cam	65339A	A2pSrCrEu	70	≲0.52	$\lesssim 3.06 \times 10^{15}$	2	15.0
β CrB	137909	F0p	31	≲0.44	$\lesssim 5.07 \times 10^{14}$	2	6.3

than the He-strong Bp stars. This could be due to their later spectral types of B7 and B8 compared to the He-strong stars (B1.5-B3), assuming the radio emission is somehow a function of the number of ionizing photons; alternatively, it could be due to the fact that the two He-weak stars observed have rather weaker magnetic fields than the He-strong stars.

2.3. The Ap Stars

If we include ω Her (B9pCr)and β CrB (F0p) in this class, we have observed and not detected 8 of these stars at 2 cm. The inferred upper limits to their radio luminosities are in the range 3.0×10^{14} - 3.0×10^{15} ergs s^{-1} Hz^{-1}. Pre-VLA radio surveys of Ap stars with sensitivities an order of magnitude poorer than the present one have also failed to detect any Ap stars as radio sources (e.g., Altenhoff _et al._ 1976).

3. DISCUSSION

It is evident from our results that the presence of a strong surface magnetic field is _not_ sufficient to produce detectable radio emission. It is possibly significant that the boundary between hot stars with and without detectable stellar winds occurs at B2 for main sequence stars (Barker and Marlborough 1984). The two stars that were detected, being B1.5 V and B2 V, should have mass loss, according to this criterion, while the remaining stars (with the exception of HD 37776), being of later spectral type, should not. Thus, this would suggest that both a strong magnetic field _and_ a stellar wind may be necessary to produce a significant radio luminosity. Since the two stars that were detected have the highest effective temperatures, it is possible that the radio emission could be optically thin thermal emission from a surrounding compact H II region. The $\nu^{-0.1}$ spectrum of σ Ori E between 2 and 6 cm is consistent with this mechanism. It can be shown that an H II region with $T_e = 10^4$ K that can produce as much radio emission as is observed _and_ remain optically thin must have an angular diameter ϕ greater than 0ʺ14. We would expect, on the other hand, that a non-thermal mechanism, such as gyroresonance or gyrosynchrotron emission, would be operative in a much smaller volume closer to the star where the magnetic field is strong, and hence be unresolvable ($\phi \ll$ 0ʺ1). In the latter case, we also anticipate that the radio emission might be somewhat circularly polarized. We hope to be able to discriminate between these alternatives by observing σ Ori E and HR 1890 in the A configuration of the VLA at 2 cm (HPBW = 0ʺ08), and at 6 cm with higher signal-to-noise than previous observations.

4. CONCLUSIONS

We have detected two stars (σ Ori E and HR 1890) as 2 cm radio continuum sources out of a sample of 14 stars with kilogauss photospheric magnetic fields that have been observed with the VLA. The two radio sources are both He-strong early B stars. None of the much nearer Ap stars were detected, implying that they have 2 cm luminosities smaller by at least 10^2 to 10^3 than the detected He-strong stars.

ACKNOWLEDGMENTS

This work has been supported by NASA grant NGL-06-003-057 through the University of Colorado.

REFERENCES

Altenhoff, W. J. et al. 1976, in Physics of Ap Stars, IAU Colloquium No. 32, p. 497.
Barker, P. K. and Marlborough, J. M. 1984, in Future of Ultraviolet Astronomy Based on Six Years of IUE Research, NASA publication, in press.
Didelon, P. 1983, Astr. Ap. Suppl., **53**, 119.
Drake, S. A., Abbott, D. C., Bieging, J. H., Churchwell, E., and Linsky, J. L. 1984, in preparation.
Groote, D. and Hunger, K. 1982, Astr. Ap., **116**, 64.
Landstreet, J. D. and Borra, E. F. 1978, Ap. J. (Letters), **224**, L5.

A VLA RADIO CONTINUUM SURVEY OF ACTIVE LATE-TYPE GIANTS IN BINARY SYSTEMS: PRELIMINARY RESULTS

S. A. Drake,[1] Theodore Simon[2] and J. L. Linsky[1,3]
[1]Joint Institute for Laboratory Astrophysics, Univ. of Colorado and National Bureau of Standards, Boulder, CO 80309
[2]Institute for Astronomy, Univ. of Hawaii, Honolulu, HI 96822
[3]Staff Member, Quantum Physics Division, National Bureau of Standards

ABSTRACT. We have made a sensitive survey at 6 cm of "active" G and K giants in binary systems, including the so-called Long-Period RS CVn stars. The systems observed have orbital periods in the range of ~10 to more than 100 days, and are judged to be "active" on the basis of their pronounced chromospheric and transition region emission lines and (where available) strong X-ray emission compared to single giants of similar spectral type. Our results to date show that strong radio continuum emission at centimeter wavelengths is a common but not universal property of this class of stars. On the basis of our VLA data and those obtained in a recent, less sensitive survey by Mutel and Lestrade, we discuss possible correlations between radio luminosity and other properties, such as X-ray luminosity, rotational period, and type of companion. Binary systems detected for the first time as radio continuum sources include 12 Cam (K0 III + ?; $P_{orb} = 80\overset{d}{.}2$), HD 185510 (K0 III-IV + sdB; $P_{rot} = 25^d$), 29 Dra (K0-2 III + wd; $P_{rot} = 31^d$), and FF Aqr (G8 III-IV + sdB; $P_{orb} = 9\overset{d}{.}21$). Sensitive upper limits are presented for five other systems including the closest Long Period RS CVn binary, α Aur (G6 III + F9 III; $P_{orb} = 104^d$).

I. INTRODUCTION

We present here the preliminary results of a 6 cm continuum survey using the NRAO VLA.* The objects observed are binary systems of 10-100 days orbital period containing an "active" giant component. Most are members of the Long Period RS Canum Venaticorum class of binaries first defined by Hall (1976). In Table 1 we summarize the data obtained

---

*The National Radio Astronomy Observatory is operated by the Associated Universities, Inc., under contract with the National Science Foundation.

R. M. Hjellming and D. M. Gibson (eds.), Radio Stars, 253–258.

Table 1. 6 cm Radio Observations of Active Late-Type Binaries

Star	HD	Spectral type	D(pc)	P_{rot}	P_{orb}	S_ν(mJy)	log$\langle L_\nu \rangle$	Comments
FF Aqr	—	G8 III: + sdOB	300	$9^d.21$	$9^d.21$	4.67 ± 0.05	17.73	Peculiar binary: First radio detection
ζ And	4502	K1 IIe + ?	27	17.8	17.8	≲0.30 (3σ)	≲14.42	
σ Gem	62044	K1 III + ?	60	19.4	19.6	{2.9 ± 0.3 0.67 ± 0.1	15.73	Known radio source
λ And	222107	G8 III-IV + ?	20	54.0	20.5	{2.9 ± 0.1 0.61 ± 0.1	14.80	Known radio source
RZ Cnc	73343	K1 III + K3-4 III	330	—	21.6	≲0.27 (3σ)	≲16.55	
—	185510	K0 III-IV + sdB	210	≃25	—	0.50 ± 0.05	16.38	First radio detection
29 Dra	160538	K0 III + wd	180	≃31.5	—	28.7 ± 0.03	18.05	First radio detection
ε Umi	153751	G5 III + A9V	74	—	39.5	≲0.17 (3σ)	≲15.06	
DK Dra	106677	K0 III + K0 III	130	63.8	64.4	{0.37 ± 0.10 1.76 ± 0.06	16.24	Independently discovered by Mutel and Lestrade (1984)
93 Leo	102509	G5 III-IV + A7 V	45	—	71.7	≲0.17 (3σ)	≲14.52	
12 Cam	32357	K0 III + ?	160	—	80.2	2.02 ± 0.04	16.79	First radio detection
α Equ†	202447	G0 III + A5 V	48	—	98.8	≲0.15 (3σ)	≲14.62	
α Aur	34029	G5 IIIe + G0 III	12.5	~7*	104.0	≲0.25 (3σ)	≲13.71	
113 Her†	175492	G4 III + A6 V	80	—	245.3	≲0.17 (3σ)	≲15.11	

*Adopted rotation period of secondary star.

†Spectroscopic binaries not considered LP RS CVn's.

in several observing runs in late 1983 and early 1984. Columns 1 and 2 list the star name and HD number, column 3 gives the spectral type, generally that quoted by Hall et al. (1984), column 4 contains our best estimate of the distance, columns 5 and 6 give the photometric (rotational) and spectroscopic (orbital) periods in days, respectively, column 7 lists the measured 6 cm flux in mJy, column 8 gives the inferred 6 cm luminosity in ergs s^{-1} Hz^{-1}, and column 9 contains additional comments. We have used these data, together with (in many cases) observed X-ray fluxes of these same systems (e.g., Walter and Bowyer 1981) to search for correlations between radio emission and other properties of these systems.

II. COMPARISON OF X-RAY AND RADIO LUMINOSITIES

For the systems observed in the present survey, and for similar systems already detected at radio wavelengths (e.g. Mutel and Lestrade 1984), there appears to be a rough correlation between the X-ray luminosity and the monochromatic 6 cm luminosity of the form $L_\nu(6\ \mathrm{cm}) \propto L_x^2$ (see Fig. 1). A linear regression for the 18 stars in Fig. 1 for which X-ray luminosities have been measured yields

$$\log L_\nu = 1.9 \log L_x - 41.9$$

as the least-squares fit, with a correlation coefficient of 0.821, assuming:

(a) mean values for the observed X-ray and radio luminosities and

(b) radio luminosities corresponding to 3σ levels for those systems that have not been detected.

A more sophisticated maximum likelihood linear regression that takes the upper limits more properly into account yields:

$$\log L_\nu = 2.4 \log L_x - 59.6$$

with an estimated uncertainty in the slope of ± 0.5.

Despite the fairly high formal statistical significance of these relations, there are clearly many possible biases influencing this result, such as the known variability of many of these systems at radio and/or X-ray wavelengths, and the treatment of nondetections. The most obvious improvement would be to increase the number of stars in the sample.

If we tentatively accept that $L_\nu \propto L_x^2$, then we can derive a relation between observed X-ray and radio fluxes, viz.:

$$S_\nu\ (6\ \mathrm{cm}) \sim 1.3 \times 10^{18}\ f_x^2 (D/\mathrm{pc})^2\ \mathrm{mJy} \quad ,$$

where f_x is the observed X-ray flux in the Einstein IPC in ergs s^{-1} cm^{-2}. This relation predicts X-ray fluxes of 3×10^{-11} and 6×10^{-12} ergs cm^{-2} s^{-1}, respectively, for the recently detected radio sources 29 Dra and FF Aqr.

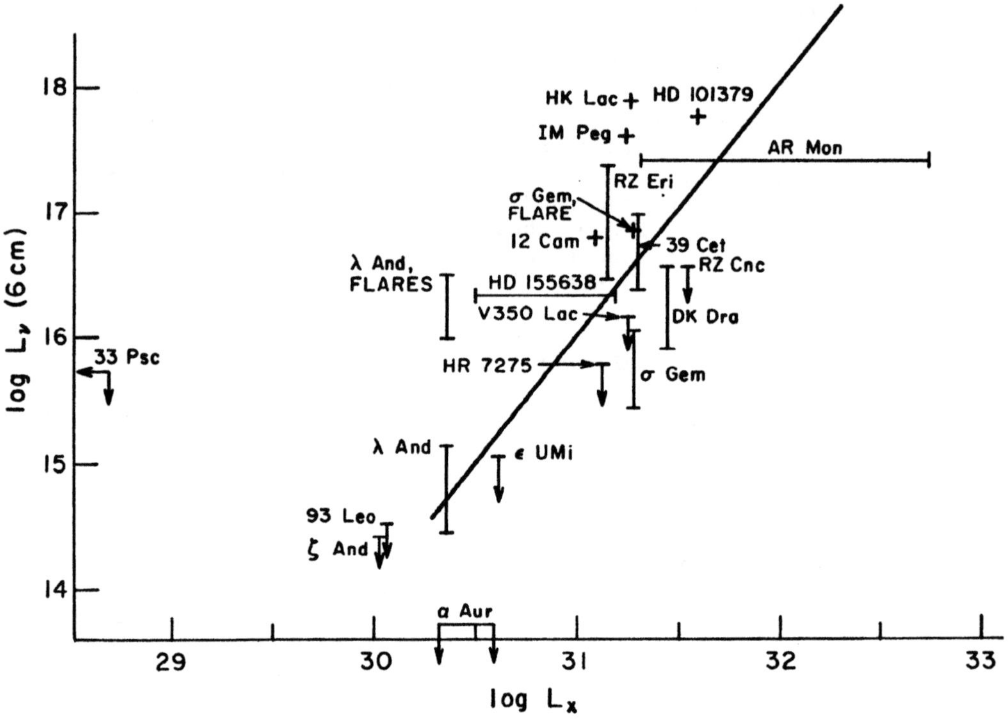

Fig. 1. The 6 cm radio luminosity in ergs s^{-1} Hz^{-1}, L_ν (6 cm), is plotted against the soft X-ray luminosity in ergs s^{-1}, L_x, in the range 0.2-4.5 keV, for all the active giant binary systems with periods between 7 and 100 days for which both quantities have been measured. The least-squares fit for the 16 binaries with $\log L_x > 30.3$ is also shown (see text).

III. SEARCH FOR OTHER CORRELATIONS WITH RADIO LUMINOSITY

For active binaries with orbital and/or rotational periods between ~10 and ~100 days, we find no significant correlation of radio luminosity with period (see Fig. 2). Using published and newly obtained data for 25 such systems, we obtain a least squares relation between the mean radio luminosity and rotational period P_{rot} (in days):

$$\log L_\nu \text{ (6 cm)} = 0.0 \log P_{rot} + 16.3$$

with the statistically negligible correlation coefficient of 0.01. Using orbital periods P_{orb} in place of rotational periods, we likewise find no statistically significant correlation between L_ν (6 cm) and P_{orb}. (Only three of these systems are significantly non-synchronous, and thus are affected by this change). Our result is in disagreement

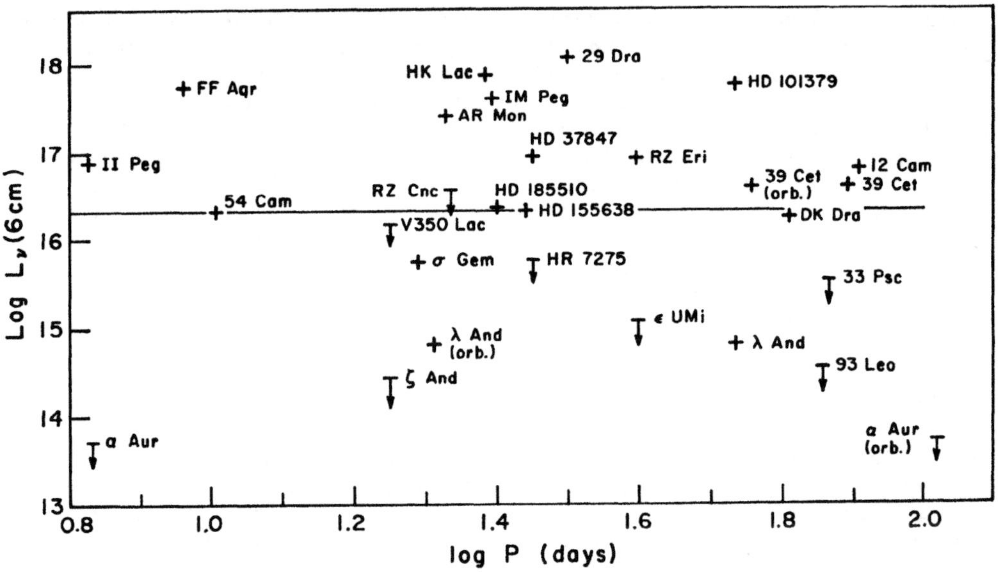

Fig. 2. L_ν (6 cm) is plotted against the (rotational) period in days for active giant binary systems with periods between 7 and 100 days. For the three systems known to be significantly non-synchronous, we also show their location if we adopt the orbital period of the system. The horizontal line is the formal linear regression found from this data set relating the rotational periods and the 6 cm luminosities (see text).

with the $L_\nu \propto P_{orb}^{-0.7}$ relation found by Mutel and Lestrade (1984). Though the latter authors observed active binaries with periods over a wider range (~1 to ~100 days) than in this study, we are skeptical that their claimed correlation is real. Their apparent correlation may be an artifact of the omission from their statistical analysis of systems for which only upper limits on the radio emission are available. Further investigation of this question is clearly required.

The only other (very tentative) correlation that is apparent in our present sample is that active binary systems with a cool giant and a hot companion (a white dwarf or OB subdwarf) tend to be stronger at radio wavelengths than the more "classical" Long Period RS CVn binaries containing cool secondaries (see Fig. 2).

IV. THE STRANGE CASE OF CAPELLA: WHY IS IT NOT A RADIO SOURCE?

The nearest active giant binary system, α Aur (≡ Capella), which is 12.5 pc away, has the highest observed X-ray flux of any "normal" cool star observed by the Einstein Observatory and yet it has never been detected at radio wavelengths. The upper limit presented in this paper is 0.25 mJy. It is perhaps also the system whose properties are most inconsistent with the tentative X-ray/radio correlation discussed above. The emission measures and temperatures of the corona(e) obtained from almost a decade of X-ray observations of this system, combined with theoretical X-ray emissivity calculations, require that the 6 cm radio flux of Capella due to bremsstrahlung alone be in the range 0.1 - 0.5 mJy. Thus if the present radio upper limit can be reduced by a factor of 3, either the system will be (at last) detected, confirming the X-ray models, or a continued nondetection at radio wavelengths will pose serious problems for these same X-ray models. We hope in the near future to make this critical reobservation of Capella at 6 cm with the lower noise level.

V. SUMMARY

Of the 12 binaries that were not known to be radio sources prior to the beginning of this survey, five were detected at 6 cm with fluxes between 0.5 and 30 mJy, and seven were not detected with 3σ upper limits of order 0.15-0.30 mJy. Given the fact that binaries at distances of 200-300 pc have been readily detected, the radio "horizon" for detectability of these systems is probably ~1000 pc. For λ And and σ Gem, which were already identified as radio sources, we have established quiescent, or at least slowly varying, levels of 6 cm radio emission. The only statistically significant correlation of radio with other properties that we have found is the relation $L_\nu(6\ \mathrm{cm}) \propto L_x^2$, which is based on 18 systems for which X-ray observations are available.

We would like to thank Eric Feigelson and Takashi Isobe for bringing to our attention the statistical techniques that are most suitable for data sets containing some upper limits, and for carrying out the maximum likelihood linear regression on the X-ray and radio luminosities of the active binaries. This work has been supported by NASA grant NGL-06-003-057 through the University of Colorado.

REFERENCES

Hall, D. S. 1976, IAU Colloq. No. 29, part 1, 287.
Hall, D. S., Zeilik, M. and Nelson, E. R. 1984, Hall Catalog of RS CVn Binary Star Systems, revised edition.
Mutel, R. L. and Lestrade, J. F. 1984, preprint.
Walter, F. and Bowyer, S. 1981, Ap. J., 245, 671.

RADIO POLARIZATION CHARACTERISTICS OF TWO RS CVn BINARIES

Robert L. Mutel
University of Iowa, Department of Physics and Astronomy,
Iowa City, IA 52242, U.S.A.

Jean-Francois Lestrade
Bureau des Longitudes, Paris, France/Jet Propulsion
Laboratory, California Institute of Technology,
4800 Oak Grove Drive, Pasadena, CA 91109, U.S.A.

D. J. Doiron
Clemson University, Laboratory of Physics,
Clemson, SC 29631, U.S.A.

ABSTRACT. We report the results of multifrequency epoch VLA observations of polarized radio emission from the nearby active RS CVn binaries UX Arietis and HR 1099. For both systems, there is an excellent correlation between handedness of circular polarization and frequency. Helicity reversal is almost always seen between 1.4 and 5.0 GHz, possibly due to optical depth effects. There may also be an anticorrelation between total intensity and fractional circular polarization, especially at 5 GHz. This is consistent with models in which intense flares are associated with compact self-absorbed synchrotron sources, while the quiescent emission arises from larger gyrosynchrotron-emitting plasma.

1. INTRODUCTION

Radio emission from RS CVn binaries is highly variable and often significantly circularly polarized. The radiation is usually interpreted as gyrosynchrotron emission from mildly relativistic (~ few MeV) electrons in magnetic fields of 10 - 100 gauss (Owen et al., 1976; Dulk, 1985). There are occasional flares of short duration which are close to 100% circularly polarized (Brown and Crane, 1978), indicative of a coherent mechanism. Direct measurement of brightness temperatures using VLB arrays (Lestrade et al., 1985) are all consistent with gyrosynchrotron models in which flares arise from a self-absorbed, compact source and expand to fill a region about equal to the binary system on a time scale of ~ 1 day (Mutel et al., 1985).

Since the discovery of radio emission from RS CVn systems in 1974 (Gibson and Hjellming, 1974), there have been more than twenty papers

R. M. Hjellming and D. M. Gibson (eds.), Radio Stars, 259–265.

reporting radio observations. However, many of these have at a single frequency, often with no polarization information. Polarization measurements have been made on a few objects (e.g., Owen et al., 1976; Spangler, 1977; Mutel and Weisberg, 1978) which indicate that the emission is often moderately circularly polarized, but not linearly polarized, presumably due to Faraday depolarization. However, there has not been a systematic investigation of the statistical properties of radio emission from RS CVn systems, except for single frequency studies of the radio luminosity function (Owen and Gibson, 1978; Feldman, 1983). In an effort to improve on this situation, we have been monitoring several nearby RS CVn systems in all four Stokes parameters at two (or more) frequencies using the VLA. In this paper, we report on polarization characteristics of two of the most active systems: UX Arietis and HR 1099.

2. CIRCULAR POLARIZATION CHARACTERISTICS

Since mid-1982, we have used the VLA to observe the strong radio emitters UX Arietis and HR 1099 for a total of sixteen sessions. The sources were originally observed during a variety of experiments, including simultaneous VLBI sessions, studies of different classes of close binary systems, and searches for new stellar radio emitters. Consequently, the time allotted for each observation and the number of frequencies observed varied. We almost always observed both sources using the full VLA in 'snap-shot' mode (typically ~ 10 min); at 1.4 GHz, 5.0 GHz, and occasionally 15 GHz. The data were reduced using standard AIPS software at the University of Iowa.

2.1. Polarization-Wavelength Correlation

The most important result of this work has been the discovery of a strong correlation between helicity of circular polarization and observing frequency. Histograms of the occurrence of fractional circular polarization at three frequencies (1.4, 4.9, and 15.0 GHz) are shown in Figure 1(a)-(b) for UX Arietis and HR 1099, respectively.

There is clearly a reversal of helicity between 1.4 and 4.9 GHz in both systems. Since the data span more than two years of observations, this indicates that the overall geometry of the magnetic field in the source region is quite stable over many orbits of the binary system. The source size, in turn, is comparable to the system size, at least during the post-flare state when the degree of circular polarization is highest (Mutel et al., 1985 and Section 3 below). This implies rather large, stable magnetic field geometry, perhaps connecting the two stellar magnetospheres. This picture is in contrast to the interacting loop models of Simon et al. (1980) and Uchida and Sakurai (1983).

2.2. Polarization-Intensity Correlation

Models of the radio emission based on VLB observations suggest that intense flares are associated with self-absorbed synchrotron sources,

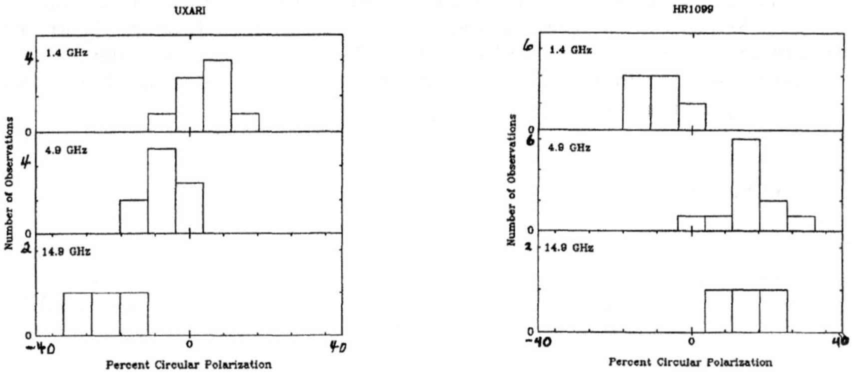

(a) (b)

Figure 1. Frequency of occurrence of fractional circular polarization at three frequencies for UX Arietis (a) and HR 1099 (b). Positive values denote net right circularly polarized flux.

while quiescent emission arises from more extended regions with less energetic electrons (~ gyrosynchrotron emission). In this scenario, one would expect the total radio luminosity to be anticorrelated with the circular polarization. Indeed, Figure 2(a)-(b) shows such an anticorrelation for UX Arietis (a) and HR 1099 (b) at 5 GHz. However, since there are only a small number of observations at high flux density, we can only conclude that the available data are consistent with such a hypothesis.

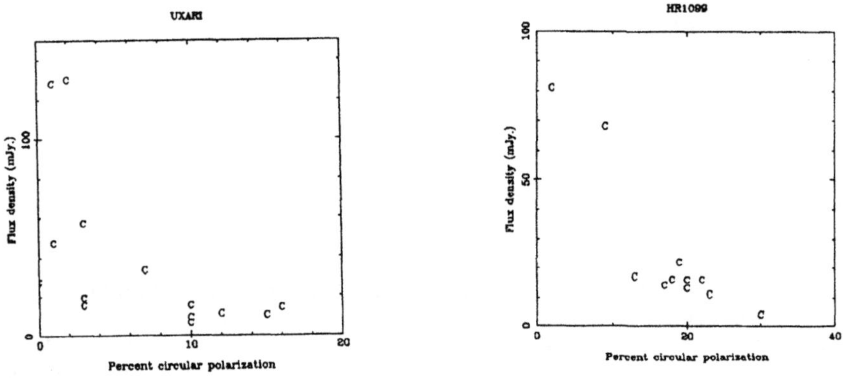

(a) (b)

Figure 2. Total flux density versus percent circular polarization at 5 GHz for UX Arietis (a) and HR 1099 (b).

We also note that some previous observations of intense flares (e.g., Brown and Crane, 1978; Fix et al., 1980) have shown rather high fractional circular polarization. These events may have been due to a coherent emission mechanism such as an electron-cyclotron maser (Melrose and Dulk, 1982).

3. COMPARISON WITH GYROSYNCHROTRON EMISSION MODELS

It is well known that synchrotron radiation from ultra-relativistic electrons (such as found in nearly all extragalactic radio sources) produces very small fractional circular polarization; roughly one has

$$\pi_c \sim \frac{1}{\gamma} << 1 \qquad \gamma = \text{Lorentz factor}$$

However, as the electron energy is reduced, the fractional circular polarization becomes substantial. Dulk and Marsh (1982) and Dulk (1985) gives approximate numerical expressions for π_c as a function of frequency for an optically thin source (Figure 3). The polarized flux is in extraordinary mode ('x-mode').

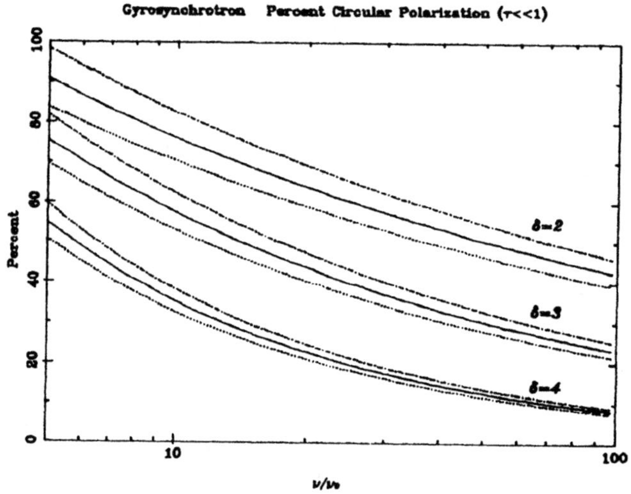

Figure 3. Percent circular polarization for gyrosynchrotron emission for power-law electrons from an optically thin source as a function of the ratio ν/ν_B (ν = observing frequency; ν_B = electron cyclotron frequency). The three sets of curves are for electron energy spectral indices δ = 1,2,3, the solid line for each curve is for a viewing angle of 45°, while the adjacent dotted lines are for 15° and 75° viewing angle. Curves are computed from equation (36) of Dulk (1985).

Comparison of Figure 3 with the observed range of percent circular polarization shown in Figure 1 shows that for $\langle\pi_c\rangle \simeq 0.10$, one requires $\delta \gtrsim 4$ and $\nu/\nu_B \sim 10^2$. For $\nu = 5$ GHz, this implies $B \sim 20$ gauss.

Synchrotron emission models predict a reversal in the handedness of circular polarization (from x-mode to o-mode) as the frequency of observation passes from optically thin to optically thick. Detailed calculations for the fractional circular polarization from ultra-relativistic electrons are available (e.g., Melrose, 1971, 1980; Jones and O'Dell, 1977), but the general case of arbitrary electron energy and optical depth has apparently not been calculated yet. Figure 4 shows the expected ratio of $\pi_c(\tau > 1)/\pi_c(\tau < 1)$ as a function of energy spectral index δ for ultra-relativistic electrons, including the effects of large Faraday depth (cf. Melrose, 1980, pp. 221 et seq.)

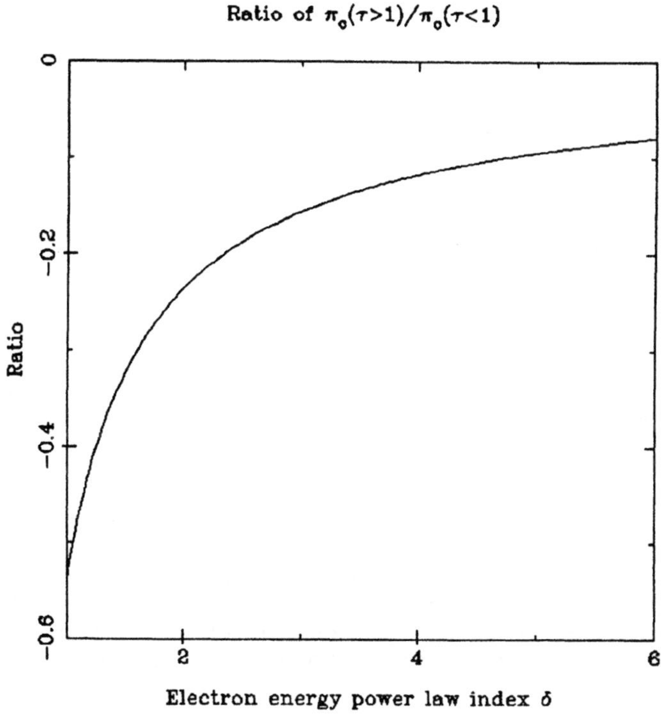

Figure 4. Ratio of optically thick to optically thin fractional circular polarization as a function of e^- energy spectral index δ for highly relativistic electrons (Melrose, 1971).

For comparison, Figure 5 shows a histogram of π_c(1.4 GHz/π_c(5.0 GHz) for all observations of UX Arietis and HR 1099, for which we have nearly simultaneous (Δt < 20 min) dual-frequency data.

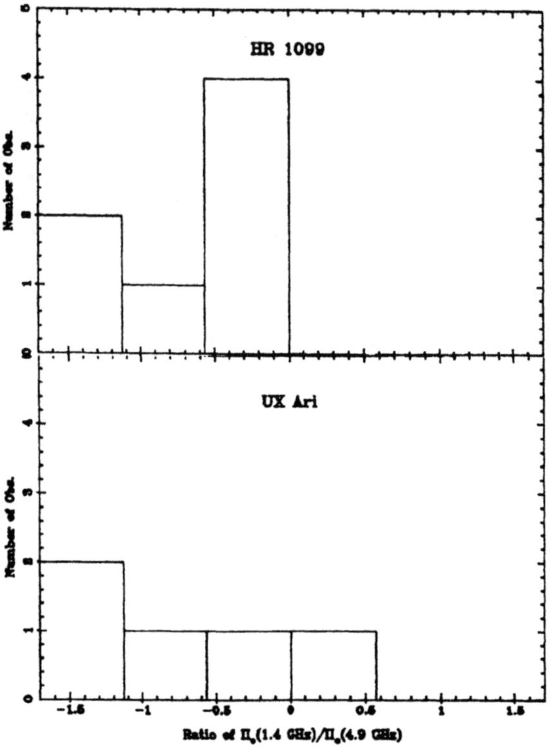

Figure 5. Histograms of ratio of fractional circular polarization at 1.4 and 5.0 GHz for UX Arietis and HR 1099.

Comparison of Figures 4 and 5 shows that while the expected helicity reversal is observed, about half of the observations for both stars have higher than predicted polarization ratios even for low values of δ (~ 'flat' e^- electron distribution). However, this disagreement is not necessarily important since the theoretical curves are valid for highly relativistic electron energies, whereas the observed high-fractional circular polarization argues strongly for gyrosynchrotron emission.

REFERENCES

Brown, R. L., and Crane, P. C., 1978, A.J., **83**, 1504.
Dulk, G. A., 1985, Ann. Rev. Astron. Astrophys., **23** (in press).
Dulk, G. A., and Marsh, K. A., 1982, Ap. J., **259**, 350.
Feldman, P. A., 1983, in Activity in Red-Dwarf Stars, ed. P. B. Byrne and M. Rodono (Reidel, Dordrecht).
Fix, J. D., Claussen, M. J., and Doiron, D. J., 1980, A.J., **85**, 1238.
Gibson, D. M., and Hjellming, R. M., 1974, P.A.S.P., **86**, 652.
Jones, T. W., and O'Dell, S. L., 1977, Ap. J., **215**, 236.

Lestrade, J.-F., Mutel, R. L., Preston, R. A., and Phillips, R. B., 1985, these proceedings.
Melrose, D. B., 1971, Astrophys. Space Sci., **12**, 172.
Melrose, D. B., 1980, Plasma Astrophysics, Vol. I (Gordon and Breach: New York).
Melrose, D. B., and Dulk, G. A., 1982, Ap. J., **259**, 844.
Mutel, R. L., and Weisberg, J. M., 1978, A.J., **83**, 1499.
Mutel, R. L., Lestrade, J. F., Preston, R. A., and Phillips, R. B., 1985, Ap. J. (in press).
Owen, F. N., and Gibson, D. M., 1978, A.J., **83**, 1488.
Owen, F. N., Jones, T. W., and Gibson, D. M., 1976, Ap. J., **210**, L27.
Simon, T., Linsky, J. L., and Schifter, F. H., 1980, Ap. J., **239**, 911.
Spangler, S. R., 1977, A.J., **82**, 169.
Uchida, Y., and Sakurai, T., 1983, in Activity in Red-Dwarf Stars, ed. P. B. Byrne and M. Rodono (Reidel, Dordrecht), p. 629.

VLA OBSERVATIONS OF LATE-TYPE STARS

K. Lang, R. Willson
Tufts University, Medford, Mass., USA

R. Pallavicini
Arcetri Observatory, Florence, Italy

ABSTRACT. We report the results of a program of observations of late-type stars at λ = 6 cm using the V.L.A. The source list includes stars with active chromospheres and coronae, UV Ceti-type flare stars, and RS CVn stars. Of the 31 objects surveyed, we have detected 6 and we have established upper limits for the remaining 25. The detected sources are all RS CVn and UV Ceti-type flare stars. No dwarf star of spectral type G and K has been detected, including the previously reported radio source χ^1 Ori.

1. OBSERVATIONS

We have used the Very Large Array (V.L.A.) in August 1983 and March 1984 to search for microwave emission from active late-type stars. Our list of sources included:

- stars with active chromospheres and coronae
- stars with detected magnetic fields
- UV Ceti-type flare stars
- RS CVn stars.

Some of the target objects were part of a coordinated program of observations of X-ray coronae with EXOSAT. We also observed the solar type star χ^1 Ori whose reported detection is unique among G and K dwarf stars, and difficult to understand on theoretical grounds.

The entire V.L.A. (27 antennas) was used in the "A" configuration. The observation frequency was 4885 MHz (λ = 6 cm) and the band width 50 MHz. The data were edited and calibrated using standard V.L.A. reduction programs. Of the 31 objects surveyed, 6 were detected. 3σ upper limits were determined for the remaining 25 objects.

In this report, we present initial results of our survey and we compare with previous observations of the same, or similar, sources by other investigators. A full account of these observations and a more extended discussion of the emission mechanisms involved will be presented elsewhere (Pallavicini, Willson and Lang 1984).

R.M. Hjellming and D.M. Gibson (eds.), Radio Stars, 267–270.

TABLE I. Detections (λ = 6 cm)

Star	Sp	D	S(mJy)	CP	$\log L_R$	Remarks
UX Ari	G5V+K0IV	50	20.7±0.24	10	16.8	RS CVn
HR 1099	G5IV+K1V	33	7.7±0.16	20	16.0	RS CVn
WW Dra	G5V+K0IV	180	3.6±0.17	$\lesssim$ 5	17.1	RS CVn
λ And	G8III-IV	24	0.84±0.12	$\lesssim$23	14.8	RS CVn
UV Cet	dM5.5e	2.7	0.95±0.25	$\lesssim$21	12.9	Flare star
EQ Peg A	dM3.5e	6.4	0.81±0.12	<20	13.6	Flare star

2. RESULTS

The results of our observations are summarized in Tables I and II. Table I gives observed fluxes S (in mJy), degree of circular polarization CP (in %) and derived radio luminosities L_R (in erg s^{-1} Hz^{-1}) for the six detected objects (four RS CVn and two UV Ceti-type flare stars). Table II gives 3-σ upper limits of the observed fluxes and upper limits of derived radio luminosities for the remaining 25 non-detected stars.

Of the detected stars, one (WW Dra) is a new detection. The derived luminosity log L_R = 17.1 erg s^{-1} Hz^{-1} falls in the range $\approx 10^{15}$- 10^{18} erg s^{-1} Hz^{-1}, typical of radio luminosities observed in RS CVn stars (Hjellming and Gibson 1980). For HR 1099 and UX Ari the derived luminosities are one to two orders of magnitude lower than the maximum radio luminosities reported previously (Hjellming and Gibson 1980). The long-period RS CVn star λ And has been seen by us at the same level reported by Bowers and Kundu (1981) for a VLA observation made three years before. We also report detection of circular polarization for UX Ari and HR 1099 (10% and 20%, respectively).

The two dMe flare stars detected by us (UV Cet and EQ Peg A) were previously detected at about the same flux level by Gary and Linsky 1981, Topka and Marsh 1982, Fisher and Gibson 1982, Linsky and Gary 1983. However, differently from previously reported observations, we have not detected EQ Peg B and YY Gem. Our upper limit for YY Gem (<0.45 mJy) is consistent with the range of variability (from 0.4 mJy to 1.8 mJy) reported by Linsky and Gary (1983) for this source.

Beside RS CVn and flare stars, we have observed 17 dwarf stars of spectral type G and K known to possess active chromospheres and coronae and/or detected magnetic fields. We did not detect any of them. The upper limits of radio luminosities for dwarf G and K stars fall in the range $\approx 10^{13}$- 10^{14} erg s^{-1} Hz^{-1}. It is interesting to note that the radio luminosities of detected dMe flare stars fall in the same range. Our negative result for G and K stars is in agreement with previous

TABLE II. Upper limits ($\lambda = 6$ cm).

Star	Sp	D	S(mJy)	$\log L_R$	Remarks
α Aur	G0III+G5III	14.0	<0.39	<14.0	RS CVn
ζ And	K1II-III	31.0	<0.45	<14.7	RS CVn
σ Gem	K1III	59.0	<0.45	<15.3	RS CVn
54 Cam	G0V+G2V	38.0	<0.42	<14.9	RS CVn
ε UMi	G5III	71.0	<0.42	<15.4	RS CVn
MM Her	G2V+K2IV	190.0	<0.51	<16.3	RS CVn
YY Gem	dM1e+dM1e	14.5	<0.45	<14.1	Flare star
EQ Vir	K5Ve	16.4	<0.42	<14.1	Flare star
κ Cet	G5V	9.5	<0.45	<13.7	
ε Eri	K2V	3.3	<0.42	<12.7	
o^2 Eri	K1V	4.8	<0.39	<13.0	
111 Tau	F8V	15.6	<0.39	<14.1	EXOSAT
β Lep	G5III	71.4	<0.42	<15.4	
χ^1 Ori	G0V	9.8	<0.23	<13.4	
π^1 UMa	G0V	15.4	<0.24	<13.8	EXOSAT
24 UMa	G2IV	26.0	<0.45	<14.6	
ξ UMa B	G0V	7.3	<0.45	<13.5	
61 UMa	G8V	8.4	<0.45	<13.6	
59 Vir	F8V	13.3	<0.21	<13.6	EXOSAT
ξ Boo A	G8V	4.7	<0.36	<13.0	
HD 131511	K2V	11.5	<0.45	<13.9	
HD 131977	K4V	5.8	<0.39	<13.2	
σ Dra	K0V	5.6	<0.36	<13.1	
HD 206860	G0V	15.0	<0.45	<14.1	EXOSAT
53 Aqr A+B	G1V+G2V	18.5	<0.17	<13.8	EXOSAT

surveys by Johnson and Cash (1980), Bowers and Kundu (1981), Gary and Linsky (1981), Linsky and Gary (1983). Contrary to Gary and Linsky (1981) who reported detection of the G0V star χ^1 Ori at a level of 0.60±0.27 mJy at 6 cm, we do not confirm quiescent radio emission from this source, as well as from a number of other stars similar to χ^1 Ori with regard to spectral type, rotation rate, chromospheric and coronal activity. Our upper limit for χ^1 Ori at 6 cm (<0.23 mJy) is a factor of 3 lower than the detection level reported by Gary and Linsky (1981). The detected radio emission likely originated from a flare on the dwarf M companion of χ^1 Ori.

3. SUMMARY AND CONCLUSIONS

Our results can be summarized as follows:

- We have detected 4 RS CVn stars and we have established upper limits for other 6. Of the detected stars, one (WW Dra) is a new detection, while the radio luminosities of the other three were one to two orders of magnitude lower than the maximum luminosities reported earlier.
- Detectable circular polarization (10% - 20%) was observed for UX Ari and HR 1099.
- We have confirmed quiescent radio emission from 2 flare stars (UV Cet and EQ Peg A) at about the same level as previous detections, and we have established upper limits for other two flare stars (YY Gem and EQ Vir).
- We have failed to detect microwave emission from any other solar type star of spectral type G and K, including rapidly rotating stars with active chromospheres and coronae. For those stars observed previously, our upper limits are consistent, and sometimes lower than those reported by Linsky and Gary (1983).

On the basis of these observations and of theoretical arguments presented elsewhere (Pallavicini, Willson and Lang 1984) we suggest that the observed quiescent radio emission from dMe stars is produced by a non-thermal process - possibly gyrosynchrotron emission by mildly relativistic electrons accelerated more or less continuously in starspots - rather than being due to thermal gyroresonance emission.

REFERENCES

Bowers, P.F. and Kundu, M.R. (1981) *Astron. J.* **86**, 569.

Fisher, P.L. and Gibson, D.M. (1982) in 'Cool Stars, Stellar Systems and the Sun' (M.S. Giampapa and L. Golub eds.), *SAO Special Rep.* 382, Vol. II, p. 109.

Gary, D.E. and Linsky, J.L. (1981) *Astrophys. J.* **250**, 284.

Hjellming, R.M. and Gibson, D.M. (1980) in 'Radio Physics of the Sun' (M.R. Kundu and T.E. Gergely eds.), p. 209.

Johnson, H.M. and Cash, W.C.Jr. (1980) in 'Cool Stars, Stellar Systems and the Sun' (A.K. Dupree ed.), *SAO Special Rep.* **389**, p. 137.

Linsky, J.L. and Gary, D.E. (1983) *Astrophys. J.* **274**, 776.

Pallavicini, R., Willson, R., Lang, K. (1984) to be submitted to *Astron. Astrophys.*

Topka, K. and Marsh, K.A. (1982) *Astrophys. J.* **254**, 641.

RADIO ACTIVITY ON W UMa SYSTEMS

V.A. Hughes and B.J. McLean
Astronomy Group, Dept. of Physics,
Queen's University,
Kingston, Ontario, K7L 3N6.

ABSTRACT. Twelve W UMa systems were observed using the VLA at 5 GHz. Radio emission was detected from two. VW Cep was seen to increase in flux density over a period of three hours from 0.6 mJy to 4.7 mJy, and could have risen further. The flare is interpreted as an event which injected $2 \times 10^{-13} M_{\odot}$ of material into the common envelope. V502 Oph was observed as a double source and could have been similar to α Sco, which is explained by the stellar wind from one component interacting with the magnetosphere or stellar wind of the second component.

1. INTRODUCTION

W UMa stars are contact eclipsing binaries with orbital periods in the range 6-15 hours. Both components are of spectral type F-K and both are overflowing their Roche lobes so that they have a common envelope. Two types have been recognized by Binnendijk (1970) depending on whether the light curve has a deeper minimum at the transit (A-type) or occultation (W-type) of the secondary, less massive component. To explain the W-type in the traditional model, it was thought that the secondary component was 5% hotter, but in the case of W UMa, UV photometry has shown that there is practically no difference in temperature between the components (Eaton, Wu, and Ruciński 1980), and the findings are interpreted as evidence for extensive areas of dark starpots (Mullan 1975). The detection of soft X-rays from VW Cep (Carroll et al. 1980) and subsequently from all the observed W-type systems, and spectroscopic observations with IUE (Dupree 1980, 1981; Dupree and Preston 1981; Hartman, Dupree, and Raymond 1982) established that these systems have an active and high temperature chromosphere and corona, similar to the RS CVn stars.

The RS CVn stars have much longer periods of 1-14 days, and are subject to intense flaring, so that if W UMa stars are in any way similar it is expected that they too would produce flares. Twelve W UMa systems were selected, principally because they were at distances of less than 150 pc. D. Florkowsky, (private communication), subsequently reported a radio flare on 44ι Boo, but no activity was detected by us.

R.M. Hjellming and D.M. Gibson (eds.), Radio Stars, 271–274.

TABLE 1
OBSERVING SESSIONS

Session I	1982 August 26	09:00-20:54 IAT	"B" Configuration
Session II	1983 January 6	03:16-09:09 IAT	"D" Configuration
Session III	1983 January 13	03:16-09:12 IAT	"D" Configuration

TABLE 2
W UMa STARS OBSERVED

STAR	DISTANCE (pc)	TYPE	STAR	DISTANCE (pc)	TYPE
SESSION I			SESSIONS II AND III		
VW Cep	24	W	VW Cep	24	W
44_1 Boo	13	W	W UMa	53	W
CW Cas	135	W	ER Ori	105	W
SW Lac	118	W	YY Eri	56	W
OO Aql	100	A			
V502 Oph	91	W			
V566 Oph	77	A			
V839 Oph	83	A			
ε CrA	38	A			

2. OBSERVATIONS

Initial observations were made using the 46-m single dish of the Algonquin Radio Observatory (ARO) to survey the nearer systems at 6.2 and 10.5 GHz. There were several marginal detections, and further observations were made using the considerably better collecting area of the Very Large Array (VLA) operated by the National Radio Astronomy Observatory*. Three sets of observations were made at 5 GHz, each of duration 6 hours. Typical angular resolutions were 3" × 1" and 14" × 10" for the "B" and "D" configurations, respectively. Dates, times and configurations are shown in Table 1; a list of systems observed, chosen because their distances were < 150 pc, together with their estimated distance and type, is given in Table 2. For further details see Hughes and McLean (1984).

3. RESULTS

For each of the sytems observed, maps were made, the CLEAN procedure used, and a best fit was made to the source using an equivalent elliptical gaussian beam. It is estimated that in most cases, any system showing a flux density greater than about 0.15 mJy would have been detected. Results were obtained on two systems. VW Cep showed a level of about 0.6 mJy during Session II from 0400 IAT to 0600 IAT when the flux density rose to 4.6 mJy at 0900 IAT and could have continued rising after this. The flare is shown in Figure 1. During Session III the flux density appears to have returned to a mean level of about 0.6 mJy. The period of the binary is 6.7 hours, but was only observed for a total of 5 hours, so it is not clear if the flare shows intrinsic time variation, or if it is the effect of rotation of the system, but since no time variations were observed during Session 3, we attribute the event as the result of major flaring during a period of smaller scale activity.

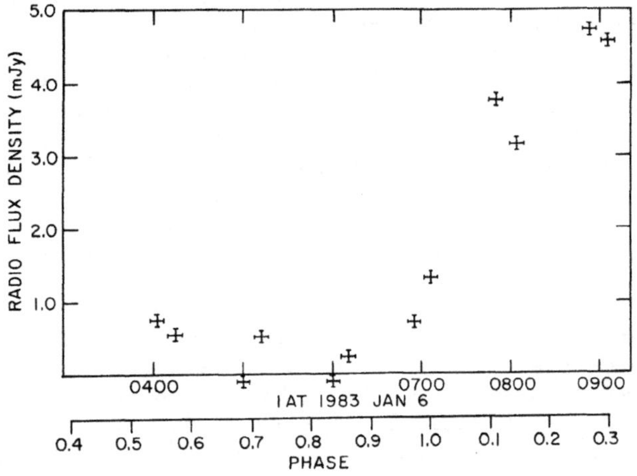

Figure 1. The radio flare on VW Cep

The other probable detection was V502 Oph which was only observed during Session 1. It was found to consist of two sources, one within the angular error of 0".6 of V502 Oph, which has a flux density of 0.4 mJy, and the other separated by 2".6 (220 au at the distance of 96 pc) with a flux density of 0.3 mJy. It is noted that if V502 Oph were at the distance of VW Cep, the flux density from each of the components would be 5 mJy, about equal to the peak observed intensity during its flare.

4. DISCUSSION

The fact that few other W UMa systems were detected leads us to the conclusion that VW Cep was exhibiting "enhanced emission" from the corona and chromosphere, which is known from the X-ray and IUE observations to be at a temperature $\sim 10^7$ K, and that the flare could be produced at the time of injection of material into the corona and chromosphere. Due to the large differential rotation that must exist in the system, any material injected must be rapidly spread out over a large area of surface. If we assume that the radius of the system is R (in units of a solar radius) and the temperature is T, then for a distance to the system of 24 pc and flux density of 0.4 mJy, we estimate that for VW Cep

$$R^2T = 2.4 \times 10^8.$$

If we assume that the radius of the emitting region is 3 $R_\odot$, then T $\sim 3 \times 10^7$ K, which is consistent with X-ray and IUE observations. We note that the radius of the outer critical Roche lobe is 3.7 $R_\odot$. If we assume that the optical depth, τ, becomes unity over a distance of 1 $R_\odot$, and that T = 3×10^7 K, then the electron density is $\sim 4 \times 10^{10}$ cm^{-3}, from which we estimate that 2×10^{-13} $M_\odot$ was injected into the common envelope. This is small in comparison with some determined values for mass-transfer between components.

The temperature of the V502 Oph system appears to be somewhat higher, at 1.4×10^8 K but is still consistent with X-ray and IUE data. It has also been shown that the system underwent a period change during 1955-1966 (Binnendijk 1969) which we interpret in terms of mass transfer of 1.5×10^{-6} $M_\odot$, perhaps at a rate as high as $10^{-6} M_\odot$ yr^{-1}. If V502 Oph has a companion, then the double source may be similar to α Sco, where in the model by Hjellming and Newell (1983), the interaction of the stellar wind from one component with the magnetosphere or stellar wind of the other produces the second source.

The other class of active binaries is the RS CVn systems which have longer periods in the range 1-14 days. They are also X-ray sources, which again is interpreted to mean that they have hot coronas with temperatures of $\sim 10^7$ K. They are also subject to intense radio flaring with an estimated luminosity at 10 GHz of $10^{17} - 10^{18}$ erg s^{-1} Hz^{-1} (Feldman 1982), which is $10^2 - 10^3$ times greater than the maximum observed from VW Cep. It is possible that VW Cep developed a more intense flare but from the rate of rise, it is unlikely that it reached the level of the RS CVn systems.

From the fact that only two systems out of 12 were detected and that of these VW Cep was not detected four months previous to the flare, suggests that the systems are not very radio active. However it may be possible to obtain more information on the mechanism of the flaring in these systems, rather than e.g. the longer period RS CVn's, since changes in the light curve, produced by mass-loss, are more easily detected with the short period light curves.

REFERENCES

Binnendijk, L. 1969, Ap.J., 74, 1031.
Binnendijk, L. 1970, Vistas Astr., 12, 217.
Carroll, R.W. et al. 1980, Ap. J. (Letters), 235, L77.
Dupree, A.K. 1980, Highlights Astr., 5, 263.
Dupree, A.K. 1981, in "Solar Phenomena in Stars and Stellar Systems", ed. R.M. Bonnet and A.K. Dupree (Dordrecht: Reidel), p. 407.
Dupree, A.K., and Preston, S. 1981, in The Universe at Ultraviolet Wavelengths, The First Two Years of the IUE Satellite", ed. R.D. Chapman (NASA Conf. Pub. No. 2171), p. 333.
Eaton, J.A., Wu, C-C., and Ruciński, S.M. 1980, Ap.J., 239, 919.
Feldman, P.A. 1982, in IAU Colloquium 71, Activity in Red-Dwarf Stars, ed. P.B. Byrne and M. Rodonò (Dordrecht: Reidel), p. 429.
Hartman, L., Dupree, A.K., and Raymond, J.C. 1982, Ap.J., 252, 214.
Hjellming, R.M., and Newell, R.T. 1983, Ap.J., 275, 704.
Hughes, V.A., and McLean, B.J. 1984, Ap.J., 278, 716.
Mullan, D.J. 1975, Ap.J., 198, 563.

* The National Radio Astronomy Observatory is operated by Associated Universities Inc., under contract with the National Science Foundation.

HIGH-ANGULAR RESOLUTION OBSERVATIONS OF STELLAR BINARY SYSTEMS

Jean-Francois Lestrade
Bureau des Longitudes, Paris, France/Jet Propulsion Laboratory, California Institute of Technology, 4800 Oak Grove Drive, Pasadena, CA 91109, U.S.A.

Robert L. Mutel
University of Iowa, Department of Physics and Astronomy, Iowa City, IA 52242, U.S.A.

Robert A. Preston
Jet Propulsion Laboratory, California Institute of Technology, 4800 Oak Grove Drive, Pasadena, CA 91109, U.S.A.

Robert B. Phillips
Haystack Observatory, Westford, MA, U.S.A.

ABSTRACT. We report on a long-term program designed to investigate the spatial structure of centimeteric radio emission from close binary systems using multi-station VLBI arrays. We have detected eleven binaries, including eight RS CVn systems, Algol, LSI 61°303, and Cyg X-1. The measured brightness temperatures vary from $T_B \sim 10^{8.5}$ K during periods of low activity to $T_B \sim 10^{10.5}$ K during flares. Extensive observation of a few sources has shown that the spatial structure is 'core-halo' with linear dimensions of about a stellar radius and the binary system, respectively. The observations are consistent with gyrosynchrotron emission of mildly relativistic electrons ($\langle E \rangle \sim 1 - 5$ MeV) in magnetic fields of $B \sim 10^{1.5 \pm 0.5}$ gauss. The core sources appear to be optically thick, while the halo component is optically thin.

1. INTRODUCTION

Radio emission from RS CVn binary systems (Hall, 1976) is a common (but not universal) property of the class. The emission is of high brightness temperature, highly variable, and often circularly polarized. The mechanism most often invoked is gyrosynchrotron radiation from mildly relativistic electrons ($E > 1$ MeV) in magnetic field of $B \sim 10 - 100$ gauss (Hjellming and Gibson, 1980).

From December 1982 to March 1984, eleven binary stars have been detected using cm-wavelength VLBI arrays with angular resolutions from

R.M. Hjellming and D.M. Gibson (eds.), Radio Stars, 275–282.

40 to 1.5 milliarcseconds (Table I). These stars are all RS CVn binary systems except for Algol, LSI 61°303, and Cyg X-1.

The scientific goals of the VLBI program are twofold: astrophysical, i.e., to measure the spatial structure of the radio emission to deduce physical properties, such as particle densities and magnetic field strengths, and astrometric, i.e., to investigate their potential in tying the future HIPPARCOS optical reference system to the VLBI extragalactic reference frame (Lestrade et al., 1982).

Table I: Summary of results of VLBI observations of stellar binary systems at 1.6, 2.3, 5., 8.4 GHz from December '82 to March '84

Star Name	Observed Ranges					Overall Size of Binary (10^{11} cm)	Stellar Diameter in Binary (10^{11} cm)	Remarks	References
	Total Flux Density (m Jy)	Angular Size of Radio Source (milliarc)	Linear Size (10^{11} cm)	Log Brightness Temperature (K)	Radio Luminosity Log L ergs/s/Hz				
HR5110	15–165	0.9–1.4	4 – 11	8.6 – 9.3	16.6–17.8	13	4	The most compact RS CVn system	a,d
UX Ari	14–145	0.4–3.2	3 – 26	8 – ≥ 10	16.6–17.6	17	4	Core-halo structure at 5 GHz	b,d
HR1099	12–400	0.8–2.1	4 – 12	8– ≥ 10.5	15.9–17.4	16	4	Possible core-halo structure at 8.4 GHz	b,c,d
Algol	5–140	0.5–2.2	1.9 – 8	8.5 – ≥ 10	15.3–17.0	14	5	Possible cyclotron-maser at 1.6 GHz	d,e
σ CrB	14	0.6	2	9.5	16.0	27	1.5		d
AR Lac	14	≤ 3	≤ 20	≤ 8	16.5	14		Size consistent with radio eclipse observations	
λ And	19	≤ 4	≤ 10	≥ 7.9	16.0	9			
II Peg	9	≤ 10	≤ 40	≥ 6.3	15.9	13			
SZ Psc	37	0.6	8	9.9	17.6	15			d
LSI 61°303	15–80	4.1	1400	9.1	20.7	120		Radio source may be broadened by interstellar scattering	f
CygX1	15	≤ 0.5	≤ 209	≥ 9.6	20.				d

a: Lestrade et al., 1984a
b: Mutel et al., 1984a
c: Lestrade et al., 1984b
d: Mutel et al., 1985
e: Lestrade et al., 1985
f: Mutel and Lestrade, 1985

2. OBSERVATIONS

High-sensitivity VLBI (Very Long Baseline Interferometry) observations were conducted using large aperture antennae and the Mark III VLBI data acquisition system recording bandwidth of up to 56 MHz (Rogers et al., 1983). Three experiments were conducted using the 64 m antennae of the Deep Space Network facilities operated by NASA/JPL at Goldstone (California) and near Madrid (Spain), the 40 m at Owens Valley (California), the 26 m at Fort Davis (Texas), and the 18 m at Westford (Massachusetts). Three experiments were conducted under the auspices of the U.S. VLB Network with the Phased-VLA (equivalent to 130 m), MPI-Effelsberg (100 m), Greenbank (40 m), Owens Valley (40 m), Haystack (36 m), and Fort Davis (26 m).

The data were correlated on the Mark III Processor at Haystack Observatory.

3. RESULTS AND INTERPRETATIONS OF THE OBSERVATIONS

3.1. The RS CVn Systems and Algol

The main result obtained from these VLBI observations is the discovery that the radio emission from RS CVn systems and Algol often exhibit variable core-halo structure with typical brightness temperatures of $\sim (1-3) \cdot 10^{10}$ K for the core and $\sim (5-10) \cdot 10^{8}$ K for the halo. For the most precise measurements we made (HR 5110, HR 1099, UX Ari, Algol), the dimension of the core component is comparable to or smaller than a stellar diameter, while the halo component is comparable to the overall size of the binary system.

Figure 1 shows a low dynamic range VLBI map of the radio emission from UX Arietis made at 5 GHz during an intense outburst in July 1983 (Mutel et al., 1985). The core-halo structure is evident.

During the July 1983 experiment, both circular polarizations were recorded simultaneously. By comparing the VLBI phase difference between the two senses of polarization for nearby calibrator sources with that of the program stars, we were able to set an upper limit of 0.05 milli-arcsecond ($4 \cdot 10^{10}$ cm at a typical distance of 50 pc) in the offset in position of the right and left circularly polarized components of the sources.

Generally, our observations in total flux density indicate that during low to moderate states of activity, the radio (at centimetric wavelengths) spectrum is often flat and the degree of circular polarization is high (> 10%), while during outbursts the spectrum is inverted ($0 < \alpha < 1$, $S \propto \nu^{\alpha}$) and the degree of circular polarization is low. These results are consistent with the observations reported by Hjellming and Gibson (1980). A simple model accounting for observed angular sizes, polarization and spectra is proposed by Mutel et al. (1985) by analogy with some types of "type-IV" radio emission observed from the sun. In this model, an outburst is initially associated with a self-absorbed (optically thick) gyrosynchrotron source, probably located on or near a starspot region on the active star. The observed source (core component) is largely unpolarized with a brightness temperature characteristic of the energetic electrons, $T_B \sim 10^{10.5}$ K, or $\langle E \rangle \sim kT_{eff} \sim$ 5 MeV. As the outburst evolves, one or more coronal loops connected with the active region expand to a size comparable with the binary system. Radio emission from the expanded loop(s) (halo component) will become circularly polarized as the source becomes optically thin. The lifetime of the halo component is much longer than the core (days vs hours) because the mean electron energy and magnetic field strength of the halo component are much lower. Hence, the model predicts that VLBI observations of binary systems during 'quiescent' periods have brightness temperatures $T_B < 10^9$ K, and hence will be strongly resolved on intercontinental baselines.

In the following section we summarize our VLBI observations of several specific binary systems.

3.2. HR 5110

The 5th magnitude RS CVn system HR 5110 (F2IV + probable K-star, P = 2.61 days, 52 pc) may be a semidetached system with the cooler component filling its Roche lobe (Dorren and Guinan, 1980). In that respect, it can be classified as an Algol-type binary system. HR 5110 is known for its strong radio outbursts (Feldman, 1979) and was the first system we detected using Mark III VLBI techniques (Lestrade et al., 1984a).
HR 5110 seems to be the most compact source we have observed so far, exhibiting a radio core always smaller than the overall size of the system (1.3 milliarcsecond) independently of its level of activity over a range of a few mJy to 165 mJy during our multiple observations. During the July 1983 outburst (140 - 165 mJy) at 5 GHz, the closure phases were all zero within 2 standard deviations indicating no asymmetric brightness distribution at a level of 0.5 milliarcsecond or more in the radio structure.

The magnetic field strength derived for the December 1982 observations (Lestrade et al., 1984a) assuming gyrosynchrotron emission yielded maximum strength B_{max} between 150 and 600 gauss depending on the electron energy spectral index. This is ten times larger than the field generally calculated for UX Ari and HR 1099 (Mutel et al., 1984a, 1985) using similar observations and assumptions.

We note that the small source size for HR 5110 measured at all levels of activity disagrees with the relatively large source size reported in Feldman (1983) during an outburst of 285 mJy.

3.3. UX Arietis

The 7th magnitude RS CVn system UX Ari (G5V + K0IV, P = 6.4 days, 50 pc), is one of the most active RS CVn systems at radio wavelength. We have detected this source using VLBI arrays in February 1983 (Mutel et al. 1984a), March 1983 (unpublished), July 1983 (Mutel et al., 1985), and October 1983 (unpublished). The most interesting data were obtained in July 1983 at 5 GHz using dual circular polarization and six telescopes. The source was observed during an intense outburst, which allowed us to detect fringes on all fifteen baselines for three scans separated by about one hour. The resulting visibility data (including closure phases) were used to map the source with the hybrid mapping algorithm (Pearson and Readhead, 1984) resulting in the core-halo morphology shown in Figure 1. The brightness temperature was twelve times higher in the core than in the halo. The map shows that the centroids of the core and halo have a position offset about equal to the projection of the major axis of the binary system at the time of observation. The halo contains 80% of the total flux density and has a size of 3.2 milliarcseconds (almost twice as large as the overall binary size), making it the most extended radio source from RS CVn systems we have detected so far. The core was unresolved on the longest baseline, indicating a size < 0.4 milliarcsecond, i.e., smaller than a stellar diameter.

Mutel et al. (1985) provide typical values for the physical quantities of both components assuming a gyrosynchrotron emission mechanism and electrons of E $\sim$ 1 - 10 MeV.

Core: $\tau > 1$, $B = 10^2$ gauss, $T_B = 10^{10\pm0.5}$ K,

$\pi_c < 0.10$, radiative lifetime $t \sim 10^4$ sec.

Halo: $\tau < 1$, $B = 10$ gauss, $T_B = 10^{9\pm0.5}$ K,

$\pi_c > 0.1$, radiative lifetime $t \sim 10^6$ sec.

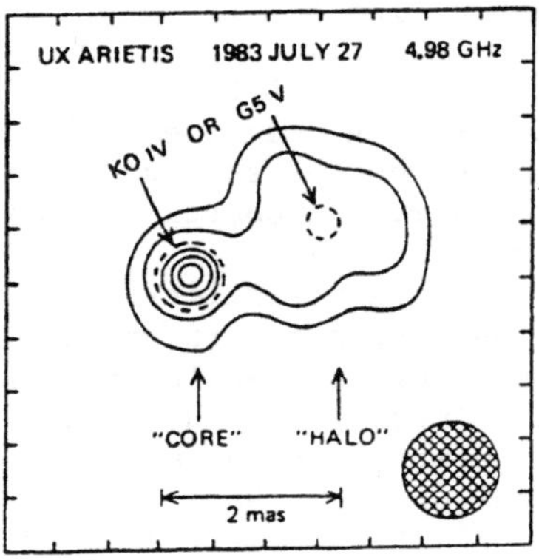

Figure 1. Hybrid map of UX Ari at 5 GHz. The location of the two stellar components relative to the radio map is conjectural. (See Mutel et al., 1985). The contour levels are at 25, 35, 50, 70, and 90% of peak brightness.

3.4. HR 1099

The RS CVn system HR 1099 consists of three stars: the close double (G5V + K0IV, P = 2.8 days, d = 35 pc) which is a RS CVn system and ADS 2644 in a long period orbit. We have detected HR 1099 using VLBI arrays during the same four epoch as UX Ari. During the March 1983 observation, the measured source size was 0.8 milliarcsecond or 75% of the stellar diameter of the chromospherically active K-star (Lestrade et al., 1984b). This outburst may have been associated with starspots, since extrapolation of the photometric data published by Dorren and Guinan (1982) shows that the starspot formation of HR 1099 was facing toward us during the time of the observations.

Lestrade et al. (1984b) derive a magnetic field strength of 30 gauss (assuming gyrosynchrotron emission) and show that it can be reconciled with a smaller value of B (6 gauss) obtained by Mutel et al. (1984a) for HR 1099 at 1.6 GHz using a scaling law deduced from Dulk and Marsh (1982).

3.5. Algol

The 3rd magnitude close binary system A-B (B8V + K0, P = 2.87 days, 27 pc) belonging to the triple system Algol is a semidetached system with the K-star filling its Roche lobe. Algol was the first radio star detected by VLBI. Clark et al. (1975) and Clark et al. (1976) obtained source sizes of ~ 4 and 1.1 milliarcseconds at 8 GHz during strong outbursts of 0.6 and 1.0 Jy, respectively. We have detected Algol with VLBI arrays during March 1983, May 1983, July 1983, and October 1983.

The May 1983 observations (Lestrade et al., 1985) utilized simultaneous observing at 2.3 GHz and 8.4 GHz. In 3.5 hours, the total flux density of Algol increased by factors of 3 and 5 at these frequencies, respectively, while the brightness temperatures increased by 3 at 2.3 GHz and 10 at 8.4 GHz. The spectral index α between these two frequencies changed from -0.2 to +0.4, suggesting that the source became optically thick during the outburst. Throughout the observations, the size of the radio source at 2.3 GHz was always smaller than the overall dimension of the binary system projected on the sky but larger than the K-star diameter. In contrast, at 8.4 GHz, the source was always significantly smaller than the diameter of the K-star.

During the July 1983 observations, the visibility data indicated that there were two components: an unpolarized core and a 5% circularly polarized halo. There was no evidence for asymmetric brightness distribution from the closure phases. The dimension of the core was smaller than the K-star diameter, but the halo was comparable with the projected separation of the binary system (Mutel et al., 1985).

During the October 1983 experiment at 1.6 GHz (recorded simultaneously in both senses of circular polarization), Algol had a very high brightness temperature, $T_B > 10^{10}$ K, associated with 53% left circular polarization (Lestrade et al., 1985). This is consistent with a coherent process, such as the cyclotron-maser described by Melrose and Dulk (1982).

3.6. Other RS CVn Systems

In general, the low signal-to-noise ratio of the detections on all but the most sensitive baselines allowed only estimates of the brightness temperature and not detailed source morphology.

Five other RS CVn systems (σ CrB, λ And, SZ Psc, II Peg, AR Lac) have been detected on VLBI baselines. During the July 1983 experiment, the sources were found to have sizes smaller than their overall binary system dimensions and high brightness temperatures (~ 10^{10} K). The eclipsing system AR Lacertae has an upper limit of $2 \cdot 10^{12}$ cm, consistent with the lower limit of $5 \cdot 10^{11}$ cm found for its (radio) size by Doiron and Mutel (1984) during an optical eclipse.

3.6.1. <u>LSI 61°303</u>. The Be star LSI 61°303 is a variable nonthermal radio source, an x-ray emitter, and may be associated with the γ-ray source CG 135 + 01. Taylor and Gregory (19823) have discovered a 26.5 day periodicity in the radio emission variation and Hutching and Crampton (1981) a 26.4 day periodicity in the radial velocity data. This strongly suggests that the Be star and an unseen companion form a binary system. The semimajor axis is likely to be 0.4 AU (Taylor and Gregory, 1982).

LSI 61°303 was observed during five VLBI sessions from December 1982 to October 1983, but because of various tecnical difficulties, only the July and October 1983 sessions provided useful data. In July 1983, during its quiescent state (15 mJy), a marginal detection on the phased-VLA-Owens Valley baseline (angular resolution = 8 mas, 5 GHz) yielded a source size < 5 milliarcseconds. In October 1983, again during its quiescent period, LSI 61°303 was clearly resolved at 1.6 GHz. A Gaussian brightness distribution was fitted to the visibility function with an angular source size (FWHM) = 4.1 milliarcseconds (Mutel and Lestrade, 1985). At 2300 pc, this corresponds to a linear size $1.4 \cdot 10^{14}$ cm (10 AU or twelve times the major axis). The large ratio of radio source to orbital size for the quiescent emission appears to rule out 'local' emission mechanisms, since radiation damping and adiabatic expansion losses would be too severe to allow sufficient numbers of relativistic electrons to propagate to such large distance. This point is emphasized by Vestrand (1983), who studied the quiescent radio emission from Cyg X-3, an x-ray binary system with similar radio properties. Vestrand proposes *in situ* acceleration of electrons by Compton collisions with γ-rays produced by the neutron star. This model appears to be applicable to LSI 61°303, since it predicts the correct angular size for the quiescent emission and since LSI 61°303 has been identified with a probable γ-ray source (Gregory et al., 1979; Pollock et al., 1981).

An alternate explanation for the large measured angular size of the radio source is interstellar scattering. The source is along the line of sight to the H II regions W3/W4/W5 (Goudis and White, 1980), which could provide a local scattering screen. Even without a local screen, the empirical formulae of Cordes (1984) predict a scattering angle of θ(ISS) = 4.5 milliarcseconds for LSI 61°303 (see Mutel, 1984b). A definitive test for interstellar scattering would be two frequency VLBI observations: if the measured angular size is due to interstellar scattering, it should scale as $\theta \propto \lambda^2$.

3.6.2. *Cyg X-1*. Cyg X-1 was detected at 5 GHz in July 1983. It was quite weak (15 mJy) and unresolved. The upper limit to its size was < 0.5 milliarcsecond, or < 10^{13} cm, at 2500 pc. We did not have sufficient signal-to-noise ratio on the weaker baselines to compute the closure phases and probe possible asymmetry in the brightness distribution.

REFERENCES

Clark, B. G., Kellermann, K. I., and Shaffer, D., 1975, *Ap. J. Letters)*, **198**, L123.
Clark, T. A., et al., 1976, *Ap. J. (Letters)*, **206**, L107.
Cordes, J. M., 1984, in 'VLBI Compact Radio Sources,' IAU Symposium 110, ed. R. Fanti, K. Kellermann, and G. Setti, p. 303.
Doiron, D., and Mutel, R. L., 1984, *A. J.*, **89**, 430.
Dorren, J. D., and Guinan, E. F., 1980, *A. J.*, **85**, 1082.
Dorren, J. D., and Guinan, E. F., 1982, *Ap. J.*, **267**, 655.
Dulk, G. A., and Marsh, K. A., 1982, *Ap. J.*, **259**, 350.
Feldman, P. A., 1979, IAU Circ. No. 3366.

Feldman, P. A., 1983, in 'Activity in Red Dwarf Stars,' ed. by P. Byrne and Rodono, p. 429.
Goudis, L., and White, N. J., 1980, Astron. Astrophys., **83**, 79.
Gregory, et al., 1979, A. J., **84**, 1030.
Hall, D. S., 1976, in 'Multiple Periodic Variable Stars,' IAU Colloquium 79, ed. W. S. Fitch, Dordrecht: Reidel, p. 287.
Hjellming, R. M., and Gibson, D. M., 1980, in 'Radiophysics of the Sun,' IAU Symposium 86, ed. M. R. Kundu and T. E. Gehrels, Dordrecht: Reidel, p. 209.
Hutching, J. B., and Crampton, D., 1981, PASP, **93**, 486.
Lestrade, J.-F., Preston, R. A., and Slade, M. A., 1982, in 'Very Long Baseline Interferometry Technique," CNES Colloquium, Toulouse, France, August 31-September 2, 1982, ed. F. Biraud, Edition Cepadues, p. 199.
Lestrade, J.-F., Mutel, R. L., Preston, R. A., Scheid, J. A., and Phillips, R. B., 1984a, Ap. J., **279**, 184.
Lestrade, J.-F., Mutel, R. L., Phillips, R. B., Niell, A. E., and Preston, R. A., 1984b, Ap. J. (Letters), **282**, L23.
Lestrade, J.-F., Mutel, R. L., Preston, R. A., and Phillips, R. B., 1985, submitted to Ap. J.
Melrose, D. B., and Dulk, G. A., 1982, Ap. J., **259**, 844.
Mutel, R. L., Doiron, D., Lestrade, J.-F., and Phillips, R. B., 1984a, Ap. J., **278**, 220.
Mutel, R. L., 1984b, these proceedings.
Mutel, R. L., Lestrade, J.-F., Preston, R. A., Phillips, R. B., 1985, Ap. J. (in press).
Mutel, R. L., and Lestrade, J.-F., 1985 (in preparation).
Pearson, T. J., and Readhead, A. C. S., 1984, Ann. Rev. Astron. Astrophys.
Pollock, A. M. T., et al., 1981, Astron. Astrophys., **94**, 116.
Rogers, A. E. E., et al., 1983, Science, **219**, 51.
Ramaty, R., 1969, Ap. J., **158**, 753.
Taylor, A. R., and Gregory, P. C., 1982, Ap. J., **225**, 210.
Vestrand, W. T., 1983, Ap. J., **271**, 304.

12CM OBSERVATIONS OF STELLAR RADIO SOURCES

K. C. Turner
Arecibo Observatory*
P. O. Box 995
Arecibo, PR 00613

During the development and operation of the Arecibo Interferometer (see Figure 1) over the last four years, a number of fields containing interesting stellar objects have been observed, both as 'filler' and as regular program sources. Classes of systems observed include RS CVn variables, dwarf, recurrent and old novae, new supernovae, and stellar wind stars.

Variations in system parameters during this period have resulted in changes in sensitivity, but typical 3 sigma non-detection upper limits are 2 to 3 mJy for a 5 minute scan. Fringe spacing is 2.4 arc seconds, so that structures smaller than one arc second are essentially unresolved.

Five sigma or better detections have been made in 17 of 76 fields containing systems of possible interest for this workshop. Half of the 14 RS CVn fields observed gave such detections, while only 11% of the dwarf novae and 'other object' fields contained sources 5 sigma above the noise. It is felt that this represents a quite conservative extimate of the confusion level for this study. Most of the detected sources showed some evidence of variability. In the tables, such sources are marked with a V in the flux column, and have the range of fluxes observed tabulated. Both upper and lower fluxes are at least 5 sigma, unless an upper limit is indicated, which is 3 sigma.

Information in the remarks column is taken from Kukarkin, et al (1969) (and supplements), or from the SKYMAP compilation of Gottlieb (1978). A more extensive table, giving dates for the individual observations, is available on request from the author.

ACKNOWLEDGEMENTS

I wish to thank the Astronomical Data Center at NASA Goddard Space Flight Center for providing magnetic tape copies of the two catalogs referred to above, together with very helpful supporting documentation.

* Arecibo Observatory is part of the National Astronomy and Ionosphere Center, which is operated by Cornell University under contract with the National Science Foundation.

R. M. Hjellming and D. M. Gibson (eds.), Radio Stars, 283–288.

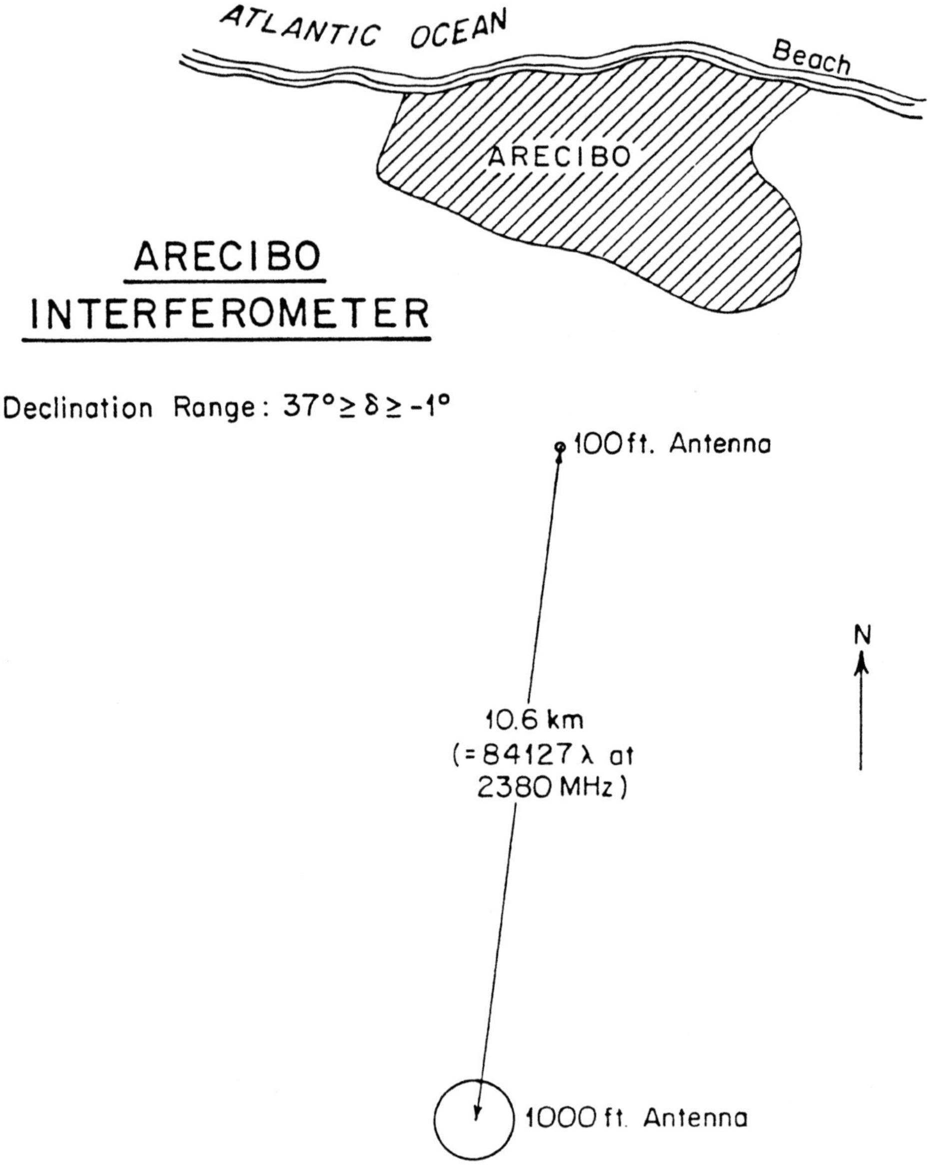

Figure 1. Geometry of the Arecibo Interferometer.

Table I. DWARF NOVAE

Source	R. A. (1950.0)	Dec.	Flux (mJy)	Remarks
TY PSC	1 22 47.93	32 7 38.8	V 10 - < 3	U GEM
UZ BOO	14 41 45.28	22 13 35.7	V? 2.4 +/- 0.2	U GEM
HW TAU	5 0 35.48	26 19 23.4	<12.	U GEM
V344 ORI	6 12 29.33	15 29 19.4	< 9.	U GEM
CZ ORI	6 13 51.24	15 25 31.4	< 2.	U GEM
IR GEM	6 45 30.75	28 8 1.0	< 3.	U GEM
RV CNC	8 6 48.24	19 35 22.7	< 2.	U GEM
CT HYA	8 48 28.57	3 18 57.1	< 2.	U GEM
SY CNC	8 58 14.46	18 6 5.1	< 1.	Z CAM
TU LEO	9 27 1.28	21 36 44.6	< 2.	U GEM
X LEO	9 48 21.40	12 6 38.9	< 2.	U GEM
AH HER	16 42 4.39	25 20 20.7	< 1.	Z CAM
V810 OPH	17 38 21.88	7 3 28.1	< 6.	U GEM
PU HER	18 7 56.23	31 57 19.8	< 3.	U GEM
CH HER	18 32 41.45	24 45 20.3	< 2.	U GEM
KW HER	18 39 5.93	12 6 23.5	< 2.	Z CAM
V550 CYG	20 3 7.65	32 13 0.3	<15.	U GEM
RU PEG	22 11 29.78	12 27 10.8	< 2.	U GEM

Table II. SYMBIOTIC NOVAE

Source	R. A. (1950.0)	Dec.	Flux (mJy)
YY HER	18 12 24.44	20 58 3.2	3.0 +/- 0.6
AG PEG	21 48 35.46	12 23 25.3	V 12 - 3

Table III. NOVAE

Source	R. A. (1950.0)	Dec.	Flux (mJy)	Remarks
T COR BOR	15 57 24.50	26 3 38.7	V 22 - 2	REC NOVA
NOVA AQL 82	19 20 50.14	2 23 35.4	V 17 - 4	NOVA
V529 ORI	5 57 8.31	20 16 25.0	< 2.	REC NOVA
CT SER	15 43 19.60	14 31 46.5	< 1.	NOVA
FH SER	18 28 16.20	2 34 29.7	< 2.	NOVA
V603 AQL	18 46 21.40	0 31 37.4	< 2.	NOVA

Table IV. RS CVn VARIABLES

Source	R. A. (1950.0)	Dec.		Flux (mJy)
UX ARI	3 23 33.18	28 32 43.1	V	96 - 15
SIGMA GEM	7 40 11.42	29 0 22.5	V	17 - < 3
HD86590	9 57 13.31	24 47 36.6		2.2 +/- .3
FK COM	13 28 26.27	24 28 41.5		5. +/- 1.
SIGMA CRB	16 12 48.27	33 59 2.2		6. +/- 1.
Z HER	17 55 51.56	15 8 31.0		3.0 +/- 0.4
HR8703	22 50 34.43	16 34 31.3	V	15 - 5
ZET AND	0 44 40.58	23 59 50.2		< 4.
UV PSC	1 14 17.95	6 32 54.1		< 4.
RU CNC	8 34 33.80	23 44 15.2		< 2.
WY CNC	8 58 57.95	26 52 43.1		< 5.
RT CRB	15 35 59.87	29 39 1.3		< 3.
MU HER	17 44 28.71	27 44 19.1		< 5.
MM HER	17 56 32.44	22 8 58.3		< 4.

Table V. OTHER OBJECTS

Source	R. A. (1950.0)	Dec.	Flux (mJy)	Remarks
RZ ARI	2 53 0.07	18 7 58.2	8.0 +/- 1.6	M6 III Irr var
SN 1979C	12 20 26.72	16 4 29.5	10. +/- 1.	
EE PEG	21 37 34.38	8 57 24.6	2.1 +/- 0.4	Algol
EQ PEG B	23 29 14.53	19 39 35.3	7. +/- 1.	Flare star
ALP AND	0 5 47.25	28 49 0.7	< 9.	A0 pec
28 AND	0 27 28.17	29 28 41.6	< 6.	Am
HD 5223	0 51 32.14	23 47 53.2	< 6.	R3
HD 5820	0 57 13.73	6 12 50.9	< 9.	M II
RR ARI	1 53 3.37	23 20 6.7	< 3.	K0 III var
14 ARI	2 6 34.27	25 42 24.4	< 3.	F2 III
XI 2 CET	2 25 30.17	8 14 17.8	< 3.	B9 III
HD 19445	3 5 29.22	26 9 19.1	< 6.	A4 pec
ETA TAU	3 44 31.23	23 57 19.1	< 5.	B7 III
BU TAU	3 46 11.74	23 54 18.0	< 6.	B8 III errupt.
BU 2 TAU	3 46 13.08	23 59 16.9	< 5.	B8 pec
X PER	3 52 15.95	30 54 13.4	< 3.	O PEC T Tau
THT 2 TAU	4 25 49.18	15 45 49.7	< 3.	A7 III
LAM ORI	5 32 23.97	9 54 15.6	< 1.	O8
ALP ORI	5 52 28.85	7 24 5.6	< 6.	M2 IAB
ETA GEM	6 11 52.71	22 31 31.5	< 2.	M3 III RR CrB E
RT AUR	6 25 22.66	30 31 40.1	< 2.	G0 Del Cep
THT GEM	6 49 31.12	34 1 31.6	< 5.	A3 III
SN NGC 3044	9 31 8.00	1 48 45.2	< 2.	Mar 1983
AD LEO	10 16 54.00	20 7 18.0	< 2.	Flare star
AM LEO	10 59 34.87	10 9 59.9	< 2.	W UMA
SN NGC 4185	12 10 53.00	28 46 33.0	< 2.	Mar 1982
NOVLKVAR	12 23 4.0	13 10 12.0	< 4.	IAU CIRC 3500
Possible SN	12 23 6.00	13 10 0.0	< 4.	
SN NGC 4753	12 49 47.23	- 0 55 54.2	< 6.	Apr 1983
SN NGC 5679	14 32 0.36	5 34 0.0	< 3.	Mar 1982
ALP CRB	15 32 34.68	26 52 51.3	< 3.	Algol
NQ HER	18 9 22.00	18 18 40.7	< 6.	Algol
GD 229	20 10 23.00	31 5 0.0	< 5.	MAG WH DW
AW PEG	21 50 2.47	23 46 38.9	< 2.	Algol
HD224808	23 58 13.99	16 42 52.2	< 6.	K0 III
SKM 154	23 58 25.37	27 36 34.3	< 6.	K0 III

REFERENCES

Gottlieb, D. M., 'SKYMAP: A New Catalog of Stellar Data', Astrophysical Journal Supplement Series, 38, pp287-308, 1978.

Kukarkin, B. V., Kholopov, P. N., Efremov, Yu. N., Kukarkina, N. P., Kurochkin, N. E., Medvedeva, G. I., Perova, N. B., Fedorovich, V. P., and Frolov, M. S., General Catolog of Variable Stars, 3rd edition, Moscow, Publishing House of the Academy of Sciences of the U.S.S.R, 1969.

DISCUSSION AFTER MULLAN PAPER IN PART III

FLORKOWSKI: I'd like to point out that HR1099 has a discrepancy of 0.3 arcsec between the optical and radio positions. This difference is not flux density dependent, and the radio source shares the same proper motion as the optical source. HR1099 is the only star, out of 26 studied in a 7-year Naval Observatory/VLA program to tie together the radio and optical (AGK3R) coordinate systems, that shows such a discrepancy.

GIBSON: Aren't the absolute errors in the two coordinate systems largest near the equator (the declination of HR1099 is $-0.5°$)?

FLORKOWSKI: I have some thoughts on that which are covered in my poster paper. However, I don't believe the effect is that strong.

LINSKY: I would like to provoke discussion by asking a number of questions raised by Dr. Mullan and the contributed papers of this session.

(1) In binary systems (RS CVn, W Uma, and Algol types) does the companion star merely spin up the active star, or does it play an active role in the flare process? In particular, can we determine where flares occur in the system?

(2) In both the binary and single flare stars, what is the time sequence of events as deduced from the relative times of the flares seen in the optical, x-ray, ultraviolet, and radio regions of the spectrum?

(3) What are the critical tests that have been, and can be, used to determine the detailed flare mechanisms and radio emission processes?

(4) What fraction of the electrons in the coronae of M dwarfs and binary flare stars are nonthermal? An important point here is that the Einstein Solid State Spectrometer data indicate a hot component ($2-6 \times 10^7$K) in these coronae, but this hot component may really be the low-energy tail of a nonthermal particle distribution.

(5) Are impulsive events ever seen in the RS CVn systems?

(6) What is the relation between "quiescent" heating and flaring? Could the "quiescent" emission detected at radio wavelengths really be "microflaring"?

(7) How are the energetic electrons accelerated?

HOLMAN: Let me make three comments concerning Mullan's talk. You commented that $E > E_{Dreicer}$ is required for electric field acceleration of electrons. In fact, electrons can be accelerated to high energies by fields that are less than the Dreicer field. The physical evolution of the system is very different for $E < E_{Dreicer}$ and for $E > E_{Dreicer}$. Also, I understood you to say that the acceleration occurs throughout the entire volume of a loop. These electric fields must, in fact, be confined to narrow current channels. Finally, it should be noted that there will always be more energy dumped into plasma heating than into particle acceleration.

Is 10-40% circular polarization consistent with the electron energies ($\gamma \approx 3-10$) that you deduce for the RS CVn systems when the

R.M. Hjellming and D.M. Gibson (eds.), Radio Stars, 289–306.

emission mechanism is incoherent synchrotron emission?

MUTEL: I would say that circular polarizations of 40% are not that common; 10% is much more common, making the constraints on the magnetic fields less severe.

HOLMAN: The numbers that you showed for the thermal and nonthermal particle densities in RS CVn stars appeared to me to be consistent with densities that are deduced for solar flares and, hence, with $n_{nonthermal} \ll n_{thermal}$. I do not see any reason for coming to your rather radical conclusion that the entire thermal plasma must be accelerated. Is there any evidence, for a single star, that $n_{nonthermal}$ must be of the order of $n_{thermal}$?

MULLAN: Electric field acceleration of some particles will occur even if $E \ll E_{crit}$ (that is the important field not the Dreicer field), but runaway of a large fraction of electrons requires $E \geq 0.1\ E_{crit}$. Local turbulent cells will provide local sites of electric current, hence the constraints of filamentation are met in a turbulent model.

With regard to your comment on the number densities. I admit that there is a range for both the thermal and nonthermal number densities. What I want to point out is that the two overlap and therefore could be the same distribution.

BROWN: If the nonthermal and thermal electron density estimates each span two orders of magnitude and overlap, it is of course possible that the acceleration efficiency is 100%. It is, however, also possible that the efficiency is the order of 0.01% - and more likely on theoretical grounds. I also doubt that the nonthermal density is known to even within two orders of magnitude, particularly in view of its E_o^{-n+1} dependence on the cut-off energy E_o, which is almost arbitrary. In the solar flare case we still only know, from joint x-ray and micro-wave data, the acceleration efficiency is in the range 0.01% to 100%.

KUIJPERS: You said that, in the case of the RS CVn stars, the coronal loops are well-tuned, i.e. L/v_A is of the order of the convective mixing time scale, while for the dMe stars, they are not. Now I understand from Ionson's work that the heating always requires that the loops be well-tuned, and that they automatically pick up the power at their eigenfrequencies. Why then is the heating less efficient for the dMe stars?

MULLAN: Two resonances are involved in any model: (1) Alfven wave resonance in a loop in order to enhance the Fresnel transmission coefficent across the density jump at the base of the corona; and (2) timing resonance to match the periods in source and sink.

MELROSE: I would like to comment on the point John Brown has just made, and make another comment on another topic.

The evidence for Ionson's model (from solar data) gives general support for the idea of a resonance between the Alfven propagation time

τ_A and the characteristic time of the turbulence τ_c. On the Sun the turbulence has peaks in its spectrum at $\tau_c \approx 300$ sec and 800 sec, and there is some evidence that loops with τ_A close to these values are preferentially heated. Also, although I agree with Holman that there are difficulties with the heating or acceleration model proposed by Mullan, in Ionson's model the circuit has a high Q, and the rate of dissipation is then independent of the dissipation mechanism. The idea of resonant heating of loops in RS CVn stars seems an attractive one which involves a plausible extension from the solar context; heating to 10 keV ($\approx 10^8$K) is not an extreme extension of the solar case (where heating up to 1 keV is common for solar loops). The important requirement is that the τ_c for the convection in RS CVn stars does coincide with the τ_A inferred for the loops.

The other topic is the suggestion that the emission may be coherent synchrotron radiation. This mechanism is implausible and encounters a serious difficulty. It requires electrons with a spectrum $N(E) \propto E^{-n}$ with $n > 2$ for $E >> mc^2$, and it also requires both $\omega \approx \gamma^2\Omega_e$ and $\omega \lesssim \gamma\omega_p$ with γ ($=E/mc^2$) in the range where the condition $n > 2$ is satisfied. Now a rising spectrum ($n > 2$) has <u>never</u> been observed, and hence it is implausible to postulate it here. Moreover, there is <u>no source</u> for which the Razin effect ($\omega < \gamma\omega_p$) is known to be important, and hence it is also implausible to invoke it here. These implausibilities are independent but compound to lead to this mechanism being intrinsically implausible. There are very strong observational arguments against the mechanism from the presence of linear polarization. The rotation measure is proportional to $n_eB \propto \omega_p^2\Omega_e$ times a length. The length must be relatively large for the optical depth to be greater than unity (required for maser action) and ω_p^2 must be large when the Razin effect is invoked. Hence there should be strong Faraday depolarization. The linear polarization in a source radiating by the coherent synchrotron mechanism should be negligible. Hence this mechanism should be rejected on observational grounds for sources with significant linear polarization.

<u>MULLAN</u>: Admittedly, special conditions are required for a synchrotron maser to work, but I think something of the nature of synchrotron emission is definitely required to explain the presence of <u>linear</u> polarization.

I would like to go back to the question John Brown raised and again point out, as Gibson has shown, that the flares in RS CVn systems seem to be extensions of the "quiescent" component of these systems. Therefore, I think it likely that the <u>same</u> mechanism is required to heat the coronae and produce flares.

<u>HJELLMING</u>: I greatly enjoyed Mullan's talk, particularly since he tried, in the face of great difficulties, to come up with fairly complete explanations for events. The strong tone of negative comments in this discussion strikes me as symptomatic of one of the major problems of this field: because it is difficult to obtain complete theoretical consistency, nearly any theoretical model can be attacked, and it is much easier to provide destructive criticism than it is to provide

constructive criticism or major advances. This is a fundamental problem that we must be careful about if we expect anyone, particularly young theoreticians, to work in this field.

BROWN: With regard to constructive suggestions, there may be scope for use of the UV continuum flare data reviewed by Mullan to constrain particle acceleration parameters. The UV continuum burst shows the flare disturbance to extend down to the chromosphere at least. If this burst is closely synchronized with the radio event, it could be that the only synchronizing mechanism which is fast enough is energy transport by energetic particles. Such a result could place useful constraints on the flux and spectrum of accelerated particles. If the flare disturbance extends still deeper, the constraints could be stronger. In these flare stars such deep (photospheric) disturbances would be in the IR. What is the state of IR data from flare stars?

RODONO: There are at least two stellar flare models which predict IR "negative" flares for $\lambda > 1$ μm. These models are due to Gurzadian and Grimin. Gurzadian's model predicts a low-amplitude, mirror-image "negative" flare in the IR coincident with major (> 2-3 magnitude) optical flares, as a consequence of a sudden injection of energetic electrons and their inverse Compton interaction with stellar (low-temperature) photons. Grimin's model predicts IR decreases due to an H^- opacity increase at or before the initial flare phase, or of any subsequent secondary peak. Only a few IR observations of stellar flares have been made. The latest observations made in March 1984 at ESO are the only ones for which concurrent UV, optical and radio data are available. We did observe a clear decrease of the IR K-band flux (0.05-0.10 magnitude) near the times of two optical flares on AD Leo and one on YZ CMi. The time scale and behavior of these IR "negative" flares appear to be consistent with Grimin's model.

MULLAN: I would like to point out that the observations of the "negative flare" that I showed were at U-band.

FEIGELSON: Most of Mullan's talk and the discussion has concentrated on dMe and RS CVn flares. I would like to comment that there is widespread evidence for flaring behavior in pre-main-sequence stars. Several stars exhibit faint flickering in the U-band on time scales of 10 sec, which is interpreted as the product of overlapping microflares. Several show rotationally modulated cool spots, and one star, SY Cha, had an extremely bright hot spot for several years. Virtually all pre-main-sequence stars show giant x-ray flares, 10^4 times more powerful than Class III solar flares. In addition, this conference contains the first reports of radio flares (V410 Tau and DoAr21). I suggest that pre-main sequence stars in general, or at least those not covered by dense winds, are rather similar to RS CVn stars. The data on the problem have been reviewed by Nuria Calvet (Rev. Mexicana Ast. Ap., 1983) and myself (Cool Stars, ed. Baliunas and Hartman, 1984).

LINSKY: I would like to point out that it is very important to obtain observations across the electromagnetic spectrum. As one who has proposed these kinds of observations, I have found it very difficult to obtain time on one telescope or satellite without the promise of time on others. This is a Catch-22 situation which I hope that those referees and schedulers at this conference will recognize and appreciate when they see similar requests in the future.

KUNDU: I would like to touch upon a new point for discussion, namely meter-wavelength observations. Dermott, you asked us to ignore type II, III, and IV moving solar analogs in flare stars, yet you talked about coronal accelerations and acceleration by shocks, which are related to type II bursts. Should we not look for them at meter wavelengths with the right kind of instruments? For example, with the Clark Lake facility we can do imaging observations and we have the flexibility of changing the frequency within a fraction of a millisecond in the range 20-25 MHz. While there is no large collecting area instrument for dynamic spectroscopy of bursts, as in the case of the Sun, we probably should use the Clark Lake facility to look for type II-like bursts in dMe flare stars.

MULLAN: I would argue that there is no evidence for frequency drifts in other stars, so there is no need to make the analogy.

LINSKY: I would like to raise the whole question of the timing of flares at different bands. Does anyone have observations?

GARY: With regard to the discussion of optical vs radio flares, the flare on YZ CMi at 6cm (on Feb. 5, 1983) was also observed optically at U-band. The timing correspondence between optical and radio (at least within the 1 minute average of the VLA data) is very close, with perhaps some small (< 1 min.) delay in radio.

GIBSON: Wait a minute Dale! Marcello and I have looked at these data and find the delay from the onset of the optical flare to the onset of the radio flare is about 2.5 minutes, while the time delay between the peaks is about 5 minutes!

RODONO: The rising portions of each light curve are coincident to within about a minute.

GARY: I would say that it is irrelevant where the peaks are, and, at most, the delay between the onsets is 1.5 minutes, because it takes some time for the radio flux to rise above the detection limit.

J. Linsky: Referring to the plot relating to time delay that Dermott showed for YZ CMi (Karpen et al., Ap.J., **216**, 479, 1977), the radio observations were made at meter wavelengths. The long time delay here could be caused by the requirement that the radio flare originate a long way from the surface of the star due to optical depth effects and the propogation time of the disturbance. Microwave flares should occur much deeper in the corona, therefore the coincidence of time in Dale Gary's observations might be expected.

TAYLOR: Dr. Mullan mentioned that there is no evidence for so-called spiky behavior in RS CVn radio flares. However, during the Ottawa workshop Dave Gibson reported on "ratty behavior", or short time scale flickering. Dave, would you bring us up to date on "ratty behavior"?

GIBSON: I'm not sure what Dermott is calling a spiky event, but I think he means events superposed on long time scale outbursts which have time scales of order 1 second. The shortest events I know of on the RS CVn stars have time scales of about 5 minutes. In virtually all cases we have determined that this behavior is correlated with the presence of highly circularly polarized emission.

MUTEL: In all of our observations at the VLA we have seen only a few cases where the variations didn't have time scales of hours. Those with time scales of 5, 10, or 20 minutes have tended to be circularly polarized up to 40%.

I would like to make another comment related to the bandwidth of RS CVn flares. Occasionally I have used the VLA in two subarrays at 18 and 21 cm and found that the detailed light curves at the two frequencies were not the same. They did not vary in any systematic way. So there may be fine-scale spectral structure superimposed on the broad band spectrum of the flare.

GIBSON: Dale Gary and I made three-frequency observations of AR Lac during October 1983 which show about as much "ratty" radio behavior as we have ever seen. This included very small "negative flares" at 2 and 6 cm. We presume the high-frequency activity is going on below the 20cm radio "horizon".

KUNDU: Dave, although 20 cm emission in AR Lac does not show any spikey features corresponding to 2 and 6 cm emission, is the 20 cm emission enhanced, that is, is there still a flare?

GIBSON: I think not, 10 mJy is typical of the quiescent level of this star. The flare is probably going on in a region lower than that from which 20 cm radiation escapes. It's interesting to note that, if one uses as typical source conditions and dimensions the ones that Fred Walter and I determined from our Einstein IPC observations in 1981, the radio source would be optically thick to thermal absorption at 20cm but optically thin at 6 and 2 cm. This goes back to the point I raised yesterday regarding absorption mechanisms - thermal absorption may be the dominant absorption mechanism.

LINSKY: Anything else about time scales for variation or the timing of flares in different bands?

BOOKBINDER: We have observed extremely short time scale activity on the flare star AD Leo, so short that it would be difficult to detect with the 3-10 second VLA integration times. I wonder whether the "smoothlyvarying" flares seen by the VLA and other interferometers could be a smoothed version of spiky events such as those seen at Arecibo. We can discuss this more tomorrow, but perhaps we should consider using large single antennas like Arecibo which are capable of high time resolution in conjunction with the VLA.

LINSKY: I would like to go on to another topic, namely the role of binarity. If a system has a period of less than 20 days or so, the components should co-rotate. Thus, the stars in short period binaries should rotate much more rapidly than single stars of the same spectral type and thus be more active. That is one "passive" aspect of being a member of a binary. Another is the local gravity. In some directions the heights corresponding to a particular gravitational potential will be very different. This could affect the coronal regions. For example, interesting things might happen at the Lagrange points. Perhaps more importantly, tidal effects may be strong in the convection zone where the dynamo operates. I would like to ask the participants to address the question of how a companion affects a star's radio emission.

BROWN: Let me respond to Linsky's request for comments on binarity effects. A complicating theoretical matter in the case of binaries is that if one or both stars exerts a high radiation pressure, the circumstellar envelope geometry can become drastically affected. For example, for very high radiation pressure the inner Lagrangian point can become stable, while the other (outer) collinear points may become "inner", and new equilibrium points can appear out of the orbital plane. The implications for envelope shape and radio diagnostics may be a fruitful area to explore.

MUTEL: I have thought about Linsky's question earlier. As far as I know no one has looked systematically for radio emission from rapidly rotating single stars. I have submitted a VLA observing request to remedy this omission. As a motivation I cite the work of Rucinski and Vilhu, who find that chromospheric activity indices appear to be independent of binarity. However, in the case of radio emission, things may be different since there is reasonable evidence that there are joint magnetospheres.

LINSKY: To rephrase my question another way, are we talking about RS CVn binaries or RS CVn stars?

KUIJPERS: If one considers the shearing of a connecting flux tube in a close binary due to differential rotation, it is not clear where the energy release eventually takes place. In the case of the Sun, with a current-carrying loop, it is unknown where the reconnection takes place (e.g. at the base or at the top) when the flux loop is sufficiently twisted. The outcome depends on the resistive effects and their distribution along the loop. Consequently, in the case of the connected stellar flux tube, it is not clear that one should expect a strong dependence of flare activity on phase. Therefore, I do not find the observed absence of such an effect to be impressive evidence for the absence of the "binary" nature of flares in such systems.

RODONO: The role of binarity in many active stars is not confined only to imposing enforced (more rapid) rotation rates. Clearly, enforced rotation triggers a more efficient $\alpha-\omega$ dynamo and consequently enhances the star's activity level. However, I would like to emphasize that active stars in close binary systems continue to experience strong tidal interactions, so that their behavior should be substantially different from that of single stars. Actually, the differential rotation regime that would occur in a single star is very likely modified by tidal coupling if the star has a close companion. This means that stars have to adjust to a new equilibrium situation which results from the competing forces internal to the stars themselves, i.e. differential rotation, and tidal interaction. I believe that indirect evidence comes from the fact that the differential rotation rates, which have been obtained from the migration rate of "photometric waves" for very active, spotted RS CVn and BY Dra stars, are one to two orders of magnitude lower than in the comparatively low-activity Sun.

This brings me to another point which I would like to address to Dermott Mullan. There seem to be very different kinds of bursts or flares on the RS CVn stars and the flare stars, not only in the radio, but at optical wavelengths too. This may be observational selection since we may not expect to observe a very spikey flare on a hotter, brighter RS CVn star. However, I think we are observing two distinct kinds of phenomena. I wonder if the binarity provides an additional source of energy responsible for the very strong flares seen in the RS CVn systems.

MULLAN: I am aware that obtaining comparable data on stars of such different luminosities is a complex problem, but I base my interpretation on the fact that I am unaware of any observation of an optical continuum flare on an RS CVn system.

GIBSON: That's not true. Mike Zeilik and collaborators have reported on a continuum flare on XY UMa.

LINSKY: I am finding that this is a most unusual session in that we have yet to hear from Anne Underhill (laughter).

UNDERHILL: That is because I have yet to be able to connect any of this to Wolf-Rayet stars. It seems to me that we have mainly been discussing possible mechanisms on a single star that will generate relativistic electrons - and what the presence of relativistic electrons in a low-beta plasma will create in the way of observable optical and radio phenomena. The nearby presence of another star acts chiefly to modify the gravitational potential well in which the low-beta plasma is confined. Thereby one may have larger volumes of low-beta plasma in which the relativistic electrons, which may be generated by either star, can interact. A second action caused by a companion star is to provide a second source of low-beta plasma, that may merge with that of the active star. Being in a binary is significant chiefly for providing more low-beta plasma that is accessible to the high energy electrons, which are provided by a process that may be relavant to only one of the stars.

Another question occurs to me as I mull over what I hear and try to think whether it has any relevance for helping us to understand the observations of WR stars. Do mechanisms exist for providing a long-lived, steady supply of relativistic electrons, or are all of the conceivable processes episodic? During this morning's discussion we have been emphasizing flares which are episodic events.

GARY: Some dMe stars do show quiescent emission that may be due to low-level flaring. This may also be the explanation for RS CVn stars.

WALTER: If we wish to investigate the role of duplicity in flares, I would ask whether there are any observed differences in flaring behaviour between close binary dMe stars (i.e. YY Gem) and single or widely separated dMe stars? If not, does this not indicate that flaring is intrinsic to the individual star - and not to the existence of the star in a close binary?

RODONO: The time scale of flares is often biased by contrast effects. In the optical region the brighter the star, the longer the average flare time scale; the small spike-like flares on the intrinsically brightest stars escape detection. On the other hand, it is also true that the most powerful long-duration flares are observed in the brightest active stars. Binarity does not seem to play a dominant role in determining the intensity and time scale of optical flares. In the UV and radio bands, binarity is, of course, essential for flares originating from interacting magnetic loops.

This brings me back to Mullan's review and the question of different time scales for RS CVn radio flares and dMe radio flares. I believe that the flare phenomena is basically the same in both types of stars, namely the conversion of stored magnetic energy into radiation. At comparable detection limits, short-duration spike flares can be detected during the course of long-duration events, but they might become undetectable during the course of powerful flares. Again, contrast effects have to be taken into account, otherwise the most active and powerful stars may appear to show only a different type of activity from faint stars with low levels of activity.

MACKINNON: The inference of electron energies at the acceleration site, based on thermodynamic ideas (adiabatic expansion and comparison of initial and final volumes) will probably yield initial energies that are much too high. These should really be inferred using the energy losses of electrons in the medium.

Given that a detector like HXRBS on SMM would detect 100 photons above 10 keV over a whole event, have there been any hard x-ray observations?

LINSKY: The answer is no.

DISCUSSION RELATED TO PART III OCCURING IN THE AFTERNOON SESSION

GIBSON: This discussion session is meant to deal primarily with quiescent stellar emission and the solar-stellar connection. I propose to ask a few individuals who presented interesting poster papers on quiescent emission from objects other than those mentioned by D. Gary to give a short summary of their work at this time. Would either Tim Bastian or George Dulk relate to us some of their work on the (WD + dMe) binary AM Her?

BASTIAN: We find that the quiescent emission from this system (0.5 mJy at 4.9 and 14.9 GHz) can be described in terms of optically thick gyrosynchrotron emission. We have observed one outburst from the M-dwarf companion star at 4.9 GHz. Due to the high degree of circular polarization, the high brightness, and the rapid variations in the flux density, we attributed the outburst to the action of an electron-cyclotron maser, implying the existence of ~1000 Gauss fields on the M-dwarf.

GIBSON: I think it worth mentioning that this system has a period of 3 hours and, as such, may cause problems for those thinking of such things as period-luminosity relations. There are two other objects which have white dwarf companions, and this raises questions of how a magnetic white dwarf might affect the radio-emitting environment of its companion. It seems in every case that systems with WD components are more active than you would otherwise expect.

I would next like to ask Steve Drake to discuss the "quiescent" emission from the newest class of stellar emitter, the Bp stars.

DRAKE: Two He-strong stars have been detected to date, one of which has a spectral index measured between 2 and 6 cm of -0.1, inconsistent with thermal emission from an optically thick wind. Possible emission mechanisms for such a radio spectrum are optically thin thermal emission, gyroresonance emission, and gyrosynchrotron emission. If the presence of circular polarization at 2 cm can be confirmed in these stars, then this would confirm the magnetic nature of the radio emission. Supporting evidence that the emission is not associated with stellar winds comes from UV-line profiles of these stars, which do not show strong P Cygni shapes that would be produced by the winds of mass loss rate 10^{-7} to 10^{-6} $M_{\odot}$/yr needed to explain the observed radio fluxes.

MUTEL: Do you have any feeling as to why you are detecting the Bp stars and not the Ap stars?

GIBSON: Certainly Steve's observations were several orders of magnitude more sensitive than the Green Bank Ap star survey. Do we have any feeling for the possible acceleration mechanism for the energetic electrons needed to radiate gyrosynchrotron emission?

UNDERHILL: Although the Bp stars have quite large (dipole?) magnetic fields, their UV spectra give very little evidence for the presence of hot outflowing plasma. This can be understood from the point of view of Ionson's (induction) theory. One needs magnetic lines of force making loops to catalyze the transfer of mechanical energy from subphotospheric layers. Strong magnetic fields hold the photosphere in a straight jacket. Bp stars tend to have sharp lines with low microturbulence and low rotation. Consequently, one can understand seeing little evidence for heated plasma. How much material leaves the star in a wind is dependent on what fraction of the surface is covered by open lines of force and what fraction is covered by the feet of magnetic loops. Large loops and many of them will lead to much gyroresonance radiation being radiated. A bountiful supply of coronal-hole-type areas will lead to much flow and bremmstrahlung radiation.

MUTEL: This is one of the age old problems: you can have 2000 Gauss fields out there but you don't get any radio emission from them unless you do something with them. What do you do with them?

UNDERHILL: You radiate gyroresonance emission there.

MUTEL: But they still won't radiate with a temperature higher than the local blackbody temperature - and that isn't enough.

TAYLOR: With regard to the radio detection of Bp stars, for a spectral index (at short wavelengths) of -0.1, I don't think you can rule out optically thin thermal emission from a wind. The detected stars are rapid rotators and have very high magnetic field strengths. It is perhaps possible that these effects conspire to enhance the mass loss rate, possibly in the equatorial plane, over that expected for a normal B main sequence star. For mass loss rates of about 10^{-7} $M_\odot$/yr and velocities of order 50 km/s the measured luminosities are in the right ball park.

ABBOTT: I'd like to relate this discussion to the talks we heard earlier and ask whether there is a solar-hot star connection? A point to keep in mind is that in the hot stars what you are seeing is nonthermal emission in a wind, whereas in the cool stars the nonthermal emission is unrelated to mass loss, although a wind may result from some kind of nonthermal process. The only point of connection that I see is in the case of V410 Tau where there is a close connection between the radio emission and the x-rays. In this case, both may result from a shock in the wind.

MUTEL: If shocks do form in stellar winds as they do in the solar wind, then I would like to ask Dermott Mullan if one could expect to see frequency drifts in the cases of these mini-bursts when the shock moves out in the stellar wind?

MULLAN: It's likely. I would also like to comment that CIRs are also good prospects for a solar-hot Star connection. All stars will have

CIRs. In particular, in Hjellming's data on α Ori, there is a breakdown in the agreement of the observationally inferred and the modeled ionized density distribution at 1-2 R_*. This occurs precisely at the radial distances where CIRs form in the wind.

GIBSON: The last objects for which we have evidence of quiescent emission are two of the dMe stars. One of the best is the detection of AU Mic (Cox and Gibson paper, this volume) . It is one of the brightness x-ray stars and has a relatively large radius. The inferred brigtness temperature at 2cm is 2 x 10^7K, about the same as the x-ray temperature. The 3-frequency observations, 2, 6, and 20 cm, show that the minimum of the spectrum is at 6 cm, so there is some question whether that is the best frequency to look at flare stars anymore.

Are there any further comments on quiescent emission from stars?

HUGHES: I would like to mention the detection of emission from the W UMa stars. The emission appears to be an extension of that seen in RS CVn binaries. The observation of a flare on VW Cep appears to be followed by continuum emission consistent with densities of 10^{10} cm^{-3} and temperatures of 10^7K. The flare seems to have involved the injection of mass into the common envelope. There is a question whether the event was thermal or nonthermal (ed. note: it was observed only at 6cm), but it appears that the flare was nonthermal and the post-flare emission was thermal.

UNKOWN: What were the time scales involved?

HUGHES: No emission was seen a few months before the flare. The flare rise time was about three hours, consistent with RS CVn events. Then a week later, the enhanced ~10^7 K emission was seen.

GIBSON: I would like to see some data on the circular polarization for that flare. Since the rise time is pretty fast compared to many RS CVn events, I'll bet it is circularly polarized. I am also interested in how your observations fit in with the upper limits for this system obtained by Andrea Dupree?

HUGHES: The event went from 0.5 to 5 mJy in three hours, but I think even at the higher level we would have trouble detecting any significant circular polarization.

MUTEL: You can do it.

COHEN: We've heard the suggestion that the quiescent emission in a star could represent a superposition of microflares. Are the very strong events in a given star due to the same physical process(es) as these microflares? In the two PMS stars that we've discussed here as having flares (V410 Tau and DoAr21) it's remarkable that each shows the same spectral index during the flare (α = +1.3) and during the quiescent phase (α ~ 0). But these two indices are so different from each other; does this argue against a common physical mechanism in the flare and quiescent phases?

GIBSON: This is something else that we should discuss. Let me say that I think, and I venture to say that people like George Dulk, Don Melrose, and Jan Kuijpers would agree with me, that there is a certain circularity in the argument regarding the determination of the emission mechanism and the physical conditions from any set of data. For example, if you make an observation at several frequencies and "determine" that the emission mechanism is synchrotron, the physical conditions you derive will be determined, to a certain extent, by the frequency at which you made the observation. Notice that this will be true from source to source as well as for subsequent observations of the same source. So I think a question which we as a workshop need to address is whether the physical conditions such as magnetic field strengths, number densities of the thermal and relativistic electrons, etc. are really the same from star to star, implying that phenomena occur under the same set of initial conditions, or whether this is just an artifact of our choice of emission mechanism. Jan, would you like to comment on this?

KUIJPERS: I think what you say is true to a certain extent, but I think that the conditions in the other solar-like stars are very different from what we see in the sun.

GIBSON: I'm not so sure. Whenever anyone has determined the peak brightness temperature of incoherent events on the Sun or other stars, that level always seems to be a few x 10^{10}°K. Thus, the only difference between flares on the sun and flares on these other stars may be one of size scale. At one time I remember calculating that the ratio in radio luminosity between RS CVn stars and the sun was about the same as ratio of their spot diameters raised to the third power. One might infer that the only way to get larger flares is to involve larger volumes.

MELROSE: Let me comment on two points. First, it has just been suggested that perhaps the quiescent level of emission may be due to unresolved outbursts. The only case I know of where such a model has been proposed in the solar context is for the Type I bursts/Type I continuum. It is now accepted that these are distinct phenomena. This provides a weak argument against the quiescent level being unresolved small flares.

Second, Dave Gibson has just commented on scaling from solar to stellar phenomena and has suggested scaling relative to the volume. However, for self-absorbed sources, or for emitting sources with a characteristic brightness temperature that comes from scatter images of much smaller sources (as is the case for most solar meter-wave bursts), the scaling is with area, not with volume.

MULLAN: I'd like to ask Dale Gary what fraction of the coronal electrons need to be superthermal in order to explain UV Ceti's quiescent emission as nonthermal gyrosynchrotron?

GARY: About 0.001.

MULLAN: Is 0.001 so high that plasma processes are important and might quickly thermalize these electrons? This could be serious if you want to have these superthermal electrons always present so as to explain quiescent emission.

GIBSON: You've raised an interesting question that is related to "filling factors." Are we talking about having "homogeneous" mixtures of thermal and superthermal electrons or can we have a situation where one loop is filled with superthermal material while the adjacent ones are essentially quiet?

MULLAN: I think it very curious that all the flare stars and all the RS CVn stars have about the same brightness temperature, 10^{10}°K. That's not a model dependent result. You have to conclude that the emission mechanism is gyrosynchrotron radiation.

GIBSON: That's probably true for all the stars, with the exception of AD Leo, if you are talking about the (relatively) slowly varying, mildly circularly polarized emission.

BOOKBINDER: For our 1983 February observations of AD Leo we had brightness temperatures > 10^{11}K for the "gradually" varying emission, assuming the source size was the size of the stellar disk.

UNDERHILL: How do you know the radius?

BOOKBINDER: You assume it. But you still have to worry about the energetics.

UNDERHILL: Well, it may be different.

GIBSON: In order to make you comfortable, and bring the brightness temperature down to 10^{10}, you would have to make the source size 3-4 R_*.

BOOKBINDER: We had the problem with this high brightness temperature for a peak flux of 50 mJy. There is a poster paper (Van den Oord et al., this volume) which shows a peak flux of 220 mJy!

VAN DEN OORD: There were two observations (Van den Oord, this volume) at 50 cm. AD Leo was observed to be above 220 mJy during both observations, made about 12 hours apart. There was a confusing source in the field that was about the same level, but it was removed by cleaning as well as possible. The AD Leo source appeared to be about 31% circularly polarized.

GARY: We observed AD Leo with the VLA on the same day as Van den Oord's 50cm observation, with a few hours overlap. His flux level was 244 mJy. We observed a flux level that was ≦ a few mJy at 20 cm, but it was somewhat variable.

GIBSON: I'm a little worried about Van den Oord's observation for two reasons. First, it is highly uncharacteristic of flare stars to be seen at exactly the same high level in observations twelve hours apart. Second, the level that he finds is about the same as that of a confusing (extragalactic) source nearby. However, Bert has assured me that he is aware of the confusion problem and has cleaned his map properly.

Let us now turn the discussion to the solar-stellar connection, in particular the period-luminosity relation for RS CVn binaries. I would like to ask Bob Mutel to summarize his results at this time.

MUTEL: I have accumulated a lot of new data on RS CVn binaries as part of our program of VLBI observations and have added to them other data from an extensive literature search. In addition, I have done a lot of reading about the connection between rotation and dynamo-related activity at other bands from which I drew a preliminary assumption that one should expect to see a relationship between the radio emission and rotation. I had the question in my own mind whether it was better to take the mean or maximum radio luminosity. I chose to plot $\log(L_{max})$ vs $\log(P)$, including upper limits and found a formal correlation of 96.5% based on 25 systems. By the same token, I wouldn't disagree with anyone who thinks it looks like a scatter diagram. Although I do not believe that the functional relation that I have found is necessarily correct, I do believe that I have probably found a correlation. I point out that the data for the long-period systems are substantially fewer than for the shorter-period sytems. I also believe that the systems with periods shorter than about 2 days are underluminous by two or three orders of magnitude, which is consistent with the picture at other wavelengths. Vilhu and Rucinski and others have suggested that this saturation of activity indices, particularly for emission lines, is related to the "fact" that the star is full of spots and there is no more surface area left. Thus, there may be a saturation of the energy deposition into the chromosphere and corona or the proximity of the two components in some way destroys the radio emission process.

KUNDU: Are your four short-period points detections or upper-limits, and, are you planning to re-observe some of the less well observed systems?

MUTEL: They are detections. We are planning to look systematically at ten systems which range in period from 2 days to 100 days, and we have another proposal to look for short-period systems.

GIBSON: Let's ask Roberto Pallavacini whether the maximum radio luminosity is the appropriate parameter to plot or whether we should be plotting some other parameter.

PALLAVACINI: To plot L_R vs P for RS CVn systems may not be appropriate. If you plot L_X vs P you find just a scatter diagram with no dependence on P. It is true that RS CVn systems as a class follow the dependence on rotation valid for other late-type stars, but there is little, if any, dependence on rotation within the class itself.

DRAKE: Drake et al. (1984) have compiled data from their own and others' radio observations of active binaries with periods between 10 and 100 days. We find no correlation between rotational periods and mean radio luminosities for these systems, unlike Mutel et al. (1984). We do however find a statistically significant correlation (about 0.8) between radio and x-ray luminosities in the sense $L_R \propto L_x^2$.

SIMON: You might perhaps consider choosing a different variable for the horizontal axis of your figure, for example the Rossby number, defined as the rotation period divided by the convective turnover time. Noyes et al. (1984) have found a good correlation between the chromospheric emission in CaII H and K lines of solar-type stars and Rossby number. George Herbig, Ann Boesgaard, and I have recently extended these kinds of empirical correlations to the high-temperature UV lines of solar-type stars, and we find that the Rossby number is a good index of activity for UV chromospheric and transition region lines as well as for the X-ray emission from these stars. For small Rossby numbers (equivalent to large dynamo numbers N_D), we find a saturation effect. Stars with $N_D >$ 10 all seem to becharacterized by flare activity, whereas stars with small dynamo numbers, $N_D \leq 0.5$, show long-term activity cycles like the Sun's magnetic cycle.

MUTEL: I think you are right, using the period is too simplistic. I know that Vilhu and Rucinski have been much more successful in getting correlations using P as a function of (B-V). I thought of doing that too, but there is a problem in finding precise spectral types.

PALLAVACINI: When plotting chromospheric and/or coronal fluxes vs Rossby number, one should be very careful not to introduce an additional dependence on (B-V) for the ordinate. For, instance, Noyes et al. find a good relation between F_{CaII} and Rossby number, but the scatter increases enormously if you take out the arbitrarily introduced dependence on T_{eff}.

JACKSON: Have you considered the integrity of your sample with regard to volume (distance)? The long period RS CVn systems may well be harder to discover and those discovered are then more likely to be the brighter and hence closer ones. These are then more likely to be detected in the radio, but yield lower luminosities.

GIBSON: You raise a very important question. We are not dealing with anything like complete samples, which makes comparing properties within a group dicey at best. The problem is even worse for inter-group comparisons. I suggest that we discuss this tomorrow when we get to the topic of future plans. How should we be conducting our surveys?

WALTER: I'm not sure that the maximum radio luminosity is the proper diagnostic to use to compare the radio luminosity-period relation with relations at other wavelengths (i.e., CaII, CIV, x-rays). The latter measure essentially the quiescent chromospheric/coronal heating, while the maximum radio luminosity is due to large flares. Perhaps a mean quiescent L_R, or a time integrated L_R, that smoothes out blips due to flares is more appropriate. If the maximum flare luminosity depends on the flare volume attainable in a loop, then one might expect an inverse correlation with P since there is more volume available in long-period systems.

GIBSON: I have done a similar analysis, and I had the same problem of not knowing what to plot. I justified using the maximum radio luminosity under the following line of reasoning. The corona exists because there are closed magnetic loops, the outer boundaries of which are probably characterized by a pretty high (relatively) plasma beta (i.e. $\beta \leq 1$). Therefore the loop can only support a certain energy density of superthermal particles before it would be blown apart. This maximum energy density in superthermal electrons would give a maximum luminosity at radio frequencies if the radiation mechanism is gyrosynchrotron. Thus I think the quiescent (in x-rays) and the flaring (in the radio) coronae are closely connected.

UNKOWN: Length scales would also be important.

MUTEL: Let me give another reason for selecting the maximum radio luminosity. The RS CVn systems simply do not exhibit quiescent emission comparable to that seen on the sun.

WALTER: I'd like to make another point concerning the downturn in radio luminosity for the W UMa systems. The W UMa's don't just saturate; their flux levels are really down by an order of magnitude in x-rays from the extrapolated F_x-P relations for the RS CVn's and other active stars. This may be due to the contact nature of these systems, which may lead to less efficient heating of the stellar atmosphere.

GIBSON: I have attempted to plot L_R vs other rotational parameters. For example, if I plot L_R vs $(v_{rot})^2$ which seems to be the appropriate parameter on the basis of Pallavacini et al.'s Einstein x-ray survey, I don't seem to get a particularly good fit. I have also attempted to plot the surface flux in radio emission (i.e. $L_R/4\pi R_*^2$) vs period analogous to what Fred Walter has done in the x-ray, which may allow us to compare giants to dwarfs. Fred's index was -2.5. I get a fairly tight relation up to the very short periods, but I mention that I have not separated the stars by spectral type as Fred has done. You can see there is still considerable question about what should be plotted. Perhaps next I should plot luminosity vs Rossby number.

Are there any final comments?

MULLAN: Thermodynamically, the existence of the chromosphere/corona in any star depends on mechanical energy. The best source of such energy is convective turbulance, which is at most 1% of L_{bol}. I believe that different stars have different coronal efficiencies because they tap into the mechanical energies with different efficiencies; but, no star has coronal and chromospheric luminosities > 0.01 L_{bol}.

BOOKBINDER: I'd like to make a quick anthropological observation. We are now reaching the point where enough data are available that we should begin to make use of some more sophisticated techniques in determining what parameters are of interest when trying to obtain correlations of physical parameters. Perhaps, we should be using some kind of principle component analysis. This applies in particular to Dave Gibson's question about what parameter is most relevant in studying the radio luminosity of RS CVn systems.

PART IV

HIGH ENERGY PHENOMENA AND STELLAR RADIO SOURCES

RADIO EMISSION FROM STRONG X-RAY SOURCES

R. M. Hjellming
National Radio Astronomy Observatory*
Socorro, NM 87801 U.S.A.

K. J. Johnston
E.O. Hulburt Center for Space Research
Naval Research Laboratory
Washington, D.C. 20375 U.S.A.

ABSTRACT. We review the properties of some of the radio sources associated with very energetic stellar systems that are also strong x-ray sources. Extended lobes of radio emission are seen in Sco X-1, SS433, and Cyg X-3, with the latter two showing regular or intermittent indications of expanding structures with velocities of the order of 0.3c. Some of the observations of these and other radio-emitting x-ray sources are summarized. A working model of accretion-disk outflows resulting in jets that expand like "cylindrical" supernovae is discussed as an explanation for the production of relativistic particles and fields in SS433. The ubiquity of jets and strong synchrotron sources in these and other mass-outflow objects is discussed.

1. STELLAR X-RAY SOURCES KNOWN TO BE RADIO SOURCES

The first galactic x-ray source found to be a radio source was Sco X-1 (Ables 1969). The Sco X-1 radio source was found to be a triple radio source with the central source that was both variable and coincident with the stellar x-ray source (Hjellming and Wade 1971a). The discovery of a variable radio source in the relatively large error box for the position of Cyg X-1 (Hjellming and Wade 1971b, Braes and Miley 1971) led to the identification of this x-ray source with the stellar binary system HDE 226868, which is now best known as a strong candidate for a stellar component that is a black hole. Since these early discoveries a large number of x-ray sources have been found associated with radio sources, making them a special class of stellar radio source. Among these are the transient x-ray sources A0620-003 and Cen X-4, which show transient radio source phenomena when x-ray flares occur.

*The National Radio Astronomy Observatory is operated by Associated Universities, Inc., under contract with the National Science Foundation.

R.M. Hjellming and D.M. Gibson (eds.), Radio Stars, 309–323.

TABLE 1 - The Major Known Radio-emitting X-ray Sources

Name	Characteristics	References
Sco X-1	Variable radio source on x-ray position, surrounded by double radio source	Ables (1969), Hjellming and Wade (1971a), Fomalont et al. (1983)
Cyg X-1	Mostly stable radio source, rare flaring, changes associated with change in x-ray source	Hjellming and Wade (1971b), Braes and Miley (1972), Hjellming (1973a)
Cyg X-2	Relatively stable radio source	Hjellming and Blankenship (1973)
Cyg X-3	Strong synchrotron flaring events, fluctuating quiescent levels, occasional expansion in N-S direction at 0.2-0.4c	Braes and Miley (1972), Gregory et al. (1972), Johnston et al. (1983)
GX17+2	Variable radio source	Hjellming and Wade (1971b)
GX5-1	Variable radio source	Braes et al. (1972), Geldzahler (1983)
A0620-003	Transient radio source associated with transient x-ray source	Owen et al. (1976) Geldzahler (1983)
Cen X-4	Transient radio & x-ray source	Hjellming (1979)
LSI +61°303	Periodic radio source source	Gregory and Taylor (1978), Taylor and Gregory (1982)
Cir X-1	Periodic and flaring radio	Whelan et al. (1977), Preston et al. (1983)
SS433	Variable radio source with "precessing" relativistic jets with 0.26c velocities	Ryle et al. (1979), Seaquist et al. (1979), Hjellming and Johnston (1981b)

In Table 1 we attempt to summarize most of the stellar systems that are both radio and x-ray sources. Some will be discussed in detail in this review, others are merely referenced in the table.

2. THE SCO X-1 TRIPLE RADIO SOURCE

Since the discovery, by Hjellming and Wade (1971a), that the Ables (1969) radio source associated with Sco X-1 was a combination of a variable radio source coincident with the stellar x-ray source and a surrounding double radio source with components 1.2' from the central object, there has been extensive work on both the variablity of the central source (Bradt et al. 1975, Mason et al. 1976) and the nature of the companion double. A recently published study of Sco X-1 is that by Fomalont et al. (1983). Figure 1 shows both their 1.465-GHz VLA

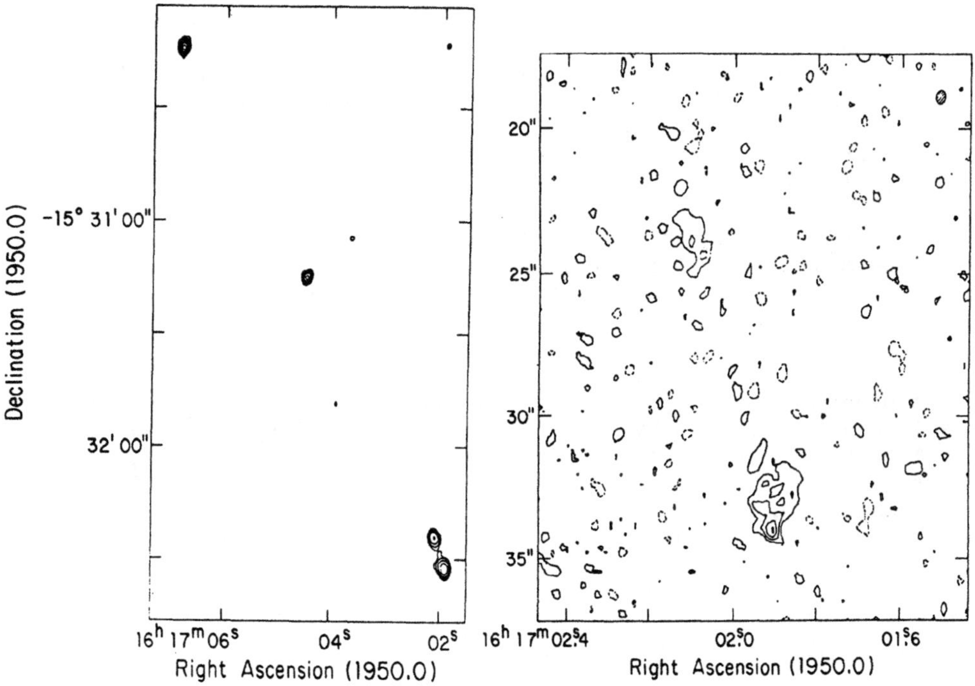

Figure 1. A 1.465-GHz map of the triple radio source associated with the x-ray source Sco X-1 (left) and a high-resolution 4.9-GHz map of the extended SW component (right) (Fomalont et al. 1983).

map of the radio triplet, and their high-resolution 4.9-GHz VLA map of the SW component, an extended and bifurcated radio source about 10" in size. The SW component contains a "hot spot" at its outer edge, which is a very common characteristic in extra-galactic double radio sources. One of the primary objectives of Fomalont et al. (1983) was the measurement of the relative proper motions of the NE and SW components. More recent results (Fomalont 1984) from the ongoing proper-motion studies indicate that the NE component may be moving away from the central radio source with a relative velocity of the order of 20 km/sec.

3. CYG X-3 AS A FLARING SOURCE

The x-ray source Cyg X-3 has an extensive history of observed radio emission (Hjellming 1973b) at flaring (> 20 Jy) and quiescent (~0.1-0.3 Jy) levels since its initial discovery as a radio source by Braes and Miley (1972). Because of the nearly unique and extensive observing of Cyg X-3 during the August-September 1972 radio flaring event, twenty-one papers appeared in the October 23, 1972 issue of Nature Phys. Sci. (**239**, No. 95). Cyg X-3 is associated with a binary system with a 4.8-hour period that is well observed at x-ray (Parsignault et al. 1972) and infra-red (Becklin et al. 1972) wavelengths. Figure 2 shows the radio behavior of Cyg X-3 during the 1982 period when it was first "caught" flaring above 20 Jy and includes the Sept. 1972 event that is one of the best known examples of an observed expanding synchrotron-radio-emitting source. Figure 3 shows multi-frequency Cyg X-3 data for a 1983 period when it was a mixture of the quiescent and erratically flaring source; this behavior is very difficult to interpret.

The model of a binary system with white dwarf and red dwarf components proposed by Davidsen and Ostriker (1974) is widely accepted, and, according to this model, the red dwarf is passing through a brief phase of extensive mass loss that produces an x-ray-emitting accretion disk around the white dwarf. The interpretation of the strong Cyg X-3 radio events as expanding synchrotron sources mixed with ionized thermal gas is best summarized by Marscher and Brown (1975), who used the early flaring observations and the linear polarization data of Seaquist et al. (1974).

Most recently the Cyg X-3 radio source has exhibited evidence that some events evolve to extended structures expanding with relativistic velocities. We will discuss this in section 6.

4. PERIODICITY OF RADIO EMISSION IN X-RAY SOURCES

The radio sources associated with Cir X-1 and LSI +61°303 exhibit a mixture of radio flaring events and modulation of low-level radio emission that is correlated with the binary period. This characteristic was established first for Cir X-1 (Whelan et al. 1977; Haynes et al. 1978, 1980; Thomas et al. 1978; Nicholson et al. 1980).

Cir X-1 as a binary system consists of a OB supergiant primary and a compact star (probably a neutron star) that orbit each other with a 16.595 day period and an eccentricity ~0.7-0.8 (Haynes et al. 1980) such that the compact star passes close to the surface of the primary for ~10^5 sec once each orbit. This passage produces an x-ray-emission cut-off and triggers radio flaring that lasts 1-3 days. The latter seems to occur because the mass transfer briefly exceeds the Eddington luminosity limit, thereby creating strongly expanding shocks that accelerate the

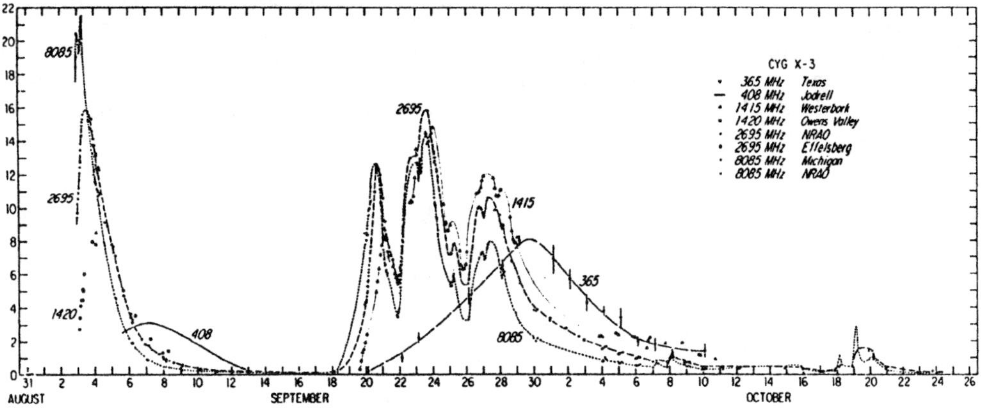

Figure 2. Radio flaring events of Cyg X-3 during 1972.

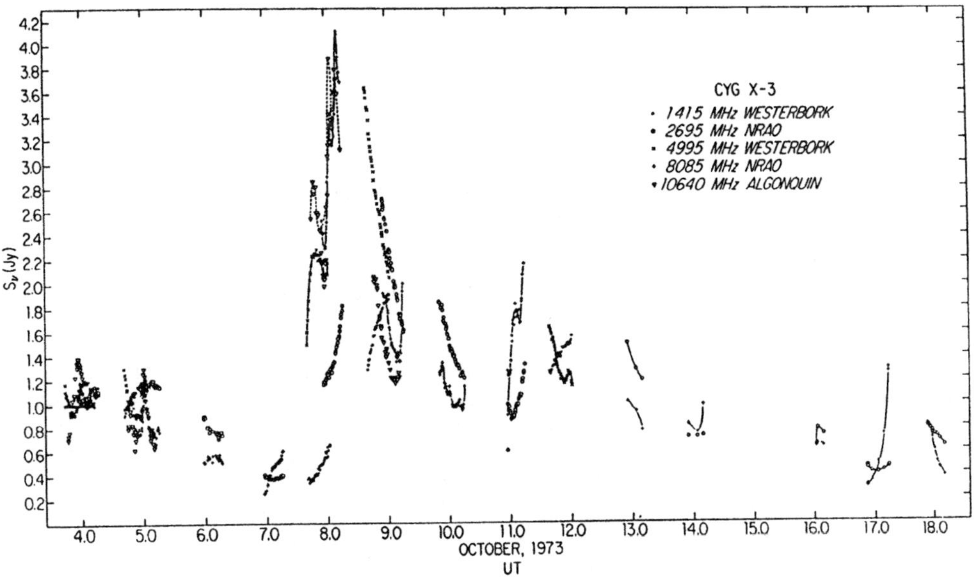

Figure 3. Quiescent and flaring events in Cyg X-3 in Oct. 1983.

relativistic particles, which are observed in the form of synchrotron radiation.

A variable radio source, first discovered by Gregory and Taylor (1978) during a survey for variable radio sources, was shown to be associated with the B0e supergiant star LSI +61°303 by Gregory et al. (1979) at a distance of 2.3 kpc, and was probably association with an x-ray source and the COS B gamma ray source CG135+1. Taylor and Gregory (1982) established a periodicity of 26.52 ± 0.04 days in the radio data as shown in Figure 4.

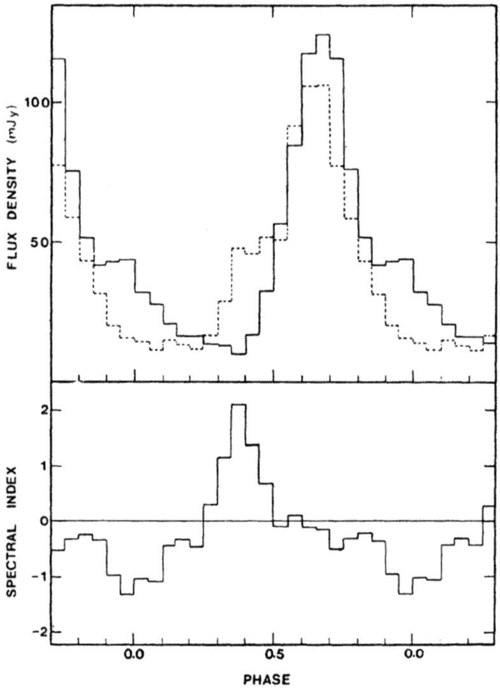

Figure 4. The periodic variations of LSI +61°303 at 5 GHz (dashed line) and 10.5 GHz (solid line) (Gregory and Taylor 1978).

The LSI +61°303 system is probably very similar to that of Cir X-1, although it is unclear whether the compact object is a white dwarf (Cyg X-3) or a neutron star (Cir X-1). Because of its long period (Taylor and Gregory 1982) and possible eccentricity, it is probably an expanding synchrotron-radiation source whose expansion and emission characteristics are affected by the modulation concomitant with movement over a range of radii in the supergiant's stellar wind.

The periodic modulation of Cir X-1 and LSI +61°303 appears cleanly in the radio data because the time scales for radio-source production (and expansion) are short compared to the binary period. Molnar et al. (1984, 1985) have argued for detection of a 4.8-hour periodicity in a few days of VLA data. The same period of modulation occurs for x-ray and IR emission. However, much more extensive data (Mason et al. 1976; Hjellming 1976; Johnston et al. 1985) indicate that, although the "quiescent" emission of Cyg X-3 tends to show fluctuations on time scales of up to a few hours, very little of the data can be phased together to show a clear 4.8-hour modulation. The October 3-7, 1983 behavior in Figure 3 is typical of this type of data.

We conclude that Cyg X-3 may occasionally show modulation of low-level radio emission that matches the 4.8-hour period, but that most of the time this periodic modulation cannot be clearly discerned. The reason is obvious - synchroton flaring events with time scales of a few hours or more are too long, compared to the binary period, for a regular modulation with period to be expected. Thus modulation of the radio emission with a 4.8-hour period will be seen only rarely.

5. SS433 AND ITS RELATIVISTIC JETS

The SS433 star system (V1343 Aql) has optical emission lines of HI and HeI (Margon 1984) that change wavelength in a manner that can be interpreted as twin doppler shifts with a range of 80,000 km/sec and a period of roughly 164 days. The twin-jet model for the optical data indicates an absolute velocity of ejection of 0.26c, a jet axis either 80° or 20° to the line of sight, and an ejection vector that rotates around the jet axis every 164 days at an angle of 20° or 80°.

The radio emission of SS433 was discovered independently by Ryle et al. (1979) and Seaquist et al. (1979) and was shown on the large scale map of Geldzahler et al. (1980) to be a compact source inside the 1° by 2° "supernova remnant" W50. Hjellming and Johnston (1981a,1981b) showed that a series of high-resolution maps made in 1979-1980 indicated that extended portions of the radio emission had oppositely directed proper motions of 3.0" per year, and that these proper motions can be used to determine all the remaining parameters in the twin-jet kinematic model that were either ambiguous or unknown from the optical data. The model that fit the 1979-1980 proper-motion data had the following characteristics: a jet axis inclined 80° to the line of sight at position angle 100°; the angle between the jets and the axis was 20°; the jets rotated in a clock-wise (left-handed) sense about the jet axis with a period of 164 days; the oppositely directed eastern and western (on the sky) jet rotation axes are on the near and far sides, respectively, of the central object; the ratio of constant jet velocity to distance to SS433 is 3.0" per year; and the measureable differences in paired features on the eastern and western sides, due to time-delay effects, allowed one to determine that the absolute jet velocity is 0.26c (where c is the speed of light). The velocity and proper-motion determinations provide a determination of the distance to SS433 to be 5.5 kpc.

Between 1979 and 1983 Hjellming and Johnston (1985) made observations at 4-6 week intervals during all the periods that the VLA was in its largest (35 km) configuration. These data provide an ongoing test of the validity of the previously published kinematic parameters that we have just summarized. Figure 5 shows an excerpt of results for a sequence of events in 1980-1981 (left) and 1982 (right). The first two maps on the left are for 15 GHz, while the bottom three maps on the left are for 4.9 GHz. The four maps on the right are made at 15 GHz. All maps have superimposed "corkscrews" (with filled circles identifying ejection intervals of 20 days) indicating the predicted proper-motion paths for the kinematic solution of Hjellming and Johnston (1981b), with the minor modification that a dP/dt term of -0.002 days per day is included. For the data in Figure 5 the additional dP/dt term only has very small effects, but it provides a significantly improved fit for the data in 1983 and thereafter. Aside from this minor correction the additional data of Dec. 1980 through Dec. 1983 confirm the prediction of the previously published kinematic model.

The SS433 radio jet data from 1979 through 1983 provide a wealth of detailed information about the evolution of ejected radio sources. One striking characteristic of the SS433 radio source is the prevalence of a spectral index of -0.6 to -0.7. In Figure 6 we show a histogram plot of the number of times a particular spectral index is found in a large number of different days of observing SS433 with the Green Bank interferometer and the VLA (Johnston et al. 1984; Hjellming and Johnston 1985). The mean spectral index from these observations is $\alpha = -0.66$. From these data we conclude that the SS433 radio source almost never exhibits any self- or external absorption effects. This behavior is contrary to the behavior of virtually all stellar radio sources, which show self- or external absorption because of the compactness of the radio sources and the high gas densities in star systems.

The 1980-1981 sequence on the left side of Figure 5 shows an apparent double radio source moving outward at a rate of 3" per year. It is one of the best examples of cases where one can derive conclusions about the evolution of the ejected material. First of all, simple measurement of the location of each component of the double reveals the asymmetry in east-west ejection that allows absolute determination of the velocity from the ratio of east-west proper motions:

$$\mu_{west}/\mu_{east} = [1 - (v/c)\cos\theta]/[1 + (v/c)\cos\theta] \qquad (1)$$

where v is the absolute velocity and θ is the angle the jets have with respect to the line of sight at the time of ejection. Since the time dependence of this angle is one of the best determined (or predicted) aspects of the model and is basically the radial velocity curve for the optical jets divided by the velocity scale (Margon 1984), measurement of the asymmetries in East-West ejection as seen in Figure 5 amount to the determination of the velocity of 0.26c. This velocity and the observed proper motion of 3.0" per year then determine the distance to be 5.5 kpc. The continuing fit of all the 1979-1983 data of Hjellming and Johnston (1985) means the quality of determination of these quantities is, in a sense, improving with time. Secondly, the data for the decay in intensity of the identifiable double in Figure 5 indicates an exponential decay with a time constant of about 80 days, with no change in the spectral index of ~ -0.6 during this observed expansion. These and other data for the extended emission indicate that the extended structure of SS433 also has a spectral index of -0.6 to -0.7.

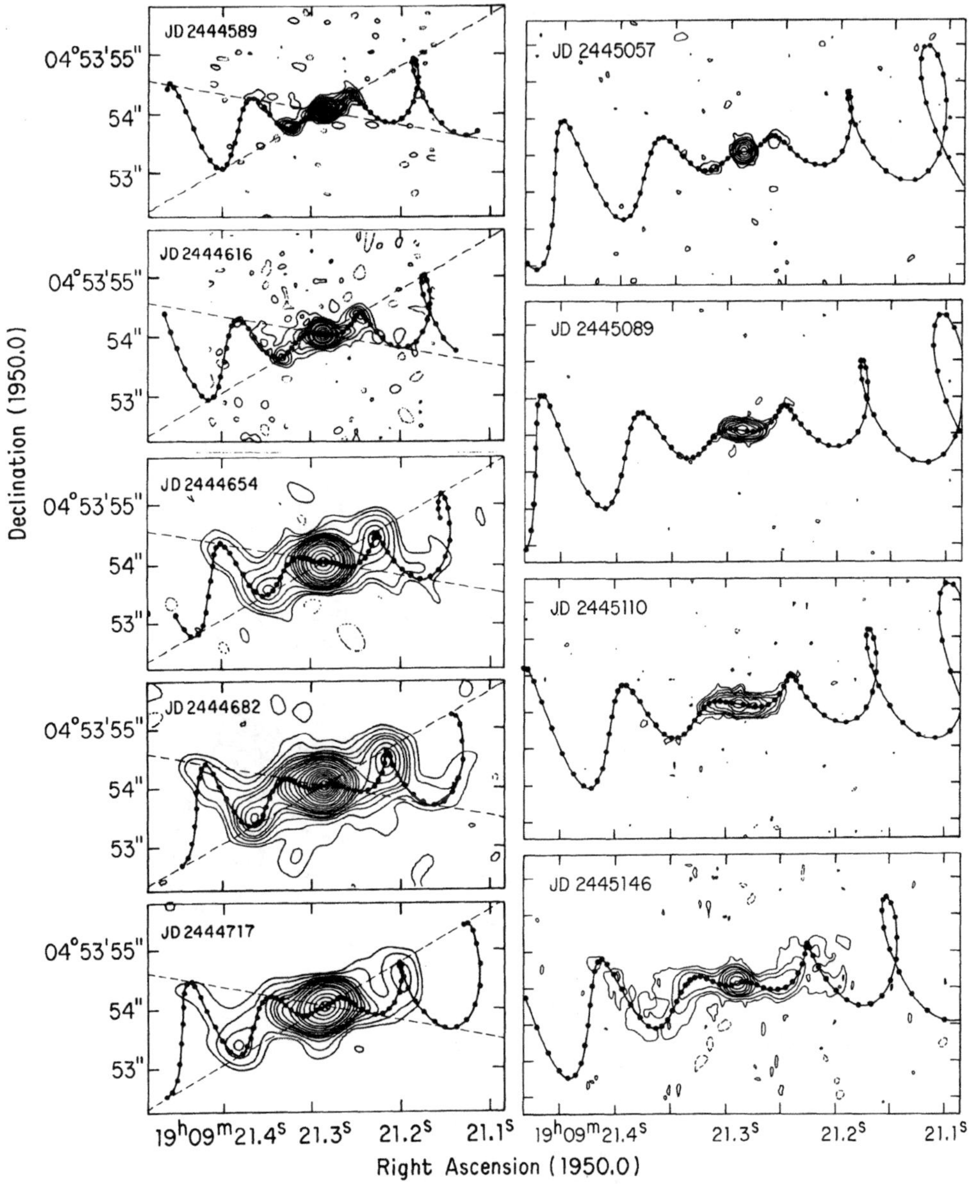

Figure 5. VLA maps of the SS433 radio jets for a sequence of observations in 1980-1981 (left, two maps at 15 GHz, then three maps at 4.9 GHz) and a sequence of observations in 1982 (right, four 15-GHz maps) together with predicted proper-motion paths.

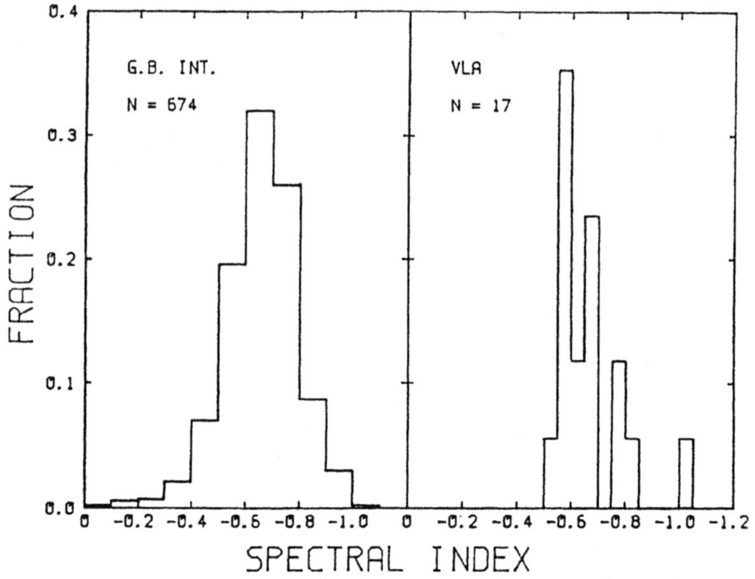

Figure 6. A histogram plot of the number of times a particular spectral index was found for 674 days of Green Bank interferometer data and 17 days of VLA data.

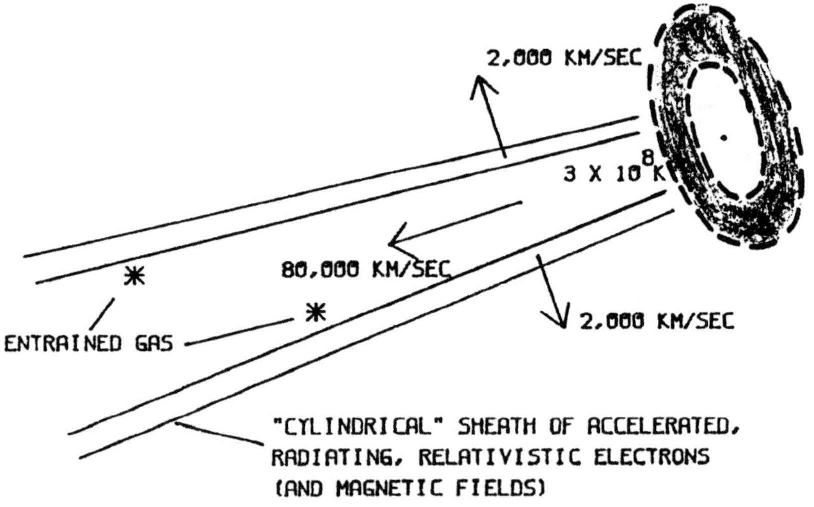

Figure 7. A schematic diagram of a model of the SS433 radio and optical jet expansion, with respect to a thick accretion-disk geometry and its "jet" axes, in which the particles are accelerated in a basically cylindrical geometry.

The reason that an identifiable double is clearly observed in the 1980-1981 sequence is the fact that its decay constant of ~80 days is much longer than the 30 days found to be typical for most ejected jet material in SS433 (Hjellming and Johnston 1985)! Thus it is survivability that makes this an evolving double radio source, not any other special feature such as "strength" at ejection.

The sequence of 15-GHz maps on the right in Figure 5 shows how a long-lived ejection of material can be seen to spread out exactly on the predicted proper-motion corkscrews, becoming resolved at 15 GHz about 100 days after ejection. The superiority in dynamic range for the map on JD 2445146 is a result of the eleven hours of observing time invested in the 4.9- and 15-GHz maps for that epoch, a result not possible for most other epochs because of the short (2-4 hours) amount of observing time scheduled for most of the observations.

Hjellming and Johnston (1985) have discussed a model for the geometry of the expanding SS433 radio jets that explains much of the data, including the relatively large and optically thin characteristics of the radio emission - and the residual "jitter" of the optical jets (Margon 1984). Figure 7 is a schematic diagram of this model in which the jets emerge from a thick accretion disk with a high temperature (and sound speed) that imparts a high initial velocity of lateral expansion, v_{exp}, to the jet material. This lateral expansion velocity means that the aspects of the jet that are involved in accelerating particles at shock-circumstellar matter interfaces are like a "cylindrical" supernova. This model was first envisioned to obtain a specific geometry in which the radio sources would always be optically thin. However, it may explain one of the puzzling features of the optical jets of SS433: the ± 2000 km/sec fluctuation in apparent "jet" velocity seen in the moving optical emission lines. If the gas emitting these lines is entrained by the expanding "cylindrical" jets shortly after they emerge from the central regions, they will share the lateral motion of the expanding jets. Since the emitting regions are small, compressed volumes of gas, they will, to first order, be located anywhere on the surface of the expanding cylinder. Hence the observed optical-emission-line velocity will be determined by the sum of the jet velocity ($\mathbf{v}$) and the peculiar velocity of expansion ($\mathbf{v}_{exp}$) at any location on the expanding cylindrical surface. If we identify $v_{exp} \approx 2000$ km/sec from the typical fluctuations of the optical-emission-line velocities about the predicted kinematic solution (Margon 1984), then this velocity can be associated with the sound speed of the jet when it emerges free from the accretion-disk environment. This velocity corresponds to a temperature of 3×10^8 K. Detailed models of x-ray sources and associated jets have in common a simple association of $kT \approx GM/R$, that is, kT is the order of the depth of the gravitational potential well from which the x-rays or gas flows are escaping; therefore it is interesting to note that 3×10^8 K corresponds to the potential well typical of a neutron star.

6. CYG X-3 AND ITS RELATIVISTIC JETS

Evidence that the strongly flaring x-ray/radio source Cyg X-3 can exhibit collimated, relativistic expansion has recently been obtained by Geldzahler et al. (1983) and Johnston et al. (1985). These papers show that in October 1982 and October 1983 strong flaring events in Cyg X-3 resulted in a radio source expanding at velocities of the order of 0.2-0.4c in the north-south direction. Figure 8 summarizes the measurements of north-south angular size. The distance to Cyg X-3 is known only from HI absorption experiments, with the assumption of a standard Schmidt model for the location of intervening spiral arms; hence it is somewhat uncertain. Furthermore, multiple outbursts, as seen in the October 1983 event, can confuse the identification of the beginning expansion. However, these data clearly indicate motions in the north-south direction of the order of 0.2-0.4c and are consistent with an SS433-like velocity of 0.27c if the distance to Cyg X-3 is 11 kpc. We can therefore conclude that both SS433 and Cyg X-3 can produce observable relativistic jets. SS433 continuously ejects resolvable radio-emitting jets while Cyg X-3 does so only rarely. The difference between them is probably due to a different degree of "cocooning," that is, different optical depth and ram-pressure-confinement effects because Cyg X-3 is surrounded by a much denser gas environment.

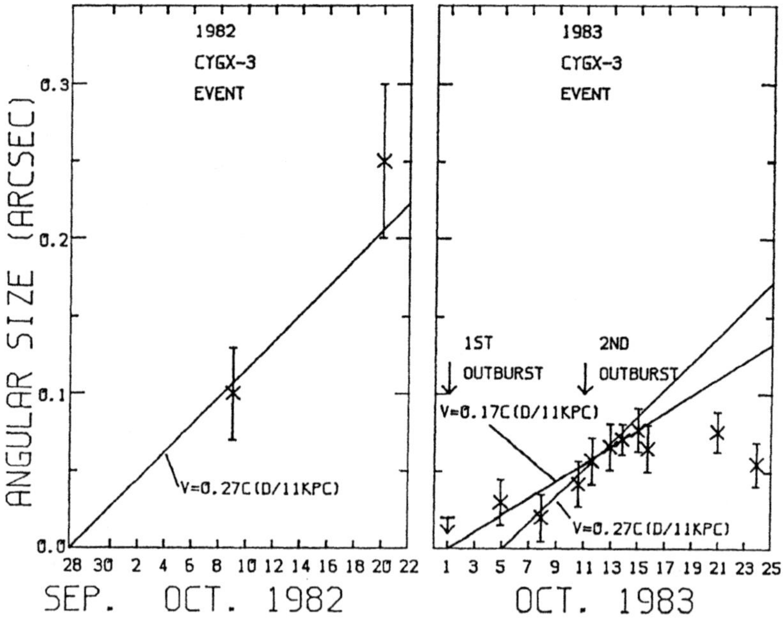

Figure 8. Measurements of the N-S angular diameter of the Cyg X-3 radio source after flaring events in 1982 and 1983.

7. THE UBIQUITY OF JETS

Rees (1982) has shown that jets, such as those found in extragalactic radio sources and SS433, have structures that scale with (dM/dt)/M, and hence are similar except for scaling in this parameter. Jet-like outflows are now found commonly associated with star-forming regions like S106 (Bally et al. 1983), "bi-polar outflows" like L151 IRS5 (Cohen et al. 1982; Bieging and Cohen 1985), and symbiotic novae like R Aqr (Sopka et. al. 1982) and AG Peg (Hjellming 1985). These facts, and recent numerical studies of supersonic jets (Norman et al. 1984), argue that jets are ubiquitous wherever there are supersonic expansions from asymmetric environments. There is a strong need for work on the simplest conditions that would suffice to make jet-like outflows.

8. CONCLUSIONS

The principal difference between radio emission associated with strong x-ray sources and radio emission from "other" radio stars is the clear dominance of strong synchrotron-emission sources. This is probably associated with the high energies of shocks expanding in spherical or cylindrical geometries. This expansive acceleration phenomena has long been used to explain the production of the radiating electrons of supernova remnants, and has recently been found to produce an extended radio source around the old nova (Nova Persei 1901) shell associated with GK Per (Reynolds and Chevalier 1984). White (1985) has discussed an equivalent phenomena for the production of nonthermal emission in stellar winds. For cases like SS433 the expansive acceleration may always occur in a basically cylindrical environment with no other phenomena, such as ram-pressure confinement seriously perturbing this geometry. Other, more "cocooned" sources like Cyg X-3 occasionally may produce observable jets in the same manner, but most of the time phenomena like ram-pressure confinement spread the expansive relativistic plasma into a more spherical, more slowly expanding radio source. This is very likely the case for most galactic x-ray sources most of the time. However, the great importance of the "occasional" Cyg X-3 jets is that 1) the SS433 jet phenomena is probably not unique to SS433 and 2) other galactic X-ray sources may exhibit transient, if not regular, jet phenomena.

REFERENCES

Ables, J.G. 1969, Proc. Astr. Soc. Australia, **1**, 237.
Bally, J., Snell, R.L., and Predmore, R. 1983, Ap.J., **272**, 154.
Becklin, E.E., Neugebauer, G., Hawkins, F.J., Mason, K.O., Sanford, P.W., Mathews, K., and Wynn-Willams, C.G. 1972, Nature Phys. Sci., **239**, 130
Bieging, J.H., and Cohen, M. 1985, Radio Stars (Reidel:Dordrecht), ed. R.M. Hjellming and D.M. Gibson, 101.
Bradt, H.V., Braes, L.L.E., Forman, W., Hesser, J.E., Hiltner, W.A.,

Hjellming, R.M., Kellogg, E., Kunkel, W.E., Miley, G.K., Moore, G., Pel, J.W., Thomas, J., Vanden Bout, P., Wade, C.M., and Warner, B. 1975, Ap.J., **197**, 443.
Braes, L.L.E., and Miley, G.K. 1971, Nature, **232**, 246.
Braes, L.L.E., and Miley, G.K. 1972, Nature, **237**, 507.
Braes, L.L.E., Miley, G.K., and Schoenmaker, A.A. 1972, Nature, **236**, 392.
Cohen, M., Bieging, J.H., and Schwartz, P.R. 1982, Ap.J., **253**, 707.
Davidsen, A., and Ostriker, J.P. 1974, Ap.J., **189**, 331.
Fomalont, E.B. 1984, private communication.
Fomalont, E.B., Geldzahler, B.J., Hjellming, R.M., and Wade, C.M. 1983, Ap.J., **275**, 802.
Geldzahler, B.J., Pauls, T., and Salter, C. 1980, Astron.Ap., **84**, 237.
Geldzahler, B.J. 1983, Ap.J.(Letters), **264**, L49.
Geldzahler, B.J., Johnston, K.J., Spencer, J.H., Klepczynski, W.J., Josties, F.J., Angerhofer, P.E., Florkowski, D.R., McCarthy, D.D., Matsakis, D.N., and Hjellming, R.M. 1983, Ap.J.(Letters), **273**, L65.
Gregory, P.C., Kronberg, P.P., Seaquist, E.R., Hughes, V.A., Woodsworth, A., Viner, M.R., Retalleck, D., Hjellming, R.M,, and Balick, B. 1972, Nature Phys. Sci., **239**, 114.
Gregory, P.C., and Taylor, A.R. 1978, Nature, **272**, 704.
Gregory, P.C., Taylor, A.R., Crampton, D., Hutchings, J.B., Hjellming, R.M., Hogg, D.E., Hvatum, H., Gottlieb, E.W., Feldman, P.A., and Kwok, S. 1979, Astron.J., **84**, 1030.
Haynes, R.F., Jauncey, D.L., Murdin, P.G., Goss, W.M., Longmore, A.J., Simons, L.W.J., Milne, D.K., and Skellern, D.J. 1978, M.N.R.A.S., **185**, 661.
Haynes, R.F., Lerche, I., and Murdin, P. 1980, Astron.Ap., **87**, 299.
Hjellming, R.M. 1973a, Ap.J.(Letters), **182**, L29.
Hjellming, R.M. 1973b, Science, **182**, 1089.
Hjellming, R.M. 1976, X-Ray Binaries, NASA SP-389, 233.
Hjellming, R.M. 1979, IAU Circular No. 3369.
Hjellming, R.M. 1985, Radio Stars (Reidel:Dordrecht), ed. R.M. Hjellming and D.M. Gibson, 97.
Hjellming, R.M., and Blankenship, L.C. 1973, Nature Phy. Sci, **243**, 81.
Hjellming, R.M., and Johnston, K.J. 1981a, Nature, **290**, 100.
Hjellming, R.M., and Johnston, K.J. 1981b, Ap.J.(Letters), **246**, L141.
Hjellming, R.M., and Johnston, K.J. 1982, Extragalactic Radio Sources (Reidel:Dordrecht), ed. D.S. Heeschen and C.M. Wade, 197.
Hjellming, R.M., and Johnston, K.J. 1985, in preparation.
Hjellming, R.M., and Wade, C.M. 1971a, Ap.J.(Letters), **164**, L1.
Hjellming, R.M., and Wade, C.M. 1971b, Ap.J.(Letters), **168**, L21.
Johnston, K.J., Geldzahler, B.J., Spencer, J.H., Waltman, E.B., Klepczynski, F.J., Josties, P.E., Angerhofer, P.E., Florkowski, D.R., McCarthy, D.D., and Matsakis, D.N. 1984, Astron.J., **89**, 509.
Johnston, K.J., Spencer, J.H., Klepczynski, F.J., Josties, P.E., Angerhofer, P.E., Florkowski, D.R., McCarthy, D.D., Matsakis, D.N., and Hjellming, R.M. 1983, Ap.J.(Letters), **273**, L65.
Johnston, K.J., et al. 1985, in preparation.
Margon, B. 1984, Ann. Rev. Astron. Ap., **22**, 507.

Marscher, A.P., and Brown, R.L. 1975, Ap.J., **200**, 719.
Mason, K.O., Becklin, E.E., Blankenship, L., Brouwn, R.L., Elias, J., Hjellming, R.M., Matthews, K., Murdin, P.G., Neugebauer, G., Sanford, P.W., and Willner, S.P. 1976 Ap.J., **207**, 78.
Molnar, L.A., Reid, M.J., and Grindlay, J.E. 1985, Radio Stars (Reidel:Dordrecht), ed. R.M. Hjellming and D.M. Gibson, 329.
Molnar, L.A., Reid, M.J., and Grindlay, J.E. 1984, Nature, **310**, 662.
Nicholson, G.D., Feast, M.W., and Glass, I.S. 1980, M.N.R.A.S., **191**, 293.
Norman, M.L., Smarr, L., and Winkler, K.A. 1984 Numerical Astrophysics: A Festschrift in Honor of James R. Wilson, ed. J. Centrella.
Owen, F.N., Balonek, T.J., Dickey, J., Terzian, Y., and Gottesman, S. 1976, Ap.J.(Letters), **293**, L15.
Parsignault, D.R., Gursky, H., Kellogg, E.M., Matilsky, T., Murray, S., Schreier, E., Tananbaum, H., Giacconi, R., and Brinkman, A.C. 1972, Nature Phys. Sci., **239**, 123.
Preston, R.A., Morabito, D.D., Wehrle, A.E., Jauncey, D.L., Batty, M.J., Haynes, R.F., and Wright, A.E. 1983, Ap.J., **268**, L23.
Rees, M. 1982, Extragalactic Radio Sources (Reidel:Dordrecht), ed. D.S. Heeschen and C.M. Wade, 211.
Reynolds, S.P. and Chevalier, R.A. 1984, Ap.J.(Letters), **281**, L33.
Ryle, M., Caswell, J.L., Hine, G., and Shakshaft, J. 1979, Nature, **276**, 571.
Seaquist, E.R., Garrison, R.E., Gregory, P.C., Taylor, A.R., and Crane, P.C. 1979, Astron.J., **84**, 1037.
Seaquist, E.R., Gregory, P.C., Perley, R.A., Becker, R.H., Carlson, J.B., Kundu, M.R., Bignell, R.C., and Dickel, J.R. 1974, Nature, **251**, 394.
Sopka, R.J., Herbig, G., Kafatos, M., and Micalitsianos, A.G. 1982, Ap.J.(Letters), **258**, L32.
Taylor, A.R., and Gregory, P.C. 1982, Ap.J., **255**, 210.
Thomas, R.M., Duldig, M.L., Haynes, R.F., Murdin, P.G. 1978, M.N.R.A.S., **185**, 29.
Whelan, J.A.J. et al. 1977, M.N.R.A.S., **181**, 259.
White, R.L. 1985, Radio Stars (Reidel:Dordrecht). ed. R.M. Hjellming and D.M. Gibson, 45.

PARAMETERS OF THE SS433 ACCRETION DISK FROM PHOTOMETRY AND POLARIMETRY

N. G. Bochkarev
Sternberg State Astronomical Institute, Moscow, USSR

E. A. Karitskaya
Astronomical Council of Academy of Science of the USSR
Moscow, USSR

ABSTRACT. Photometric and polarimetric data for SS433 show the presence of a geometrically and optically thick slaved disk oriented normally to the jets. The observed matter is a hot, outflowing gas spreading mainly along the accretion disk and forming a photosphere at a distance of several times 10^{11} cm. The ratio of the effective diameter of the disk to the thickness is 2 to 3. The disk is surrounded by a corona of outflowing matter that forms a large gas cloud around the binary and that is elongated in the orbital plane.

In order to interpret the non-jet radio emission of SS433 it is necessary to know the geometry of the system and the distribution of the ionized gas. The most reliable results for the geometrical parameters can be determined from analysis of photometric and polarimetric data obtained mainly by Cherepashchuk, Gladyshev and Kemp (see references in Margon 1984 and also paper by Gladyshev et al. 1983). These data show not only clear primary and secondary eclipses, but also strong variations of all parameters of the light curve with the precession period. There are changes in the amplitude of variations of brightness in both maximum and minimum regions of the light curve. In any orbital phase SS433 is brightest near the times of the largest separation of the moving lines and faintest near the time of their crossing (Gladyshev et al. 1980, 1983). When the moving lines cross, the photometric amplitude modulation is also at its lowest, whereas at the time of the largest line separation, the photometric amplitude modulation is greatest.

Polarimetric data (McLean and Tapia 1980, Shakhovskoj, Efimov and Piirola 1984) are not as extensive, but show variations with precession period (more than 1%, McLean and Tapia 1980) and possibly with orbital period (Bochkarev et al. 1980).

The photometric and polarimetric data show that the secondary component of SS433 cannot be a star or any nearly spherical body. The large widths of the minima show that the secondary component is very large (0.3 - 0.4 of the distance between mass centers of components), probably larger than the Roche lobe (Bochkarev et al. 1980; Bochkarev and Karitskaya 1983a), and has a ratio between thickness H and diameter D such that H/D = 1/2 to 1/3. Our analysis of the observational data shows

R. M. Hjellming and D. M. Gibson (eds.), Radio Stars, 325–328.

definite precession effects in the system (Bochkarev et al. 1980). The secondary has a high temperature T = (4-5) X 10^4 K (Cherepashchuk 1981; Cherepashchuk et al. 1982), which is approximately twice that of the primary component.

Only for the case of a partially eclipsing, geometrically and optically thick hot slaved disk, which has no full eclipses, is it possible to understand the photometrical behavior of the system (Bochkarev et al. 1980). Dependence of the polarization on the precession period, found by averaging the model polarization through the orbital period, shows good agreement with the polarimetric data of McLean and Tapia (1980) and Shakhovskoj et al. (1984). Comparison of models with the variations of polarization with time shows that the orientation of the orbital plane of SS433 is perpendicular to the radio (Walker et al. 1980) and X-ray (Seward et al., 1980) jets in W50 (Bochkarev et al. 1980) and also shows the direction of motion of the precession disk (Bochkarev and Karitskaya 1983b), which corresponds to the radio data on the jets' rotation (Hjellming and Johnston 1981). A constant amount of polarization, p ≈ 1.5%, was found that is the sum of interstellar polarization (p_{is}) and the local polarization (p_o). Multicolor polarimetric measurements (Shakhovgskoj et al. 1984) show that probably p_{is} ≈ 5% and p_o = 3-4%. The last fact strongly suggests a non-spherical distribution of matter in SS433.

These data correspond to a model of a geometrically and optically thick disk in the supercritical regime of accretion. Very strong mass transfer results in a large radius for the accretion disk, which can exceed the diameter of its Roche lobe, and one can have rapid inflow of matter from the outer parts to the inner parts of the disk. A schematic drawing of the accretion disk geometry together with major inflows and outflows is shown in Figure 1.

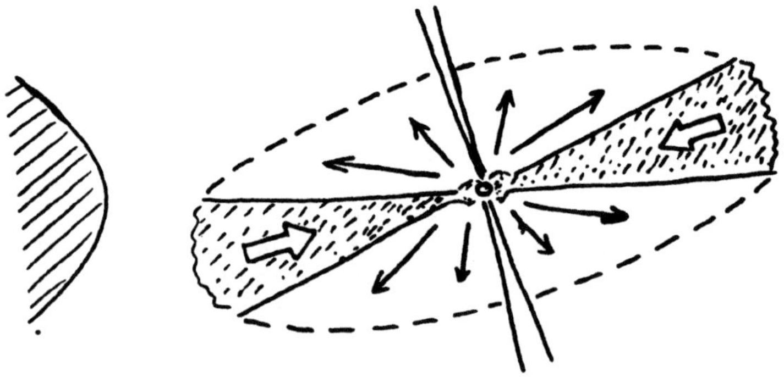

Figure 1. A schematic drawing of the geometry of the SS433 accretion disk and its major outflows and inflows.

Only a small part of the accreting matter can reach the central body, and most of the matter flows out from the central parts of the disk. The outflowing gas is hot, highly ionized, and optically thick to electron scattering. It forms an effective photosphere at a distance of several times 10^{11} cm from the central body and has a density $\rho \sim r^{-2}$.

Centrifugal force results in outflow predominately along the accretion disk. Therefore the effective photosphere is located farther from the central body, in directions closer to the boundary of the accretion disk, rather than in the perpendicular direction. Therefore the effective photosphere of the disk may be quasiflat with a lower density and a higher albedo of scattering than what is normal for stable stars. In contrast to stars with stable radiative atmospheres, where the albedo can never be close to unity in the optical regime (Bochkarev and Karitskaya 1983b, Bochkarev et al., 1984), the albedo of electron scattering of the disk can be of order 90 to 95%.

The photometric data, as well as the polarimetric data, show clearly that the disk-like structure precesses with the jets, which are at all times perpendicular to the disk plane (Bochkarev et al. 1980; Bochkarev and Karitskaya 1983b). The outer boundary of the accretion disk is the coldest part of the disk surface and may be responsible for the FeII lines in the SS433 spectrum. This could account for the spectroscopic data on variation of FeII lines with precession period (Margon 1984). The radio observations of SS433 deal with the outer part of the outflowing corona of the disk, which must be also strongly elongated in the orbital plane of the system.

REFERENCES

Bochkarev, N.G., Karitskaya, E.A., Kurochkin, N.E., Cherepashchuk, A.M. 1980, Astron. Tsircular Acad. Sci. USSR, No. 1147, 1.

Bochkarev, N.G., Karitskaya, E.A., 1983a, Sov. Astron. Lett., **9**, 6.

Bochkarev, N.G., Karitskaya, E.A. 1983b, Astron. Tsircular Acad. Sci. USSR, No. 1255, 1.

Bochkarev, N.G., Karitskaya, E.A., Sakhubullin, N.A. 1985, Ap. Space Sci., **108**, 15.

Cherepashchuk, A.M. 1981, M.N.R.A.S., **194**, 761.

Cherepashchuk, A.M., Aslanov, A.A., Kornilov, V.G. 1982, Sov. Astron., **26**, 697.

Gladyshev, S.A., Goranskij, V.P., Kurochkin, A.M., Cherepashchuk, A.M. 1980, Astron. Tsircular Acad. Sci. USSR, No. 1146.

Gladyshev, S.A., Goranskij, V.P., Cherepashchuk, A.M. 1983, Sov. Astron. Lett., **9**, 1.

Hjellming, R.M., Johnston, K.J. 1981, Ap.J. Lett., **246**, L141.

Margon, B. 1984, Ann. Rev. Astron. Ap., **22**, 507.

McLean, I.S., Tapia, S. 1980, Nature, **287**, 703.

Seward, F., Grindlay, J., Seaquist, E., Gilmore, W. 1980, Nature, **287**, 806.

Shakhovskoj, N.M., Efimov, Yu., Piirola, V.E. 1984, in press.

Walker, R.C., Readhead, A.C.S., Seielstad, G.A., Preston, R.A., Niell, A.E., Resch, G.M., Crane, P.C., Shaffer, D.B., Geldzahler, B.J., Neff, S.G., Shapiro, I.I., Jauncey, D.L., and Nicholson, G.D. 1981, Ap. J., **243**, 589.

CONFIRMATION OF RADIO PERIODICITY IN CYGNUS X-3

Lawrence A. Molnar, Mark J. Reid, and Jonathan E. Grindlay
Center for Astrophysics
60 Garden Street
Cambridge, MA 02138

ABSTRACT. Simultaneous observations of Cyg X-3 in its low state were made on June 11, 12, 14 and 16 of 1984 at radio, infrared, and X-ray wavelengths. The radio observations were made with the VLA at 1.3, 2, 6, and 20 cm wavelengths. These data confirm the tentative results reported in Molnar et al. (1983, 1984); flux density variations at 1.3, 2, and 6 cm wavelengths are consistent with a period of 4.95$\pm$0.04 hr, a value significantly longer than the 4.79 hr X-ray period. (Another possible period is 5.49$\pm$0.04 hr, a sampling alias of the 4.95 hr period.) The amplitude of the flux density variations decreases with wavelength. Further there is a well-defined wavelength dependent time delay in the sense that flux density variations at longer wavelengths lag behind those at shorter wavelengths. A model of the radio emission in which the variations are a series of flares due to the repeated injection of relativistic particles into a small volume which subsequently expands at constant velocity is qualitatively consistent with the data.

1. INTRODUCTION

After many years of study, Cyg X-3 remains a poorly understood, but yet an extremely interesting galactic X-ray source. The source is unusual in that it shows both variable low state (S_ν < 1 Jy) radio emission (Braes and Miley 1972) as well as spectacular radio flares ($S_\nu \sim$ 20 Jy). The large radio flares have long been interpreted as synchrotron emission from an expanding source (Gregory et al. 1972); more recent work by Geldzahler et al. (1983) implies that Cyg X-3 may be one of the few identified radio jets in the Galaxy. Flux density variations with a period of 4.8 hr have been discovered at X-ray (Brinkman et al. 1972) and infrared (Becklin et al. 1973) wavelengths, and may have been detected at gamma ray energies from 10^8 to 10^{16} eV (Lamb et al. 1977, Samorski and Stamm 1983). It is likely that the X-ray emission is produced in a binary system composed of a low mass star orbiting a compact object with an accretion disk and an accretion disk corona (White and Holt 1982). As interstellar absorption (Av = 19, Weekes and Geary 1982) precludes study of the optical emission, the applicability

R. M. Hjellming and D. M. Gibson (eds.), Radio Stars, 329–334.

of this model to Cyg X-3 rests heavily on analogy with other sources with similar X-ray properties (4U1822-37 and 4U2129+47).

There is little agreement on the location and mechanism of the low-state radio emission. Seaquist and Gregory (1977) have suggested synchrotron emission from particles embedded in a stellar wind. Vestrand (1983) has suggested that gamma rays accelerate the relativistic electrons through Compton collisions and pair production. The low flux density of Cyg X-3 in its low state and source confusion make observations difficult, and relatively few observations have been reported. The published data (Hjellming, Hermann, and Webster 1972; Hjellming and Balick 1972; Mason et al. 1976; and Geldzahler et al. 1979) do not define either the shape or the time evolution of the radio spectrum well.

2. OBSERVATIONS

We have initiated a program to study the evolution of the radio spectrum of Cyg X-3 in its low state and to study the relationship between the low-state radio emission and the emission at other wavelengths. Molnar et al. (1984) reports the first phase of this program, in which we monitored the flux density of Cyg X-3 at the VLA for six hours on both 17 and 18 September 1983 and for eleven hours on 3-4 December 1983 at 1.3, 2, 6, and 20 cm wavelengths. We observed a series of flares with properties such that the longer the wavelength, the lesser the amplitude and the later the time of maximum. We concluded that the time between flares is consistent with the 4.8 hr X-ray period although favoring a value about 0.1 hr longer. Here we present follow up observations made with the VLA on June 11, 12, 14 and 16 of 1984. The source was observed for 11 hours each day with a measurement at each wavelength every 15 minutes. Coordinated X-ray and infrared observations were also made and will be reported elsewhere. These new data confirm the wavelength dependent characteristics and the periodic nature of the flares. The longer time baseline of the new data distinguishes the 4.8 hr X-ray period from the longer radio period.

The June observations again showed a series of low-level flares with wavelength dependent amplitudes and time delays as in the 1983 data. The time delays as calculated for each day in December and June are listed in table I. The values are consistent with a single set of time delays applying to all of the December and June data.

The 2 cm wavelength data from June are plotted in figure 1. The flux density variations on each day show patterns repeating with separations of about 5 hr (and somewhat longer on June 11). We analyzed the data for periodicity with the following chi-square test. We binned the data and folded the data at all the periods our sampling would allow. For each period we compared the variance of the points within the bins to that among the bins. We computed a chi-square for the scatter of the binned fluxes, weighted by the square root of the number of points in a bin times the variance within the bins, about their mean. The resulting function is similar to a power spectrum, but has the advantage that it is sensitive to light curves of arbitrary shape,

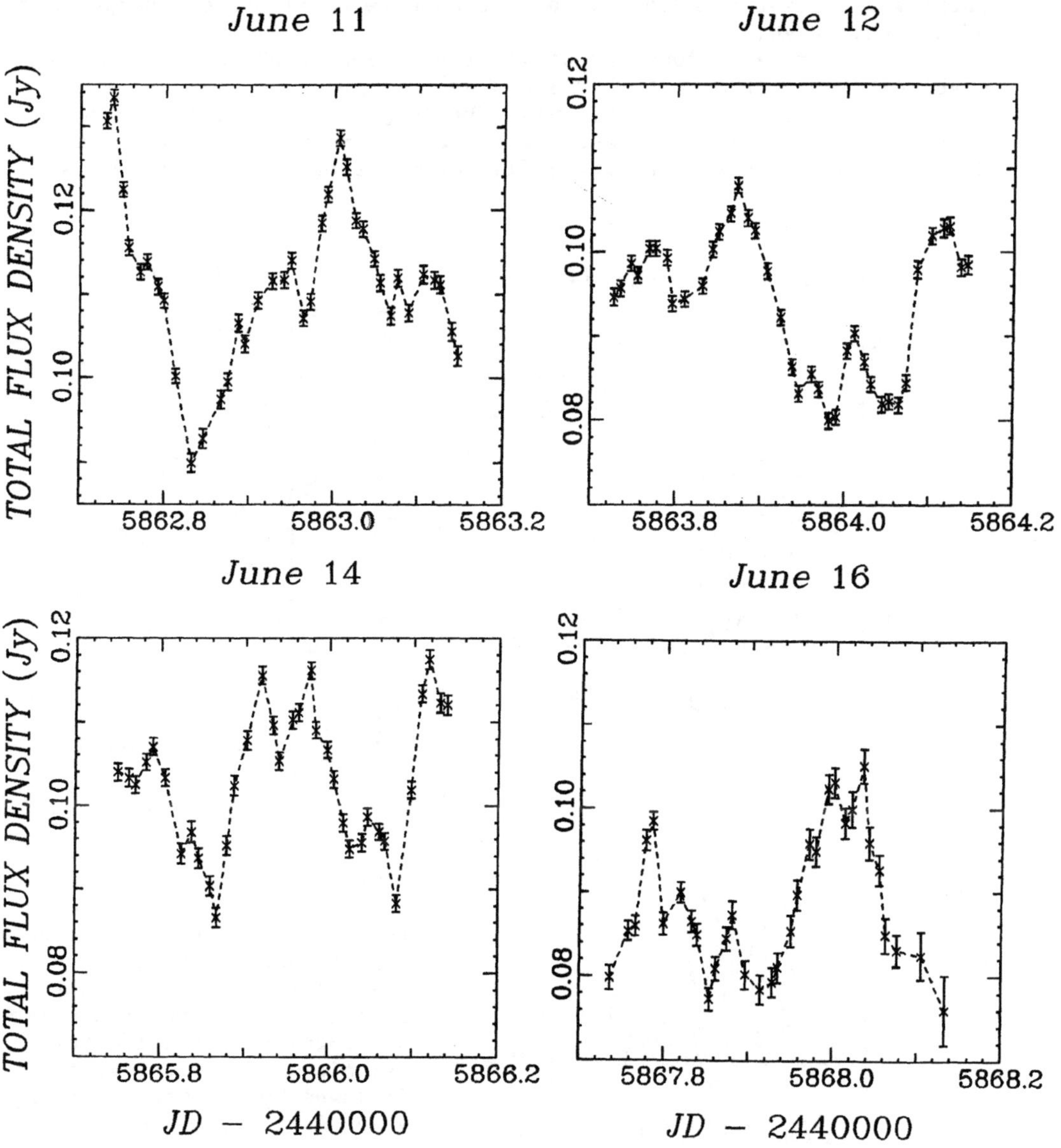

Figure 1. Flux density at 2 cm wavelength plotted as a function of time for the June 1984 data. The time axis is labeled with Julian days minus 2440000.

whereas a Fourier transform is only sensitive to sinusoidal components of a light curve.

Chi-square as a function of inverse period is shown for the 2 cm wavelength data from September and June in figure 2a. The other wavelengths show similar patterns. The larger size of the June data base is particularly important for searching frequencies lower than 4 $days^{-1}$. Both data sets show significant peaks between 4 and 5 $days^{-1}$

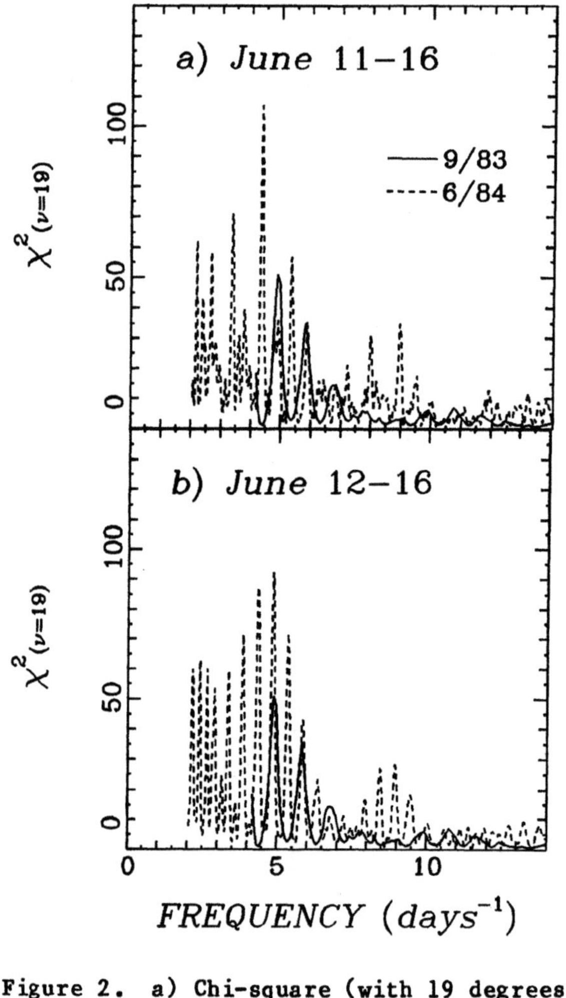

Figure 2. a) Chi-square (with 19 degrees of freedom) plotted as a function of inverse period for the 2 cm data from September (solid line) and June (dashed line). See the text for the calculation of chi-square. b) The same as above omitting the data from June 11.

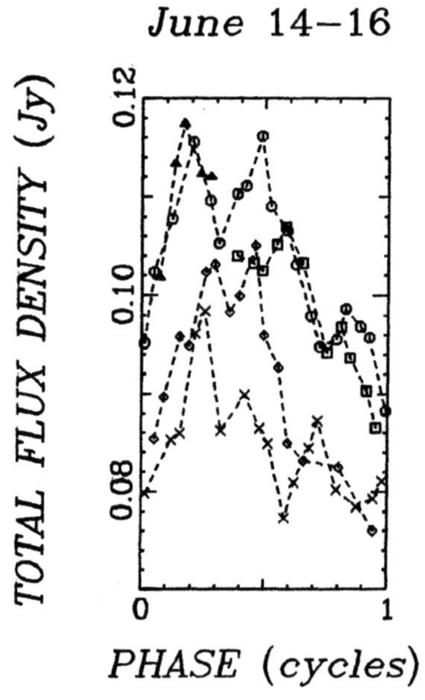

Figure 3. Flux density at 2 cm wavelength plotted as a function of phase with period 4.95 hr and arbitrary zero point for the data of June 14-16. Successive cycles are distinguished by differing symbols as follows (dates are followed by a letter denoting wrap--A first, B second, C third):

---- June 14 A
---- June 14 B
---- June 14 C
---- June 16 A
---- June 16 B

(.20 to .25 day periods), indicating the system is periodic. Furthermore, this behavior is probably typical for Cyg X-3 as all of our observing dates were chosen at random.

The spacing of secondary peaks (i.e., aliases) is determined by the spacing of gaps in the sampling. The interpretation of the small difference between the periods of the two peaks is ambiguous. It turns out that the distinction between the heights of the peaks at 5.49 hr and 4.95 hr (4.4 and 4.8 days^{-1}) in the June data is mostly determined by the data from the first half of June 11. Hence if we call the first half of June 11 abnormal behavior (a beat was skipped), then the data are consistent with a good clock with a period of 4.95 hr. Figure 2b shows chi-square versus inverse period omitting data from June 11. (The June 14 and 16 data, folded at 4.95 hours in figure 3, agree particularly well.) On the other hand, it may be that the period wanders somewhat over long time scales. Additional observations, preferably without sampling gaps, will be required to distinguish these two cases.

TABLE I

Date	A_{2-6}(min)	$A_{1.3-2}$(min)
12/3/83	52.2 (1.1)	15.1 (5.6)
6/11/84	53.4 (1.0)	6.3 (2.1)
6/12/84	46.9 (2.7)	9.2 (1.6)
6/14/84	49.5 (1.4)	12.0 (4.0)
6/16/84	51.8 (2.6)	----
mean	51.8 (0.6)	8.8 (1.2)

Table I. Cross-correlations among the wavelengths and their 1-sigma errors. They are calculated for each day with 11 hours of coverage, and a weighted mean of their values is calculated.

3. DISCUSSION

Periodic nonthermal radio emission has been observed from the much longer period binaries Cir X-1 (Nicolson et al. 1980) LS I+61°303 (Taylor and Gregory 1982), and SS 433 (Grindlay et al. 1984 and Band and Grindlay 1984), although the flaring mechanisms in those more widely separated systems are probably different. In the case of Cyg X-3, periodic production of relativistic particles may be a natural result of a surge in mass transfer at periastron passage of an eccentric orbit. In this scenario, the difference between the X-ray and radio periods would be due to the precession of the periapsis and would be expected to be constant. Another inviting analogy is with the superhumps in SU UMa stars, which have periods 1 to 7% longer than the binary orbital period.

Vogt (1982) suggested that this was due to a precessing, eccentric ring of material orbiting the compact object. In this scenario, the period should wander slightly over long time scales as the ring evolves.

The VLA is a facility of the National Radio Astronomy Observatory, which is operated by Associated Universities, Inc., under contract with the National Science Foundation.

REFERENCES

Band, D. L., and Grindlay, J. E. 1984, submitted to Ap. J..
Becklin, E. E., Neugebauer, G., Hawkins, F. J., Mason, K. O., Sanford, P. W., Matthews, K., and Wynn-Williams, C. G. 1973, Nature, 245, 302.
Braes, L. L. E. and Miley, G. K. 1972, Nature, 237, 506.
Brinkman, A., Parsignault, D., Giacconi, R., Gursky, H., Kellogg, E., Schreier, E., and Tananbaum, H. 1972, IAU Circ., No. 2446.
Geldzahler, B. J., Johnston, K. J., Spencer, J. H., Klepczynski, W. J., Josties, F., J., Angerhofer, P., E., Florkowski, D. R., McCarthy, D. D., Matsakis, D. N., and Hjellming, R. M. 1983, Ap. J. Lett., 273, L65.
Geldzahler, B. J., Kellermann, K. I., and Shaffer, D. B. 1979, A. J., 84, 186.
Gregory, P. C., Kronberg, P. P., Seaquist, E. R., Hughes, V. A., Woodsworth, A., Viner, M. R., Retallack, D., Hjellming, R. M., and Balick, B. 1972, Nature Phys. Sci., 239, 114.
Grindlay, J. E., Band, D., Seward, F., Leahy, D., Weisskopf, M. C., and Marshall, F. E. 1984 Ap. J., 277, 286.
Hjellming, R. M., Hermann, M., and Webster, E. 1972, Nature, 237, 507.
Hjellming, R. M., and Balick, B. 1972, Nature, 239, 443.
Lamb, R. C., Fichtel, C. E., Hartman, R. C., Kniffen, D. A., and Thompson, D. J. 1977, Ap. J. Lett., 212, L63.
Mason, K. O., Becklin, E. E., Blankenship, L., Brown, R. L., Elias, J., Hjellming, R. M., Matthews, K., Murdin, P. G., Neugebauer, G., Sanford, P. W., Willner, S. P. 1976, Ap. J., 207, 78.
Molnar, L. A., Reid, M. J., and Grindlay, J. E. 1983, IAU Circ., No. 3885.
Molnar, L. A., Reid, M. J., and Grindlay, J. E. 1984, Nature, 310, 662.
Nicolson, G. D., Feast, M. W., and Glass, I. S. 1980, M. N. R. A. S., 191, 293.
Samorski, M., and Stamm, W. 1983, Ap. J. Lett., 268, L17.
Seaquist, E. R., and Gregory, P. C. 1977, Ap. Lett., 18, 65.
Taylor, A. R., and Gregory, P. C. 1982, Ap. J., 255, 210.
van der Klis, M. and Bonnet-Bidaud, J. M. 1981, A. A., 95, L5.
Vestrand, W. T. 1983, Ap. J., 271, 304.
Vogt, N. 1982, Ap. J., 252, 653.
Weekes, T. C. and Geary, J. C. 1982, P. A. S. P., 94, 708.
White, N. E., and Holt, S. S. 1982, Ap. J., 257, 318.

AN EXTREMELY VARIABLE RADIO STAR IN THE RHO OPHIUCHI CLOUD

Eric D. Feigelson
Department of Astronomy
The Pennsylvania State University
University Park, PA 16802 USA

and

Thierry Montmerle
Service d'Astrophysique
Centre d'Etudes Nucleaires de Saclay
91191 Gif-sur-Yvette, Cedex, FRANCE

ABSTRACT. During a multi-epoch radio continuum survey of pre-main sequence stars in the ρ Ophiuchi cloud made with the VLA, rapid radio variations were seen in the young star DoAr 21. Typically present at levels of 2-5 mJy with a flat spectral index, it rose on timescales of hours to 48 mJy at 5 GHz on 18 February 1983, during which the spectrum was steeply inverted. Interpretation in terms of mass loss, which is usually invoked to explain T Tauri radio emission, encounters difficulties. (Gyro)synchrotron radiation from energetic electrons in a magnetic loop several times larger than the star provides a better model. The event is very similar to radio flares seen on RS CVn stars. Along with an event seen in V410 Tau (Bieging et al., this volume) this is the first report of a microwave flare on a pre-main sequence star, and adds to the growing body of evidence that young stars exhibit very high levels of magnetically induced surface activity.

1. INTRODUCTION

Radio continuum surveys of low mass pre-main sequence (PMS) stars have revealed about a dozen T Tauri stars with detectable emission (e.g., Cohen, Bieging and Schwartz 1982). The radio emission is generally interpreted to be free-free emission from ionized portions of stellar winds with $\dot{M} \gtrsim 10^{-7}\ M_\odot$/yr. There is a growing body of evidence that PMS stars possess enhanced nonthermal activity compared to main sequence stars. Indications of flares and starspots from optical photometry and spectroscopy have been strongly confirmed in X-ray studies of PMS stars (see review by Feigelson 1984). The X-ray evidence for powerful flares motivated us to conduct a search for variable nonthermal

R. M. Hjellming and D. M. Gibson (eds.), Radio Stars, 335–338.

radio emission from PMS stars. We chose to study the ρ Ophiuchi star formation cloud (Barnard 42) because of its concentration of 50-70 X-ray variable PMS stars within a few square degrees (Montmerle et al. 1983, henceforth MKFG). A number of radio sources, interpreted as H II regions around embedded B stars, had previously been found in the cloud (Falgarone and Gilmore 1981). The results of our extensive survey of the cloud will be given elsewhere (Montmerle et al., in preparation). This paper presents findings concerning a particular PMS star that exhibited extremely variable radio intensity and spectrum. A more detailed version of this work has been submitted to Ap.J. Letters.

The star under investigation, known as DoAr 21 (Dolidze and Arakelyan 1959), is one of the brightest in the cloud at infrared and X-ray wavelengths, though is faint in the blue due to strong interstellar reddening. It is unusual in that it exhibited strong Hα emission in 1949, weak emission around 1960, and no emission lines during the past decade. Recent optical spectra show no clear absorption lines (except a DIB feature) and near-IR photometry suggests a G0 spectral type with $L_{Bol} \simeq 25\ L_{\odot}$. DoAr 21 thus appears to be a classical T Tauri star that has (temporarily?) lost its emission line wind. In X-rays, DoAr 21 = ROX-8 is brighter that almost all of the X-ray emitting PMS stars in the cloud, with L_x varying between $1\text{-}2.6 \times 10^{31}$ erg s^{-1} on timescales of days or shorter.

2. RADIO OBSERVATIONS AND FINDINGS

Snapshots of DoAr 21 were obtained on six days in February, April, June, and September 1983 with the VLA. The array was in various configurations and combinations of 1.4, 4.9 and 15 GHz data were obtained at each epoch (see Figure 1). On 18 February 1983 the source was dramatically brighter than during our other observations, including 17 February. If 18 February is excluded from consideration, the source still exhibits variability at 1.4 and 5 GHz, though at a low level. The spectral index is also clearly changing. The slope of the spectrum was steeply positive, with $\beta = +1.23 \pm 0.05$ ($S_\nu \propto \nu^\beta$) on 18 February, but is flat or negative other times. DoAr 21 exhibited short timescale variability on February 18 when scans spanning several hours were obtained (Figure 2). A large and definitive increase in flux occurred at 5 GHz: from 33.6 ± 0.4 mJy to 48.3 ± 0.3 mJy during 2 1/2 hours. The change can not be attributed to instrumental gain variations since the calibrator 1622-297 was observed during this interval and exhibited random variations of only $\pm$ 0.3%. No polarized flux was seen during the event. Linear and circular polarization fractions are less than 1.4% and 9% at 5 and 1.4 GHz respectively.

3. INTERPRETATION OF THE 18 FEBRUARY EVENT: WIND OR FLARE?

The radio emission from late-type PMS stars like DoAr 21 is usually attributed to free-free emission from a spherically symmetric, ionized stellar wind. If the wind has a constant velocity, the observed

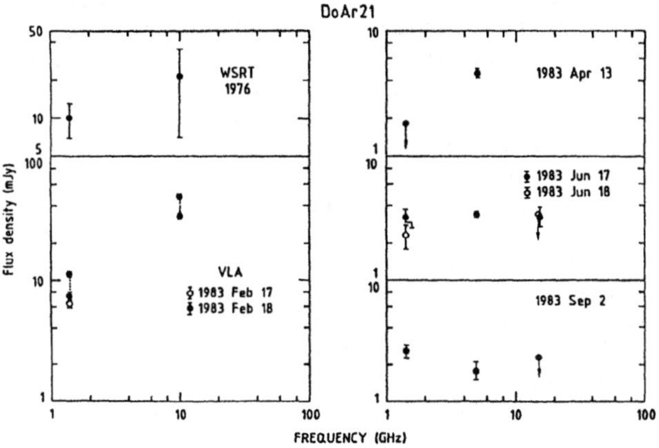

Figure 1 (above). Long-term history of radio emission from DoAr 21.

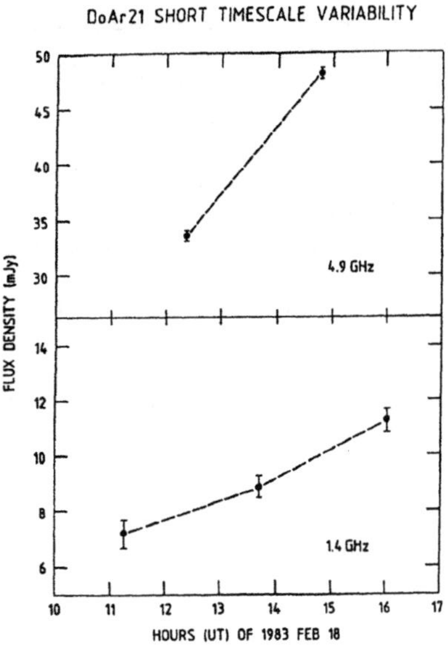

Figure 2 (right). Short-term light curve showing event on 18 February 1983.

spectral index will be $\beta = 0.6$ (Wright and Barlow 1975). Steeper spectra, such as $\beta = 1.2$ seen on DoAr 21, can be produced if the wind is accelerated in the radio emitting region, if it is "truncated" at an appropriate distance from the star, or if the source has a composite structure. In the case of an accelerated spherical outflow, reasonable values for the wind temperature ($T = 1 \times 10^4$ °K) and terminal velocity ($v_\infty = 300$ km/s), and given S(1.4 GHz) = 11 mJy and $\beta = 1.2$ from the 18 February data, we find $v(r) \sim r^{1.2}$ out to $R \approx 7 \times 10^{14}$ cm with $\dot{M} \sim 1 \times 10^{-6}\ M_\odot$/yr. Such a wind model has several serious problems: mass loss rate exceeds values estimated for even the most extreme T Tauri stars, whereas no Hα emission has been seen on DoAr 21 for many years; the radio emission from such a wind should vary on timescales $t \gtrsim R/v_\infty \sim 2 \times 10^7$ s, far longer than the observed timescale $t \lesssim 10^4$ s; and there is no clear way the star can inject energy to accelerate the wind 10^4 radii from its surface.

The principal alternative to wind models is that DoAr 21 undergoes solar-type microwave flares such as those seen on dMe and RS CVn stars. The spectral index of the 18 February event, $\beta = 1.2$, is similar to 1-10 GHz indices seen on the Sun. The peak observed power, 1.5×10^{18} erg/s Hz at 5 GHz, is 10^7 greater than that produced by the most powerful solar flares, and the ratio of radio to X-ray emission in DoAr 21 is 10^3 times higher. The DoAr 21 event most closely resembles radio flares seen in some RS CVn stars. They exhibit peak luminosities of $\sim 1 \times 10^{18}$ erg/s Hz (though levels of 10^{16}–10^{17} erg/s Hz, like the quiescent level of DoAr 21, are more typical), spectral indices $-0.6 \lesssim \beta \lesssim 1.0$, and variation timescales of 10^4 s (Feldman *et al*. 1978). The main inconsistency is that RS CVn flares nearly always show circular polarization at levels higher than the limits found for DoAr 21.

We have made rough estimates of the physical conditions implied by the 18 February event by considering a simple model of (gyro)synchrotron emission from non-thermal electrons, radiating in a homogeneous plasma of density n and magnetic field B filling a spherical volume of radius R. Such conditions will produce a rising spectrum up to a frequency ν_{max} due to various absorption processes (e.g. Kundu and Vlahos 1982). Applying the expected effects of gyroresonance, gyrosynchrotron and Razin-Tsytovitch absorption, we estimate $B \lesssim 500$ G and $n \leq 3 \times 10^{10}$ cm^{-3} in an emitting region $\sim 1 \times 10^{12}$ cm in size. The electron energies are probably more energetic (MeV compared to 10^{-2} MeV) than in solar flares. The duration of the flare, $\sim 10^4$s or longer, implies that *in situ* particle acceleration takes place in the loop. Since the observed radio power, spectrum, and risetime are identical to flares occasionally seen in RS CVn systems, the physical parameters are likely to be similar as well.

4. CONCLUSIONS

The radio behavior of DoAr 21, along with the similar event seen in V410 Tau (Bieging *et al.*, this volume), constitutes the first evidence for microwave flares in pre-main sequence stars. It directly supports the accumulating evidence, based on X-ray and other data, that PMS stars are characterized by extremely high levels of magnetically induced surface activity. An important question is whether or not DoAr 21 is unusual in its radio variability perhaps because, like V410 Tau, it is more rapidly rotating than most PMS stars. Alternatively, all ρ Oph stars may exhibit occasional radio flares. Our failure to detect a second flare from DoAr 21 in several tries, combined with the X-ray evidence that at least 50 ρ Oph stars are flaring (MKFG), suggests other radio stars may be found. Our VLA survey of the ρ Oph cloud has uncovered as many as 11 possible additional radio emitting PMS stars.

T.M. wishes to thank G. Dulk, M. Pick, and especially L. Klein and C. Trottet for enlightening discussions. We appreciate the assistance of the NRAO staff, and particularly the VLA scheduling committee, in performing these observations. E.D.F. is partially supported by an NSF Presidential Young Investigator Award.

REFERENCES

Cohen, M., Bieging, J.H., Schwartz, P.R. 1982, Ap.J. 253, 707.
Dolidze, M.V., Arakelyan, M.A. 1959, Sov. Astr.-AJ 3, 434.
Falgarone, E., Gilmore, W. 1981, Astr. Ap. 95, 32.
Feigelson, E.D. 1984, in *Cool Stars, Stellar Systems, and the Sun*, (S.L. Baliunas and L. Hartmann, eds.) Springer-Verlag, p. 27.
Feldman, P.A., *et al*. 1978, Astron. J. 83, 1471.
Kundu, M.R., Vlahos, L. 1982, Sp. Sci. Rev. 32, 405.
Montmerle, T., Koch-Miramond, L., Falgarone, E., Grindlay, J.E. 1983, Ap.J. 269, 182. (MKFG)
Wright, A.E. and Barlow, M.J. 1975, M.N.R.A.S. 170, 41.

DISCUSSION RELATED TO PART IV

MUTEL: There are about 20 x-ray binaries similar to SS433 (i.e., normal star + neutron star). What is different about SS433 that allows us to see the jets?

HJELLMING: My prejudices are that SS433 is a unique combination of super-critical accretion and relatively little "cocooning" by gas external to the binary system. Unlike most x-ray binaries, SS433 appears to be in a temporary state where the mass transfer to the compact object is super-critical in the sense that the x-ray luminosity exceeds the Eddington luminosity. In the case of Cyg X-3, which has symptoms of relativistic jets, external ram pressure or some other "cocooning" factor may disturb the collimation of the jet outflows in all but the most favorable circumstances.

BEGELMAN: The SS433 jets may be accelerated by radiation pressure. While this is not the case for extragalactic sources or even bipolar outflows, one can conceive of the possibility that the flow may be accelerated within a few Schwarzschild radii of the compact object where the radiation pressure is high enough to accelerate the flow to 0.26c. However, one cannot argue for this on the basis of the directly observed radiation fields. In SS433 the observed radiation is insufficient to accelerate the jets since SS433 is a "weak" x-ray source and a "weak" optical source. The x-ray luminosity is down by about four orders of magnitude from what is needed; and the optical luminosity is down by about two orders of magnitude. If the radiation density needed to accelerate the jets by radiation pressure is coming out of the central regions of SS433, it must be doing it in the far UV where we can't detect it.

MUTEL: I'm still not satisfied. What's different about SS433?

BEGELMAN: The optical depth. There is a lot more junk around the compact object than for typical massive x-ray binaries.

MUTEL: The mass loss rate for the primary is larger than in other cases?

BEGELMAN: Not necessarily, but the mass accretion rate onto the compact object must be higher.

FLORKOWSKI: The companion star of SS433 may not be a normal star but rather a very massive O-star with a high mass loss rate.

HJELLMING: The work that Mark Wagner, now at Arizona State, did for his Ph.D. thesis made a good case for the central, nonmoving emission lines in SS433 optical spectra being due to the wind from a massive OB star with a high mass loss rate. Since there is only a little variation in intensity of the nonmoving line in the optical emission spectrum of SS433, the very high mass loss rate seems to be a constant feature of SS433 in its current phase of evolution.

R.M. Hjellming and D.M. Gibson (eds.), Radio Stars, 339–348.

KUNDU: What is the lowest frequency at which SS433 has been observed?

HJELLMING: About 400 MHz. They see similar corkscrews at Jodrell Bank using the MERLIN array.

HEINRICHS: Let me go back to the question of why SS433 is different than other x-ray binaries. In the evolution of a massive x-ray binary the relatively long-lasting wind-powered x-ray phase (low accretion rate) is followed by a 10^4 yr (or shorter) Roche-lobe overflow-(RLO-)powered X-ray phase (higher accretion rate) simply because of the expansion of the massive star. Toward the end of the RLO x-ray phase the mass transfer continues to increase, becoming higher and higher until the x-rays are completely smothered by the infalling matter ($\lesssim 10^3$ yrs). Ultimately, the compact star might be swallowed in the very rapidly expanding envelope of the massive star. SS433 fits this picture if it is near the very short end phase. This is in accordance with the statistics of the roughly two dozen known massive x-ray binaries.

In other words, the very steep increase in the accretion rate, as a function of time, allows any such binary to be in an SS433 stage only during a very short time. ("Accretion Driven Stellar x-ray Sources", 1983 (Cambridge U. Press), ed. Lewin and Van den Heuvel).

BEGELMAN: I think we should discuss some of the other points that Bob Hjellming raised in his talk. In particular, I'm interested in the ubiquity of jets or bipolar flows in a wide variety of objects. It seems possible to get them if you have but two ingredients, an axis of asymmetry (presumably due to rotation) and a source of free energy (at the origin of this axis) which makes it possible for the gas to escape. You then get some kind of bipolar outflow.

How do you confine it? In the case of bipolar flows radiation pressure won't work, but it seems reasonable that they could be channeled inertially. Radiation pressure doesn't seem to work for extragalactic radio sources either because you don't observe enough radiation (I don't think it is hiding in the EUV band as mentioned earlier). In this case people have proposed exotic mechanisms to get the material out from the deep potential well - such as squeezing the material out of a magnetic funnel like a toothpaste tube.

In intermediate cases like SS433 it seems that radiation pressure is a plausible acceleration mechanism. It has nevertheless the same basic ingredients I spoke of earlier.

KWOK: When you say bipolar flow do you mean a molecular bipolar flow?

BEGELMAN: Yes, but they could also be observed in the form of reflection nebulae.

KWOK: Many observers have argued that molecular bipolar flows cannot be driven by radiation pressure because the observed momentum in the flow is greater than L/c. In fact, the observed flow may represent swept-up mass rather than matter directly ejected from the star. Imagine that a radiatively driven stellar wind (low dM/dt, high v) comes out of the star and runs into a circumstellar molecular medium; a high density region will form at the interface. A strong adiabatic shock propagates backward to the star and the high-temperature shocked region begins to push the swept-up gas by gas pressure. Since the interaction is adiabatic, momentum is not conserved. What we should be comparing is the mechanical luminosity of the bipolar flow to the stellar luminosity; the latter is always larger.

BEGELMAN: That's right. However, the question is how much of the energy in optical radiation can be tapped. You need to tap ~v/c to radiatively drive the wind, but this seems to be precluded by the optical depth measurements and the local reddening.

People like Ben Zuckerman have said similar things. Their arguments are that there isn't much extinction, and thus you get pretty good (optical) spectra. In order to drive a bipolar flow by radiation pressure the fraction of the energy which must be scattered is about half. If you have photons coming out that are only scattered a few times they won't be able to transfer enough energy since the fraction exchanged is of order v/c where v is the flow velocity.

UNDERHILL: The picture you suggest supposes that some energy and momentum becomes "free" at the base of the jet and that what we observe in radio, optical, and x-ray wavelengths is a result of this "free energy and momentum" being transferred to the plasma. In order to relate this to the theory of evolution of objects having a prescribed mass, it is essential to describe how the energy and momentum become "free" at the base of the jet. We can only measure how "observable factors" evolve; how the energy and momentum becomes "free" at the base of the jet, as time passes, is the crucial link to the theory of evolution of stars and other massive objects. Which of these questions are we discussing?

BEGELMAN: The evolutionary sequence which leads to these strong winds may have something to do with the axial nature of the flows. It may be misleading to distinguish between the star and the material just outside the star which may have a different form. In this way the production of the jet and the collimation of the jet may all be part of the same process. However, I think it is an unsolved problem whether this is the proper scenario or whether bipolar flows involve the interaction of a source of stellar energy and a surrounding but unrelated molecular cloud.

I was under the impression that some of the central stars in these nebulae were of quite low luminosity, and this is one of the reasons that the resonance-line-driven wind model that works for O stars doesn't work for cool stars.

COHEN: You have to remember that pre-main-sequence objects can evolve within the dynamical time scales of Herbig-Haro (HH) flows. Therefore, L/c can only be assessed now, not at the time that the HH object was ejected by the exciting stars. Worse than that, you can only assess the present luminosity from our perspective and we know that we are often looking at the HH exciting stars through edge-on disks. Also, bipolar flows are of two quite different kinds. The higher luminosity OB stars drive bipolar flows that are very crudely confined. The low luminosity T Tau and pre-T Tau stars produce much more highly confined jets. The low mass protostars may operate by totally different principles than the high mass ones.

BEGELMAN: Is there any way to assess how much kinetic energy the star is losing now, for example by observing how it impacts on the disk?

COHEN: To observe a shock interface is very difficult. When we do, they appear to be elliptical and may be due to inflow.

BEGELMAN: Do you see any evidence for steady ionization in the inner regions of the disk?

COHEN: If you look at IRS5 in L1551, you certainly see changes in ionization.

BEGELMAN: What about the collimation? How does it come about?

COHEN: I think that one of the most important questions is whether you have a spherical wind that is inertially confined or a genuinely anisotropic flow. The L1551 IRS5 jet can be estimated to have a length-to-thickness ratio of somewhere between >10 (from the observational limit to the jet width) to ~400. The latter comes from assuming thermal emission from the jet, with $T_e \sim 10^4$ K, and the $\nu^{0.13}$ spectrum (which indicates perhaps that the jet is just becoming optically thick at 6 cm). My feeling is that the jets are truly anisotropic.

BEGELMAN: Is that so obvious? In some extragalactic sources the initial flow diverges with large opening angles, but then over long length scales the flow can become cylindrical. In these sources inertial confinement won't work.

FEIGELSON: Perhaps we have to think about magnetic collimation in the jets associated with Herbig-Haro objects, just as we consider it in highly collimated extragalactic jets.

COHEN: John Beiging and I hope to use our current VLA data on IRS5 to try to determine the polarization of the radio jet. This may give us a clue to the magnetic field strength in the circumstellar environment. Optical polarization measurements in L1551 show that the cloud magnetic field is parallel to the flow direction from IRS5, and perpendicular to the circumstellar dusty toroid around IRS5.

CAMPBELL: Let me mention the correlation of the position angle of roughly a dozen star-forming-region bipolar outflows observed in CO with the position angle of the local interstellar B-field obtained from stellar polarization measurements. In all but two cases, those axes are aligned (± 10-15°); the two exceptions have position angles essentially perpendicular to the local field.

BEGELMAN: Perhaps the magnetic field helps in the collimation of the jets or, more likely, the magnetic field controls the collapse of the protostellar cloud such that the rotation axis is parallel to the interstellar magnetic field.

COHEN: The alternative to magnetic confinement is, as mentioned before, inertial confinement. We believe it to be more likely. We find that very young low-mass objects like L1551 IRS5 are surrounded by sizeable dusty disks which become thinner and less radially extended as these protostars evolve. Thus a visible T Tau star like DG Tau, which is optically polarized, has only a rather thin disk. In comparing the 6 cm continuum emission from L1551 IRS5 and DG Tau we find both are spatially extended, but the collimation of IRS5's jet is much tighter than the biconical ionized zone about DG Tau. Perhaps there really is some inertial confinement mechanism.

BEGELMAN: So you would argue that these jets arise in a similar way to "extragalactic funnels." This mechanism works if you have an accretion disk that can't cool, but there is still viscosity, so it is losing angular momentum leading to differential rotation and the material spiraling inward. The material gains gravitational potential energy which goes into internal energy. Usually this energy is radiated away, as in the disks of compact x-ray sources. However, if it can't be radiated away, the gravitational energy goes into heating and the inner part of the disk thickens such that it resembles a funnel. However there are some calculations by Pringle and collaborators which suggest that such a structure may be unstable.

FEIGELSON: Researchers concerned with highly collimated stellar flows might be interested in a summary of recent thinking on extragalactic jets in Astrophysical Jets (1984, ed. A. Ferrari and A. Pacholcyzk) and in Rev. Mod. Phys. by Begelman et al.

BOCHKAREV: I think that at least in the case of SS433 there is hydrodynamical collimation of the jets. In SS433 we clearly see precessing jets and a slaved disk. There exist a lot of photometric light curves for SS433 which show strong variations with phase of the precession period. It is so strong an effect that it is possible to determine the precession period, not only from the radial velocities, but from the photometry. Detailed analyses of light curves together with data on radial velocities show that the jets are permanently perpendicular to the disk to within several degrees. This is easy to understand in the case of hydrodynamical collimation by a thick disk, but difficult in the case of magnetic collimation.

BEGELMAN: Can you determine the thickness of the disk?

BOCHKAREV: Photometric analyses show that the disk is geometrically thick. Only if this is the case can we understand the variations in brightness of SS433 in different orbital phases with phases of the precession period. But we see the disk as an outflow disk, not an inflowing one, because the optical thickness of outflowing gas due to Thompson scattering drops to 1 at only a distance of a few x 10^{11}cm.

FLORKOWSKI: The radio behavior of SS433 on time scales ~ 100 days shows intervals of both flaring and quiescence. This could be caused by instability in the inner region of the disk.

There is a possible correlation between the 13 day photometric amplitude and whether SS433 is quiescent or flaring in the radio regime. The changes in amplitude may be measuring the changes in the disk structure. More optical data are needed to confirm this correlation.

BEGELMAN: A question for Bob Hjellming. You made a case for treating the jets from SS433 like cylindrical supernovae remnants but you didn't mention the relationship to the small structures found with VLBI observations. Also, do you have a handle on the time scales for emission and the behavior of the optical emission lines?

HJELLMING: We know that the structures observed with VLBI resolution are relatively large, because they are seen only in the shorter VLBI spacings. Spacings corresponding to a few milliarcsec completely resolve out the radio sources. Therefore we know that the size scale of radio-emitting structures in the SS433 jets is ≥ 0.01". If the "cylindrical" supernova approach is correct, the structures seen with VLBI resolution occur only on the outer parts of the cylindrically expanding jets.

The optical emission lines typically show 1-3 components, each of which is present for only a brief period of 1-3 days. If the optical emission comes from gas entrained by the cylindrically expanding jets, it will exist for only a brief time at temperatures of 10^4 K because it is gas (originally from the cooler circumstellar material) being compressed and heated. Further compression heats the gas to temperatures too high to produce significant emission. Qualitatively this behavior is helpful in explaining the transient nature of the optical emission lines.

BEGELMAN: There is some evidence in the optical for line changes that imply the ejection of material every few hours, but this would be washed out by the time the clouds got to distances where they could emit in the radio. It's not clear to me whether we should be discussing bullets or jets.

UNDERHILL: It seems that we are discussing the evolution of the plasma which generates the observable jets, not the evolution of the central disk, and how it sets "free" the energy and momentum which go into the jets. This is a problem with its own interest, but not the one I thought was to be discussed.

BEGELMAN: Unfortunately that's right. In the case of observed jets you cannot get close to the central object. In fact, you can probably get closer in the case of the bipolar objects than any others. In the case of SS433 you probably have a relativistic object with a scale of a few kilometers. The point at which you can begin mapping the object is at about 10^{11}cm, so you have five orders of magnitude between the size scale where you know what is happening and where the jets originates. The same is true for extragalactic jets. Anything can happen over five orders of magnitude of distance.

BOCHKAREV: We cannot see the central sources in any range of size scale as a result of the very dense outflowing matter around the secondary which has Thompson scattering optical depths ~ 1 at distances ~ 3 x 10^{11}cm perpendicular to the disk plane. Nevertheless it is possible to get some information on the mass of the secondary from optical data. The variations of radial velocities with orbital period show the mass ratio $q \approx 0.2-0.5$. The optical component has a temperature 2 - 2.5 x 10^4 and a size which corresponds to a star with ~ $20M_\odot$. Thus, the secondary component has a large mass, at least close to the limit for neutron stars. Most probably the secondary is a black hole.

BEGELMAN: To paraphrase Greg Benford, when you are looking at a jet you are not looking at the channel where the material is actually flowing, but at radiation pulses which are travelling through some other region. In extra-galactic objects you see these jets coming into a hot spot and turning by 90°. It seems they are continuous, but can go in any direction with very little advanced notice. This is not something that you would expect from a supersonic flow. I would expect that the models run by Norman, Smarr, and Winkler fortuitously look like some of the sources that we now see, but they may be entirely on the wrong track if you try to explain sources like Cyg A. There the jet comes into the lobe, wiggles around a bit, hits a hot spot, and goes off in some other direction. Of course Norman et al. can't handle that type of phenomena with their current limitation to two-dimensional code, but I will still defy them to do it when they have a three dimensional code.

Is there further discussion? Perhaps on Sco X-1.

GIBSON: Bob Hjellming made the statement that a skeptic could say that there was no obvious reason to associate the outlying sources of Sco X-1 with the central object, but I recall there was a proper motion study which showed them to have the same proper motion.

HJELLMING: That's right. The NE component and the central component have common proper motions corresponding to about 80 km/sec, while the possible proper motion of the NE component with respect to the central source is 20-30 km/sec.

MOLNAR: To clarify the last point, I noticed that you did not say the SW component of Sco X-1 shares the motion of the core and the NE component.

HJELLMING: That is correct. The structures inside the 20" SW component are large enough to make a precise determination of proper motion much more difficult than for the other two components.

MUTEL: I would like to go back and try to make a connection between the jets seen in Cyg A and those in R Aqr and IRS5. One way that extragalactic jets are characterized, especially the VLBI jets, is that they expand with velocities ≥ 0.1 c. This is true for SS433 as well. I wonder whether this is a universal characteristic of jets? Is it possible to rule out jet velocities of this order for galactic objects like IRS5 since it is very difficult to observe inside the jet itself?

COHEN: It's very hard to find an optical, infrared, or radio probe of the inside of a jet that would be observable. Recombination lines would be too weak for velocity studies. All of the radial velocities from shocked emission lines are lower than about 750 km/s.

MUTEL: That's probably all right. I'm thinking about the basic energetics and acceleration properties, and therefore predicting velocities ~0.25 c in the region of the inner core.

BEGELMAN: That would imply that the jets have a very low density since the lobes move very slowly.

MUTEL: I am talking about well-collimated jets, not bipolar outflows. In particular, what does a $\sim M_{\odot}$ object like SS433 have in common with the $10^8 M_{\odot}$ cores of extragalactic jets?

UNDERHILL: Your remark that all jets seem to supply electrons moving with velocities which are a significant fraction of c means that the process which sets "free" energy and momentum occurs in objects of a wide range of masses: neutron star, pre-main-sequence star, and galaxy. This question is related to the evolution of accretion disks. What physical process is implied? Why does it occur in such diverse objects?

BEGELMAN: Any neutron star or black hole is going to give you that kind of velocity. Anything that escapes from within a few "radii" of the compact object will automatically come out with that kind of velocity unless you "tune" things.

No one has yet brought up what I consider the most disturbing thing about Sco X-1, i.e. if it is a galactic object, then the radio lobes are moving out with what I would have thought to be implausibly slow velocities. Using standard synchrotron theory you can estimate the minimum energy density of the hot spots, associate a pressure with it, associate a confining ram pressure with that, and ask what kind of velocities do those ram pressures correspond to. One finds it is much larger than 20 - 30 km/s. Such low velocities would seem to be achievable in the dense parts of molecular clouds. However, there appears to be no evidence of that sort of density from other observations. A dense molecular cloud would be expected to obscure the x-ray source itself. Has anyone addressed this problem?

FEIGELSON: The lobes of Sco X-1 present a real problem: the compact object has shot out two bullets, but the bullets aren't moving! You need a brick wall to stop bullets. The difficulties are (1) that there is no reason to believe that a wall (i.e. molecular cloud) is present around Sco X-1, and (2) that walls don't survive long in the vicinity of x-ray sources. Lepp, McCray, and colleagues have shown that such an x-ray source will heat its environment to 10^4 to 10^8 K on rapid (10^2 yrs) time scales. One must resort to less appealing alternatives. Perhaps the source is pointed towards us, or the beam itself recollimates and forms an apparently stationary lobe in a rapidly flowing medium.

BEGELMAN: I find the nodal point approach hard to swallow because laboratory experiments of the same type show that the nodes oscillate. Thus, we would have to be observing Sco X-1 at a special time.

GIBSON: Likewise, I find an end-on geometry hard to swallow because you already have a problem with the size-to-separation ratio even if the outlying sources are ejected perpendicular to our line of sight.

I would like to ask whether Sco X-1 could be an intermittent source?

MOLNAR: Dave Gibson's point about the axial ratio being overly large if Sco X-1 is viewed end-on is a point against using that idea to explain the low apparent expansion velocity. I don't think that being end on would prevent us from seeing the source expand.

HJELLMING: It is quite possible that the dynamic range limitations of the observations may prevent us from seeing the source expand. One can imagine, for example that all of the observed emission is due to the equivalent of a hot spot, and that the "real lobes" are so low in surface brightness as to be unobservable.

HOGG: Even assuming the outer components were once associated with the central source, there is little evidence that the components and the central object are still connected by, for example, a beam. It is troublesome that there is no sign that variations in the flux of the central source appear later in the fluxes of the outer components.

MUTEL: Have people looked at Sco X-1 with other VLA arrays, such as the compact D-array, for evidence of large-scale diffuse structure?

HJELLMING: Campbell Wade and I have observed Sco X-1 with the VLA C- and D-arrays. However, no additional flux was found except what exists in the three compact components.

WADE: Bob, I don't think you have mentioned there is one potential observation of an even larger size scale for the Sco X-1 system.

HJELLMING: Oh yes! Fomalont et al. have found a source several arcmin NE of the NE component of Sco X-1, and noted that it lies on the same line as the Sco X-1 triplet. Our C- and D-array observations showed this source very well, and a map that shows these four sources dominating the field is tacked to Wade's bulletin board. Our results show that on the SW side there are lots of sources in line with Sco X-1 several arcmin away. However, there are real questions about probabilities when trying to associate them with Sco X-1.

BEGELMAN: What is the possibility that the "associated" sources are in fact unrelated extragalactic sources?

HJELLMING: Anything is possible given our current state of information. The common proper motion of 80 km/sec is the strongest additional argument to association of the triplet, beyond the improbability of the triplet being a coincidence.

UNKNOWN: Has anyone looked for HI-absorption towards the outlying sources?

HJELLMING: No. The NE and SW components are too weak, being only several mJy. Occasionally the central source flares to levels at which HI could be detected in absorption. The central source appears to be more active now, so one could wait until flaring to 0.1-0.3 Jy levels becomes common once again.

HOGG: It is unfortunate that, even if an HI absorption experiment can be done against the central component, it will not be possible to do it against the outer components. This means that absorption experiments won't help settle whether the components are really associated.

PART V

TECHNIQUES AND PROBLEMS IN STELLAR RADIO ASTRONOMY

THEORETICAL PROBLEMS RELATED TO STELLAR RADIO EMISSION

D. B. Melrose
School of Physics
University of Sydney
Sydney N.S.W. 2006
Australia

ABSTRACT. Some outstanding research problems relating to stellar radio flares are discussed. Emphasis is placed on "coherent" emission mechanisms for the very brightest bursts and on magnetic coupling between stars in a binary system.

1. INTRODUCTION

Formulating a theory for radio stars or for any other radioastronomical objects involves three stages: (i) the identification of the emission mechanism or mechanisms, (ii) the construction of a mathematical model for the source, and (iii) the use of the model to infer information about the source from the radio data. In identifying the emission mechanism, the brightness temperature T_B is particularly important; it allows us to estimate whether the source is thermal ($T_B \lesssim 10^6$ K for stellar coronae and winds), nonthermal and incoherent ($\kappa T_B \lesssim \varepsilon$, κ = Boltzmann's constant, ε = energy of radiating particles), and "coherent" ($\kappa T_B \gg \varepsilon$). Thermal emission from radio stars is attributed to stellar winds; in this case we are at stage (iii) in our theoretical interpretation. The emission from most flare stars has $T_B \lesssim 10^{10}$ K and so can be interpreted as incoherent gyrosynchrotron emission from electrons with energy $\lesssim$ 1 MeV. Accepting the gyrosynchrotron interpretation we are at stage (ii) in this case. However it is not certain that emission with $T_B \cong 10^{10}$ K is incoherent rather than coherent. This point may be decided from the polarization; a source with $T_B \cong 10^{10}$ K would be optically thick if it were a gyrosynchrotron source and then it should be weakly polarized whereas a cyclotron maser source should be highly polarized. The very brightest emission ($T_B \gg 10^{10}$ K) from some flare stars, notably the M dwarfs, certainly requires a coherent emission mechanism. Although cyclotron maser emission seems the most plausible mechanism, there is one serious difficulty with it.

My brief for this paper is to point out some outstanding research problems in the theoretical interpretation of radio stars. The problems concerning stellar winds involve details of the models, specifically the

R. M. Hjellming and D.M. Gibson (eds.), Radio Stars, 351–358.

matching of the predicted and observed frequency spectra. I do not propose to discuss these problems further. The standard problem connected with the gyrosynchrotron sources is the modelling of the magnetic structure and the particle spectra to account for the observed radio spectra. A particularly challenging problem with the radio stars in binary systems involves the magnetic coupling between the stars; this coupling is likely to provide the free energy for the acceleration of the gyrosynchrotron emitting electrons. I comment further on this problem in Section 4. My main interest is in the coherent emission mechanism: cyclotron maser emission is discussed in Section 2 and some qualitative comments on possible alternative plasma emission mechanisms are made in Section 3.

2. ELECTRON CYCLOTRON MASER EMISSION

Electron cyclotron maser emission has had a long history as a possible solar radio emission mechanism, e.g. the discussion by Hewitt et al. (1982), but it has become accepted as a plausible mechanism only over the past few years. The first major success was a theory by Wu and Lee (1979) for the auroral kilometric radiation (AKR). In discussing the theory here, I start by comparing and contrasting Wu and Lee's theory with an earlier one of my own (Melrose 1976) which had many but not all the attractive features of the later theory. I then discuss problems with the application of this theory to solar and stellar sources.

(a) The Parallel Driven Cyclotron Maser

A spiralling electron and a wave are said to resonate at the sth harmonic when the resonance condition

$$\omega - s\Omega - k_\parallel v_\parallel = 0 \tag{1}$$

is satisfied. Here $\Omega = \Omega_e(1-\beta^2)^{\frac{1}{2}}$ is the electron's gyrofrequency, with Ω_e the nonrelativistic electron gyrofrequency, $v_\parallel = \beta_\parallel c$ and $v_\perp = \beta_\perp c$ are the velocity components and $k_\parallel = (N\omega/c)\cos\theta$ and $k_\perp = (N\omega/c)\sin\theta$ are the wavenumber components parallel and perpendicular to the magnetic field respectively, and N is the refractive index. For waves in a mode M(= x, o or z for the x-mode, o-mode or z-mode) the dispersion relation $N = N_M(\omega,\theta)$ is assumed to be satisfied. Let $f(\underset{\sim}{p})$ be a distribution of electrons. Cyclotron maser emission at the sth harmonic occurs when the absorption coefficient $\gamma(\underset{\sim}{k})$ is negative. The contribution from resonance at the sth harmonic to $\gamma_s(\underset{\sim}{k})$ is of the form

$$\gamma_s(\underset{\sim}{k}) = \int d^3\underset{\sim}{p}\, A_s(\underset{\sim}{k},\underset{\sim}{p}) \left\{\frac{s\Omega}{v_\perp}\frac{\partial f(\underset{\sim}{p})}{\partial p_\perp} + k_\parallel \frac{\partial f(\underset{\sim}{p})}{\partial p_\parallel}\right\} \delta(\omega - s\Omega - k_\parallel v_\parallel), \tag{2}$$

where $A_s(\underset{\sim}{k},\underset{\sim}{p})$ involves Bessel functions of order s and argument $k_\perp v_\perp/\Omega$. We may classify cyclotron masers as <u>parallel-driven</u> or <u>perpendicular-driven</u> depending on whether the dominant negative contribution comes from the terms $k_\parallel \partial f/\partial p_\parallel$ or $(s\Omega/v_\perp)\ \partial f/\partial p_\perp$ respectively.

If one makes the nonrelativistic approximation $p_\perp = m_e v_\perp$, $p_\parallel = m_e v_\parallel$, $\Omega = \Omega_e$, then the resonance condition (1) does not depend on $v_\perp$, the integral in (2) may be written in terms of $dv_\perp v_\perp dv_\parallel$ and the $v_\parallel$- integral performed over the δ-function. In this case there is no restriction on the $v_\perp$-integral, and one may perform a partial integration over the $\partial f/\partial p_\perp$ term

$$-\int_0^\infty dv_\perp v_\perp\ A_s(\underset{\sim}{k},\underset{\sim}{p})\ \frac{s\Omega_e}{v_\perp}\frac{\partial f}{\partial v_\perp} = s\Omega_e \int_0^\infty dv_\perp f\ \frac{\partial A_s(\underset{\sim}{k},\underset{\sim}{p})}{\partial v_\perp} .$$

For waves with refractive index $N \lesssim 1$, only $s > 0$ is allowed and only $k_\perp v_\perp/\Omega << 1$ is relevant, then $\partial A_s/\partial v_\perp$ is positive and one concludes that the $p_\perp$-derivative cannot lead to maser emission. Thus in the nonrelativistic approximation the only possibility is a parallel-driven maser.

This nonrelativistic case was analysed by Melrose (1973,1976) using a bi-Maxwellian streaming distribution, which is a special case ($j = 0$) of the distribution

$$f(\underset{\sim}{p}) \propto \left(\frac{v_\perp^2}{V_\perp^2}\right)^j \exp\left[-\frac{v_\perp^2}{2V_\perp^2} - \frac{(v_\parallel - U)^2}{2V_\parallel^2}\right] . \qquad (3)$$

In (3), U is a streaming speed, $V_\perp^2/V_\parallel^2 \neq 1$ describes a temperature anisotropy, and $j = 1,2,3,...$ simulate loss-cone anisotropies. It was found that growth in the x-mode at $s = 1$ can occur when the following conditions are satisfied. (i) Negative absorption is restricted to frequencies below the center of the line at $\omega - \Omega_e - k_\parallel U = 0$, and $k_\parallel U$ must be large enough so that the center of the line is above the cutoff frequency for the x-mode. This requires $\omega_p^2/\Omega_e^2 << 1$, where ω_p is the plasma frequency, and

$$N_x \frac{U}{c} \cos\theta > \frac{\omega_p^2}{\Omega_e^2} . \qquad (4)$$

(ii) For the driving term ($p_\parallel$-derivative) to overcome the stabilizing contribution from the $p_\perp$-derivative, one requires

$$\frac{V_\perp^2}{V_\parallel^2} > \frac{c}{\sqrt{2} N V_\parallel |\cos\theta|} . \qquad (5)$$

It is the condition (5) which is difficult to satisfy under plausible astrophysical conditions. Inclusion of a loss cone anisotropy leads to the left-hand side of (5) being replaced by $(j+1)\ V_\perp^2/V_\parallel^2$. Thus inclusion of a loss-cone anisotrophy ($j > 0$) is effectively equivalent to increasing $V_\perp^2/V_\parallel^2$ by the factor $(j+1)$. (Revision is required to the existing discussion of the nonrelativistic loss-cone instability given by

Goldstein and Eviatar (1979). These authors started from a formula for the absorption coefficient in which the sign of the $p_\perp$-derivative was erroneously taken to be opposite that in (2).)

(b) The Perpendicular-Driven Cyclotron Maser

The major advance made by Wu and Lee (1979) was recognizing that when Ω in (1) is approximated by $\Omega_e(1-\beta^2/2)$ the resonance condition (1) becomes a quadratic rather than a linear equation for $v_\parallel$. The resonance condition may be interpreted in terms of a resonance ellipse in velocity space (Hewitt et al. 1981, 1982, Omidi and Gurnett 1982, Melrose et al. 1982). The integral in (2) is then around a semicircle $(v_\parallel-v_0)^2 + v_\perp^2 = v_R^2$ with $v_0 = k_\parallel c^2/s\Omega_e$ and $v_R^2 = v_0^2 - 2(\omega-s\Omega_e)\, c^2/s\Omega_e$. The important qualitative points are: (i) the $v_\perp$-integration is then limited and for a loss-cone distribution (which has $\partial f/\partial v_\perp > 0$ at small $v_\perp$) the $p_\perp$-derivative in (2) can be destabilizing at every point around the contour of integration, (ii) $|k_\parallel c/\omega|$ is necessarily small for the semicircle to be in the region of velocity space populated by electrons, and this implies (a) emission at large angles to the magnetic field, and (b) that the term $k_\parallel \partial f/\partial p_\parallel$ in (2) is intrinsically small and relatively unimportant. This instability is driven by the $p_\perp$-derivative and so may be called perpendicular-driven.

In the application to AKR the instability is attributed to precipitating electrons which reflect as B increases, leading to a one-sided loss-cone distribution. Upward emission ($k_\parallel v_\parallel > 0$) is implied for these upward propagating electrons. It is this version of the cyclotron maser theory which has been applied to solar (Holman et al. 1980) and stellar (Melrose and Dulk 1982, Dulk et al. 1983) sources.

(c) Difficulties with the Cyclotron Maser Mechanism

A difficulty with the cyclotron maser theory for solar and stellar applications is that in order to escape radiation generated at $s = 1$ must pass through layers where ω is equal to $2\Omega_e$, $3\Omega_e$ etc. Three ways of overcoming this difficulty have been suggested: (i) the radiation may escape through the "window" at small $\sin\theta$ (Holman et al. 1980) (the optical depth is proportional to $\sin^2\theta$ and so is necessarily < 1 for sufficiently small $\sin\theta$), (ii) the escaping radiation is generated by the maser at $s = 2$ (Melrose and Dulk 1982), (iii) the maser generates z-mode waves at $\omega > \Omega_e$ and the escaping radiation results from the coalescence of two z-mode waves into o-mode or x-mode radiation at $\omega \gtrsim 2\Omega_e$ (Melrose et al. 1984).

Suggestion (i) encounters the difficulty that it would require a very special source structure for radiation emitted at $\theta \cong \pi/2$ at $\omega \cong \Omega_e$ to arrive at the layer $\omega \cong 2\Omega_e$ at $\sin\theta \cong 0$. It might appear that this difficulty can be overcome by appealing to the parallel-driven maser which does generate radiation at small $\sin\theta$. However, if the requirement (5) can ever be satisfied it is likely to be for downgoing (into increasing B) electrons and then the maser emission is initially directed downwards; thus a change in θ by $\cong 180^\circ$, rather than $\cong 90^\circ$ for the perpendicular-driven maser, would seem to be required. No plausible source model based on either maser mechanism has been developed.

The difficulty with suggestion (ii) is twofold. The maximum growth rate at the sth harmonic for either the parallel-driven or the perpendicular-driven maser is of the form

$$|\gamma_{max}| \cong \xi_s(\theta) \frac{n_1}{n_e} \frac{\omega_p^2}{\Omega_e} \langle\beta_\perp^2\rangle^{s-2} , \tag{6}$$

where $\xi_s(\theta)$ is a factor of order unity, n_1/n_e is the ratio of the precipitating to the thermal electrons and $\langle\beta_\perp^2\rangle$ is the average of $\beta_\perp^2$ over the distribution function. Effective growth at s = 2 requires (a) that the much faster growth at s = 1 be suppressed, and (b) that $|\gamma_{max}|$ at s = 2 be sufficiently large for growth at s = 2 to be optically thick. Requirement (a) can be satisfied for $\omega_p/\Omega_e \gtrsim 0.3$ (Hewitt et al. 1982) and there is then a range of ω_p/Ω_e where growth of either the z-mode or the o-mode at s = 1 or the x-mode at s = 2 is favored. Although this leads to hope that either suggestions (ii) or (iii) might overcome the difficulty, to date there is no plausible model which clearly demonstrates that this is the case.

Besides this difficulty with thermal absorption, there are other basic problems in formulating a model based on the cyclotron maser mechanism. One concerns the relation between the large and the small scales. Maser emission saturates quickly and so is restricted to localized regions (Melrose and Dulk 1984). In view of this it is by no means clear how one incorporates the basic physics (calculations of growth rates etc.) into a large scale model for an astrophysical maser. A related problem is that maser emission must be driven continuously; it requires a "pump" just as does a laboratory maser or laser. The relatively long duration of stellar flares requires that this deriving mechanism operate continuously over the observed timescale. For the perpendicular-driven maser one requires a continuous supply of precipitating electrons. Moreover these electrons must be forced to precipitate, as in AKR; a dribble of precipitating particles from a trapped electron distribution is inadequate. This leads to a third problem in the formulation of a model: an acceleration mechanism and a source of free energy is required. The coherent radio emission is not restricted to stars in binary systems and so the source of free energy is not necessarily associated with magnetic coupling between stars. Flaring correlates with rapid rotation in convective stars (e.g. Bopp 1980), and presumably the driving mechanism for the maser emission is similar to that in solar spike bursts and is associated with a solar-type flare.

One may conclude that the formulation of an acceptable model based on the cyclotron maser mechanism involves some formidable problems.

3. PLASMA EMISSION AND STELLAR FLARES

The dominant radio emission mechanisms for solar radio bursts are gyrosynchrotron emission in the microwave range and plasma emission at decimeter, meter and longer wavelengths. I define "plasma emission" to include any mechanism which involves conversion of plasma microturbulence into escaping radiation. In solar decimeter-wave bursts the effect of the magnetic field on the microturbulence seems to be important and upper-hybrid and Bernstein modes are probably involved, e.g. the reviews by Kuijpers (1980) and elsewhere in these Proceedings. At meter and longer wavelengths the magnetic field is weak, in the sense $\Omega_e \ll \omega_p$, and the microturbulence involves Langmuir waves.

Consideration of possible analogs for solar bursts leads to two alternatives to the cyclotron maser mechanisms for stellar flares. First, the emission could be due to many unresolved type III-like bursts. The solar bursts would need to be scaled up to higher frequencies. This scaling in itself presents a difficulty because collisional damping becomes stronger at higher plasma frequencies. Also the brightness temperature T_B would need to be scaled up: the maximum T_B observed for type III bursts in the solar corona is $\cong 10^{13}$ K compared with $\cong 10^{15}$ K for some stellar flares. Also type III bursts are only weakly polarized and one would need to invoke some additional mechanisms, such as thermal absorption at s = 3 which eliminates the x-mode component to produce the observed high polarization of some stellar flares. A final difficulty is energetic: it is estimated that only $\cong 10^{-3}$ of the solar energetic electrons escape to produce type III bursts with the net conversion efficiency $<10^{-10}$ for the radio power compared to the power going into electron acceleration. These difficulties seem formidable, but in view of the difficulties with the cyclotron maser theory, it would be worthwhile exploring such an alternative model.

Another variant on plasma emission involves Langmuir (or other) waves generated by a loss-cone anisotropy, as has been proposed for decimeter-wave bursts (e.g. Kuijpers 1980), type I bursts (Melrose 1980, Benz and Wentzel 1981) and microwave flare kernels (Zaitsev and Stepanov 1983). A detailed discussion of the growth of the Langmuir waves has been presented recently by Hewitt and Melrose (1984). A further alternative model could be based on this mechanism. The difficulties are similar to those with the type III model mentioned above. However the energetic requirements are less severe.

4. MAGNETIC COUPLING AND ELECTRON ACCELERATION

It seems that the radio emission from most flare stars can be explained in terms of gyrosynchrotron radiation. For the RS CVn stars, for example, the radio brightness temperature may not exceed about 10^{10} K, and if this is the case then it is probably unnecessary to invoke a coherent emission mechanism. If we accept the gyrosynchrotron hypothesis, the question arises as to how the electrons are accelerated, and this leads to the underlying question of the source of free energy. At least for the RS CVn stars and the AM Her stars it is plausible that the ultimate source of the free energy is from magnetic coupling between the two stars in a binary system.

The magnetic coupling between stars is not understood. Three types of model have been envisaged. The central problem is that in the ideal MHD approximation, as the stars rotate and move around their orbits, the magnetic field lines would be wound up indefinitely. One way of overcoming this is to invoke dissipation or magnetic reconnection (Bahcall et al. 1973, Treves 1978, Joss et al. 1979, Lamb et al. 1983). The build up of magnetic stresses is minimized, but not eliminated, for stars in synchronous rotation. The idea is that the system finds a magnetic configuration in which dissipation in a localized region or regions allows the build up in magnetic stresses to be continuously relaxed. The details are ill-defined. The second type of model is based on an assumed analogy with the Io-Jupiter system (Dulk et al. 1983, Chanmugam and Dulk 1983) in that one star is regarded as a conductor moving through the magnetic field of the other. However, again the details are somewhat uncertain, especially in view of recent changes in ideas concerning the Io-Jupiter system (e.g. Neubauer 1980, Goertz 1983). In the third approach (Uchida and Sakurai 1983) the distribution of field lines threading between the two stars is calculated under specified boundary conditions.

A much more systematic and thorough investigation of the magnetic coupling is required. One interesting point arising from the existing literature is that the rate of energy dissipation may be only weakly dependent on the model. Typically it seems that when a conducting object (one star) of area A moves with speed v through an ambient field B (due to the other star) free energy becomes available at a rate of order $(B^2/2\mu_0)$ Av. It would be desirable to confirm this and determine the condition under which it applies.

Finally there is the problem of electron acceleration. Again this is poorly understood. Magnetic dissipation, double layers and small-amplitude Fermi acceleration are all possible mechanisms. Realistically the best one can hope to achieve at our present level of understanding is to modify proposed models for acceleration in the solar corona (e.g. Chapter 4 of Sturrock 1980) to the stellar sources. I know of two such models currently in press (Kuijpers and van der Hulst 1984, Bogdan and Schlickeiser 1984).

REFERENCES

Bahcall, J.M., Rosenbluth, M.N., and Kulsrud, R.M.: 1973, Nature 243, 27.

Benz, A.O., and Wentzel, D.G.: 1981, Astron. Astrophys. 94, 100.

Bogdan, T.J., and Schlickeiser, R.: 1984, Astron. Astrophys. (in press).

Bopp, B.N.: 1980, Highlights of Astronomy 5, 847.

Chanmugam, G., and Dulk, G.A.: 1983, in M. Livio and G. Shaviv (eds) Cataclysmic Variables and Related Objects D. Reidel (Dordrecht) p. 223.

Dulk, G.A., Bastion, T.S., and Chanmugam, G.: 1983, Astrophys. J. 273, 249.

Goertz, C.K.: 1983, Adv. Space Res. 3, 59.

Goldstein, M.L., and Eviatar, A.: 1979, Astrophys. J. 230, 261.

Hewitt, R.G., and Melrose, D.B.: 1984 'The Loss-Cone Driven Instability

for Langmuir Waves in an Unmagnetized Plasma' preprint.
Hewitt, R.G., Melrose, D.B., and Rönnmark, K.G.: 1981, Proc. Astron. Soc. Australia 4, 221.
Hewitt, R.G., Melrose, D.B., and Rönnmark, K.G.: 1982, Aust. J. Phys. 35, 447.
Holman, G.D., Eichler, D., and Kundu, M.R.: 1980, in M.R. Kundu and T.E. Gergeley (eds) Radio Physics of the Sun D. Reidel (Dordrecht) p. 457.
Joss, P.C., Katz, J.I., and Rappaport, S.A.: 1979, Astrophys. J. 230, 176.
Kuijpers, J.: 1980, in M.R. Kundu and T.E. Gergeley (eds) Radio Physics of the Sun D. Reidel (Dordrecht) P. 341.
Kuijpers, J., and van der Hulst, J.M.: 1984, Astron. Astrophys. (in press).
Lamb, F.K., Aly, J.-J., Cook, M.C., and Lamb, D.Q.: 1983, Astrophys. J. (Letters) 274, L71.
Melrose, D.B.: 1973, Aust. J. Phys. 26, 229.
Melrose, D.B.: 1976, Astrophys. J. 207, 651.
Melrose, D.B.: 1980, Solar Phys. 67, 357.
Melrose, D.B., and Dulk, G.A.: 1982, Astrophys. J. 259, 844.
Melrose, D.B., and Dulk, G.A.: 1984, Astrophys. J. 282, 308.
Melrose, D.B., Hewitt, R.G., and Dulk, G.A.: 1984, J. Geophys. Res. 89, 897.
Melrose, D.B., Rönnmark, K.G., and Hewitt, R.G.: 1982, J. Geophys. Res. 87, 5140.
Neubauer, F.M.: 1980, J. Geophys. Res. 85, 1171.
Omidi, N., and Gurnett, D.A.: 1982, J. Geophys. Res. 87, 2377.
Sturrock, P.A.: 1980, Solar Flares Colorado Associated Press (Boulder).
Treves, A.: 1978, Astron. Astrophys. 67, 441.
Uchida, Y., and Sakurai, T.: 1983, in P.B. Byrne and M. Rodono (eds) Activity in Red Dwarf Stars D. Reidel (Dordrecht) P. 629.
Wu, C.S., and Lee, L.C.: 1979, Astrophys. J. 230, 621.
Zaitsev, V.V., and Stepanov, A.K.: 1983, Solar Phys. 88, 297.

HIGH-ANGULAR RESOLUTION STUDIES OF STELLAR RADIO SOURCES

Robert L. Mutel
University of Iowa, Department of Physics and Astronomy
Iowa City, IA 52242, U.S.A.

ABSTRACT. Scientific goals and instrumentation for high-angular resolution studies of stellar radio sources are discussed. By plotting the sampling range of various instruments in luminosity-brightness temperature coordinates, an appropriate instrument can be selected for the study of a given emission process independent of observing wavelength.

1. INTRODUCTION

The combination of small angular size (typically << 1") and low flux levels (typically << 1 Jy) has made observations of the spatial structure of stellar radio sources nearly impossible until quite recently. During the past few years, it has been possible to map the thermal emission from extended nebulosities surrounding a few luminous stars using the VLA. However, most stellar radio sources have angular sizes of a few milliarcseconds or less, so that high-sensitivity VLBI techniques are required. A few pioneering stellar VLBI observations were made of intense flares from Algol in the mid-1970's (Clark et al., 1975, 1976), but the video recorder bandwidths were relatively narrow-band, which limited the minimum detectable flux density to $S \sim 100$ mJy. With the advent of the wide-band MkIII VLBI system (Rogers et al., 1983), it became possible to investigate the spatial structure of stellar sources with flux densities $S \gtrsim 20$ mJy. This system has been used recently to study several nearby RS CVn binary systems (Lestrade et al., 1984).

Unfortunately, current VLBI arrays are not sensitive enough to map the radio continuum emission from most of the stars which are being studied in total power at the VLA. During the next five years, the proposed dedicated VLB array (Kellermann, 1984) should be completed in the U.S., allowing observations of stellar radio emission at flux densities at least as low as the VLA, but with an angular resolution $\sim 10^3$ times higher. Also planned is a VLB telescope in high eccentricity orbit (the 'QUASAT' project, Schilizzi et al., 1984) which will extend the angular resolution by a factor of nearly 3, but with a large reduction in sensitivity. These new instruments will allow detailed studies of a

R. M. Hjellming and D. M. Gibson (eds.), Radio Stars, 359–369.

wide range of stellar phenomena, from determining the geometry of stellar winds to measuring the angular expansion of novae.

2. SCIENTIFIC GOALS OF HIGH-ANGULAR RESOLUTION STELLAR OBSERVATIONS

2.1. Thermal Emitters* (Stellar Winds, Novae, Chromospheres and Coronae)

Thermal radio emission has been detected from a variety of stellar environments: extended chromospheres (e.g., α Ori: Newell and Hjellming, 1982), winds from late-type (e.g., α' Her: Drake and Linsky, 1983), and early-type (e.g., Cyg OB2#12: White and Becker, 1983) stars, nova shells, and even interacting winds in binary systems (e.g., HM Sge: Kwok et al., 1984). From these observations, estimates can be made of mass-loss rates, radial dependence of electron density and temperature, and fractional ionization. However, the estimates are quite model dependent. Typically, a spherically symmetric, constant velocity wind model (Wright and Barlow, 1975) is assumed. However, in some stellar winds, a combination of thermal bremsstrahlung and (nonthermal) gyroresonance radiation may be responsible for the observed emission (e.g., Linsky and Gary, 1983). Underhill (1984) has recently pointed out that mass-loss estimates in early-type stars may have been overestimated by a factor of ~ 10 if gyroresonance radiation dominates the centimeter-wavelength radio emission.

What information can high-angular resolution observations of these objects provide? Some examples include:

(a) Direct measurement of the angular size of stellar winds at several frequencies, which effectively measures the surface of ~ unity optical depth at each observing frequency. This will allow much more straightforward calculation of mass-loss rates than estimates based only on flux density measurements.

(b) Two-dimensional imaging of winds will check the assumption of spherical symmetry, which is not strictly valid even for the sun (and is probably grossly incorrect for rapid rotators).

(c) The effective angular size, or equivalently the brightness temperature, is an unambiguous discriminator between thermal processes ($T_B \lesssim 10^4$ K) and nonthermal processes ($T_B \gg 10^4$ K).

(d) Regular monitoring of the angular size of galactic novae would provide a determination of the temporal dependence of expansion velocity, ionization fraction, and morphology of the ionized nebulosity. The angular expansion velocity could be combined with measurements of the linear velocity (using emission lines) to determine the nova's distance. With the VLBA it should be possible to study these phenomena in nearby extragalactic supernovae. For example, a supernova in the galaxy M31

---

*The term 'thermal emission' is used to mean free-free emission from a nonrelativistic Maxwellian electron distribution.

would attain an angular size of $\theta \sim 1$ mas a few months after the initial outburst, if it expands at $v \sim 10^4$ km s^{-1}.

2.2. Flare Stars

The term 'flare star' has historically been applied to certain nearby M-type dwarfs which exhibit highly variable radio emission. These were the first true 'radio stars' and were detected over 20 years ago at meter wavelengths (Lovell et al., 1963). The emission mechanism for the low-level emission is thought to be either gyroresonance or gyrosynchrotron emission, while at least some intense, highly polarized flares (e.g., Lang et al., 1983) appear to arise from a coherent process.

No high-angular resolution studies of flare stars have been attempted. This is probably because the flux levels of even the burst emission is almost always below the minimum detectable flux of present VLB arrays, at least at centimeter wavelengths. Bursts of up to 100 mJy have been occasionally observed from some of the most nearby systems (Lang et al., 1983), but the probability of such flares occurring during a prescheduled VLBI Network observation is fairly low. For example, Gary, Linsky, and Dulk (1982) observed the dMe system UV Ceti for 6 hours and detected only one intense burst of duration ~ 10 minutes.

With the completion of the VLBA, it should be possible to map the quiescent flux from nearby ($\lesssim$ 30 pc) flare stars and, of course, flare emission when it occurs (cf. Section 4).

2.3. Close Binary Stars (RS CVn, Algols)

The nearby ($d \lesssim 50$ pc) RS CVn systems are some of the few stellar radio sources which can profitably be studied using current VLBI arrays. Their characteristic angular size ($\sim 0.5 \lesssim \theta \lesssim 5$ mas) and flux levels during flares (often $S > 50$ mJy) make them mappable, but only using the MkIII system and the world's largest fully steerable radio telescopes (Lestrade et al., these proceedings). Even with the most sensitive VLB array presently available, there are only ~ 10 RS CVn systems which exceed the minimum detectable flux density of $S_{min} \sim 10$ mJy often enough to allow a reasonable chance for detection during prescheduled observations (Mutel and Lestrade, 1985). The VLBA will lower this limit to $S_{min} \sim 0.1$ mJy, increasing the number of detectable systems by $\sim 10^3$, or $\sim 10^4$ sources.

Some of the scientific questions which future VLBI observations can address are:

(a) What is the characteristic size and shape of the emission region during quiescent and active states? Is the core-halo geometry reported by Mutel et al. (1985) representative?

(b) Is coherent emission ever seen? This possibility has been suggested for RS CVn systems (Brown and Crane, 1978; Melrose and Dulk, 1982) to explain highly polarized bursts with time scales of a few minutes. A definitive test for coherent emission would be a VLBI measurement of brightness temperature in excess of the inverse Compton limit ($T_B \gtrsim 10^{12}$ K). In practice, earth-based interferometers are limited to measurements of $T_B \lesssim 10^{8.5} \cdot S$ (mJy), so that only extremely intense flares could even be tested for coherent emission (cf. Section 3.4).

(c) Do compact core sources move? One expects that flare events which originate near the stellar surface may propagate radially outward with speeds the order of the local Alfvén velocity (~ 10^2 -10^3 km s^{-1}). This phenomenon is seen in moving type IV bursts on the sun. At a distance of 50 pc, a source moving at v ~ 10^3 km s^{-1} transverse to the line of sight would have an apparent motion of θ ~ 0.45 mas/hour, or ~ 180° of fringe phase per hour on transcontinental baselines at 5 GHz. This is easily detectable at moderate flux levels, particularly on the sensitive Bonn-phased VLA (MkIII) baseline, where the RMS phase noise for 60 sec integration is σ_ϕ ~ 65°/S, where S (mJy) is the correlated flux. However, the phase measurement must be done differentially, i.e., using a nearby calibrator as a phase reference, because the short coherence time of VLBI interferometers.

2.4. Other Systems (Magnetic Binaries, X-Ray Binaries, etc.)

Radio emission has now been observed from several other classes of stars, including magnetic cataclysmic variables (e.g., AM Her, Chanmugan and Dulk, 1982) and X-ray binaries (e.g., LSI 61°303, Taylor and Gregory, 1982). High-resolution VLBI observations have been made of several x-ray binaries: Cyg X-1 (Clark et al., 1975, 1976; Lestrade et al., these proceedings); Cyg X-3 (Geldzahler, Kellermann, and Shaffer, 1979); Circ X-1 (Preston et al., 1983); LSI 61°303 (Lestrade et al., 1984). For the latter three systems, the derived linear size of the radio emission was $\gtrsim$ 100x the semimajor axis of the system (Lestrade et al., 1984). The radio size of Cyg X-1 varies with observing epoch, but appears to be considerably smaller (~ 10^{13} cm). There have been no VLBI observations of AM Her binaries because they are a little too weak to be reliably detected using present VLBI arrays. The dedicated VLBA (Section 3.3) will allow the determination of the spatial structure of both quiescent and flare emission in these objects.

3. EXISTING AND PLANNED INSTRUMENTS

3.1. Phase-Connected Arrays

The VLA (Napier, Thompson and Ekers, 1983) is currently the most powerful radio telescope available for the study of stellar radio emission. Its unprecedented sensitivity, polarization diversity, and ability to rapidly change frequencies make it ideal for measurements of spectral index, polarization, and time variability. Unfortunately, the maximum angular resolution (θ ~ 140 mas at 15 GHz) is too low to resolve most stellar sources. There is, however, one class of stellar radio source which is extended at VLA resolutions: extended thermal nebulosities surrounding very luminous stars. Some examples of such observations include maps of the supergiant Antares (Hjellming and Newell, 1984) and the peculiar variable HM Sge (Kwok et al., 1984).

Table I. Centimeter Arrays

Array	F (GHz)	B (km)	Smin* (mJy)	Log T_B† (K)	FWHM(5 GHz) (mas)	Availability
VLA	0.3-22	0.04-36	0.4	$\lesssim$ 4.8	350	Now
U.S. VLBN	0.3-22	500-8000	20(Mk III)	$\lesssim$ 2.6	1	Now
Eur. VLBN	0.6-22	200-2000	20(Mk III)	7.0-9.5	4	Now
VLA+1VLBA‡	0.3-22	0.04-90	0.4	6.2-8.2	140×350	1987.5?
U.S. VLBA	0.3-43	200-8000	0.4	6.2-9.5	1	1988.5?
QUASAT+VLBA	1.4-22	200-20000	50.	6.2-12.2	0.4	> 1990?

*Smin referes to 8 hr integration, 10:1 dynamic range.
†Normalized to 10 mJy source.
‡VLA+1VLBA refers to VLA and Pie-town, NM VLBA telescope.

Inspection of Figure 1 shows that the largest VLA baselines are only a factor of ~ 3 short of those required to study several other classes of stellar sources, such as optically thick chromospheres (e.g., α Ori, α Her) and stellar winds. These baselines will be provided by the nearby Pie-Town, NM VLBA site, which is currently scheduled to be completed in late 1986 as the first VLBA telescope. The 'VLA + Pie-Town' configuration will suffer from nonuniform (u,v) plane coverage, so that the angular resolution will be position-angle dependent. The ideal array would provide continuous baseline coverage from the longest VLA baseline (~ 30 km) to the smallest VLBA baseline (~ 200 km). These intermediate baselines will be important for all classes of stellar radio sources with brightness temperatures in the range $10^4 < T_B < 10^6$ K.

3.2. Current VLBI Arrays

There are presently two VLBI Networks available: the U.S. Network (USVLBN) and the European VLBI Network (EVN). Both Networks consist of 6-8 telescopes (the exact number available depends on the observing frequency) which are made available approximately 10 days every two months for coordinated VLBI observations.

For stellar observations, only the wide-band MkIII recorder system (56 MHz maximum bandwidth) and large telescope apertures provide adequate sensitivity for all but the most intense stellar flares ($S \gtrsim 100$ mJy). Also, the angular sizes of stellar sources are often only a few milliarcseconds, so that transcontinental baselines are necessary. These requirements currently limit the number of available telescopes to a maximum of ~ 6. Furthermore, the joint visibility between U.S. and European telescopes is sharply reduced at low elevations (e.g., there is only ~ 1.5 hours joint visibility between Bonn and California for the active RS CVn source HR 1099 ($\delta = 0°$). The consequent restricted transfer plane (u,v) coverage and low signal-to-noise on baselines to smaller telescopes has limited the maximum dynamic range of stellar VLBI maps to $\lesssim$ 10:1.

3.3. Dedicated VLBI Array (VLBA)

The VLBA is a proposed array of ten 25-meter telescopes located in the continental U.S., but with elements in Puerto Rico and Hawaii (Kellermann, 1984). As of October 1984, funding for the array (about $60M) has been approved by NSF, and the first $9M for FY 1985 has also been approved by Congress. If funding continues at the expected levels, the array should be operational by ~ 1988.5.

The VLBA will be the principal instrument used for high-angular resolution studies of stellar radio emission in the next decade. Its greatly improved sensitivity (0.4 mJy detection limit in 8 hours), frequency coverage (0.3 to 43 GHz) and polarization diversity (all four Stokes prameters routinely available) will allow detailed studies of radio emission from several classes of stars, including close binaries, M-dwarfs, novae, x-ray binaries, and stellar winds. The instrument will be controlled by a central facility with real-time communication links to each element, so that telescope operation and data integrity can be continuously monitored. The user will 'operate' the telescope in a fashion very similar to the VLA.

3.4. Orbiting VLBI

Observations of many nonthermal stellar sources with low flux densities will require VLB arrays with extremely long baselines in order to map the spatial structure. For example, mapping a source with S = 3 mJy and $T_B \approx 10^{9.5}$ K (typical of flares from gyrosynchrotron emitters) requires baselines $B \gtrsim 15{,}000$ km. This implies an array with one or more large-aperture VLB elements in high eccentricity orbits. Additional advantages of auch a system include denser, more uniform (u,v) plane coverage and removal of degrading atmospheric and ionospheric effects, especially at short wavelengths.

Several detailed designs for an orbiting VLB telescope have been proposed during the past decade (Burke, 1984). At the present time, however, most of the planned programs incorporate relatively small apertures (15 m - 30 m class), so that their use for stellar radio astronomy may be limited. An example is the joint NASA-ESA project QUASAT (Schilizzi et al., 1984), which consists of a single 15 m paraboloid in a 45° inclined orbit with an apoapsis of ~ 15,000 km. For a 24^h integration, the expected noise level is ~ 50 mJy (1.6 and 5.0 GHz). A somewhat more promising design for low flux density observations is the 50 m telescope devised by the Marshall Space Center in 1982. Unfortunately, the telescope was fixed to the Space Shuttle (i.e., low-earth orbit) and hence could not provide long enough baselines. Given the extremely high cost and lead-times involved in major new space instrumentation, detailed maps of weak, high-brightness radio sources (of all kinds) will probably not be possible during the next decade.

3.5. Millimeter Arrays

Table II summarizes some characteristics of the major mm-arrays either currently available or in the planning stage.

Table II. Millimeter Arrays

	Max Baseline (km)	S_{min} (mJy)	F_{max} (GHz)	θ_{FWHM}[e] (mas)	Available
Phase-connected arrays (e.g., Hat Creek, OVRO, Nobeyama	$\lesssim$ 5	20-200[b]	≈ 300	100	Now
mm-VLBI (OVRO (10 m), Hat Creek (6 m), KPNO (10 m), FCRAO (15 m), Onsala (20 m))	8000	1200[c]	90	0.075	Now
NRAO mm-array (≈ 21 × 10 m dishes)	2 (36)[a]	0.5[d]	230	100	$\gtrsim$ 1990's (no funding committment)

[a]Could be extended to maximum VLA baseline if sited 'piggyback' on VLA rail system.
[b]Depends on array, 20 mJy refers to Nobeyama array.
[c]Assumes 500^s coherent integration, 56 MHz MkIII, OVRO-Hat Creek baseline, 5 σ detection.
[d]Assumes 8 hr integration, 5 σ detection.
[e]Atmospheric seeing of the limits resolution to $\gtrsim$ 0.5" for long integrations.

A basic limitation of present phase-connected mm-wave arrays is that the maximum baseline lengths are insufficient to map high-brightness temperature sources (cf. Section 4.2). Even the planned NRAO mm-array will only be capable of mapping sources with $T_B \lesssim 10^5$ K (assuming S = 10 mJy) even with the maximum 36 km baseline. The mm-VLBI array now in operation can map much brighter sources, but it is much too insensitive to detect most stellar sources.

3.5.1 Advanced mm-VLB Array: Imaging Stellar Disks

An intriguing possibility for future mm-VLBI arrays would be the ability to map thermal emission from stellar surfaces, starspots and plages. In principle, one could take advantage of the $S \propto \nu^2$ flux density increase for the optically thick stellar disk, along with the $\theta \propto \nu^{-1}$ resolution improvement to design a mm-VLBI array capable of stellar disk mapping. Table III gives examples of expected angular sizes and flux densities for a few representative stars at 300 GHz. Note the last two columns in particular, which assume a minimum of nine resolution elements (3 × 3 grid) on each stellar disk.

Table III. Mapping Thermal Disks at 300 GHz (1 mm)

Star	Spectral Type	Dist. (pc)	T_{eff} (K)	R/R_{sun}	Size (mas)	S_{tot}(mJy)	B_{max}(km)	S(mJy/beam)
Sirius	A1V	2.7	10,000	1.8	5.7	16.0	100	1.6
UV Ceti	M6e	2.7	2,500	0.3	0.9	0.1	630	0.01
Capella	G0III	14.0	6,000	6.3	3.9	4.5	150	0.4
HR1099	K0IV	35.0	5,500	3.0	0.7	0.1	800	0.01
α Ori	M2I	150.0	3,000	800	46.0	320.0	1.8	32.0

Note: Last two columns refer to minimum mapping requirements (3 × 3 resolution elements/disk).

It is obvious that a mm-array with considerably improved sensitivity will be needed to map even nearby solar-type stars. The ideal array would have a sensitivity $\lesssim$ 10 μJy (5 σ) and a maximum baseline of ~ 1000 km.

4. CHOOSING AN APPROPRIATE INSTRUMENT

4.1. Flux Density-Angular Resolution Analysis

In order to determine which of the arrays discussed in Section 3 is appropriate for mapping a given class of radio emission, observers typically think in terms of resolving power and minimum detectable flux. An example of such a plot is shown in Figure 1 below, which has been calculated assuming an observing frequency of 5 GHz. Such plots are useful at a single frequency, but since nearly all of the existing and proposed arrays are frequently agile, it is more convenient to choose frequency-independent coordinates which depend only on the physics of the emission, as described below.

4.2. Luminosity-Brightness Temperature Analysis

If we describe an emission process by its intrinsic properties (i.e., brightness temperature and luminosity), then at least for optically thick sources we have a set of parameters which are frequency and distance independent. To see how these parameters are related to the properties of interferometer arrays, we note that the range of brightness temperatures which can be resolved (not simply detected) is a simple function of the baseline lengths only:

$$(T_B)_{\substack{min \\ max}} \simeq 4.5 \times S \times D^2_{\substack{min \\ max}} \ \text{°K}$$

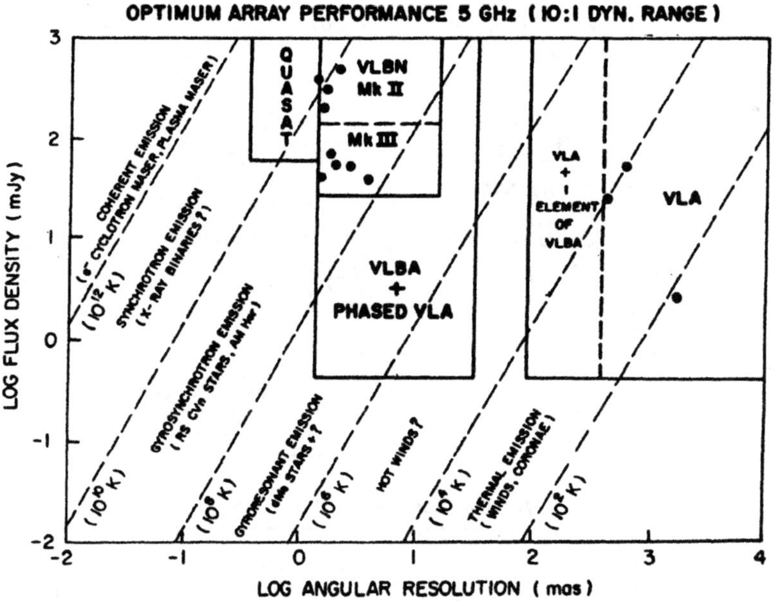

Figure 1. Sensitivity and angular resolution ranges of existing and planned arrays at 5 GHz. Dashed lines are loci of constant brightness temperature. Dots represent published measurements of individual objects.

where D (km) is the baseline length, S (mJy) is the flux density and I have used $\theta = \lambda/3D$ as the resolution criterion. An interesting example is the present VLA, which has sufficient sensitivity but insufficient baseline length to map weak sources ($S \lesssim 1$ mJy) with $T_B \gtrsim 6000$ °K at any wavelength. This eliminates virtually all stellar radio sources except very luminous thermal sources.

Since instruments are flux-density limited, overlays of instrument response on a brightness temperature-luminosity plot will be distance dependent. Figure 2 is a plot of all reported classes of stellar radio emission in luminosity-brightness coordinates. Overlays of the current VLBN (MkIII mode) the planned VLBA, and are shown overlayed in Figure 2(a) and (b), respectively.

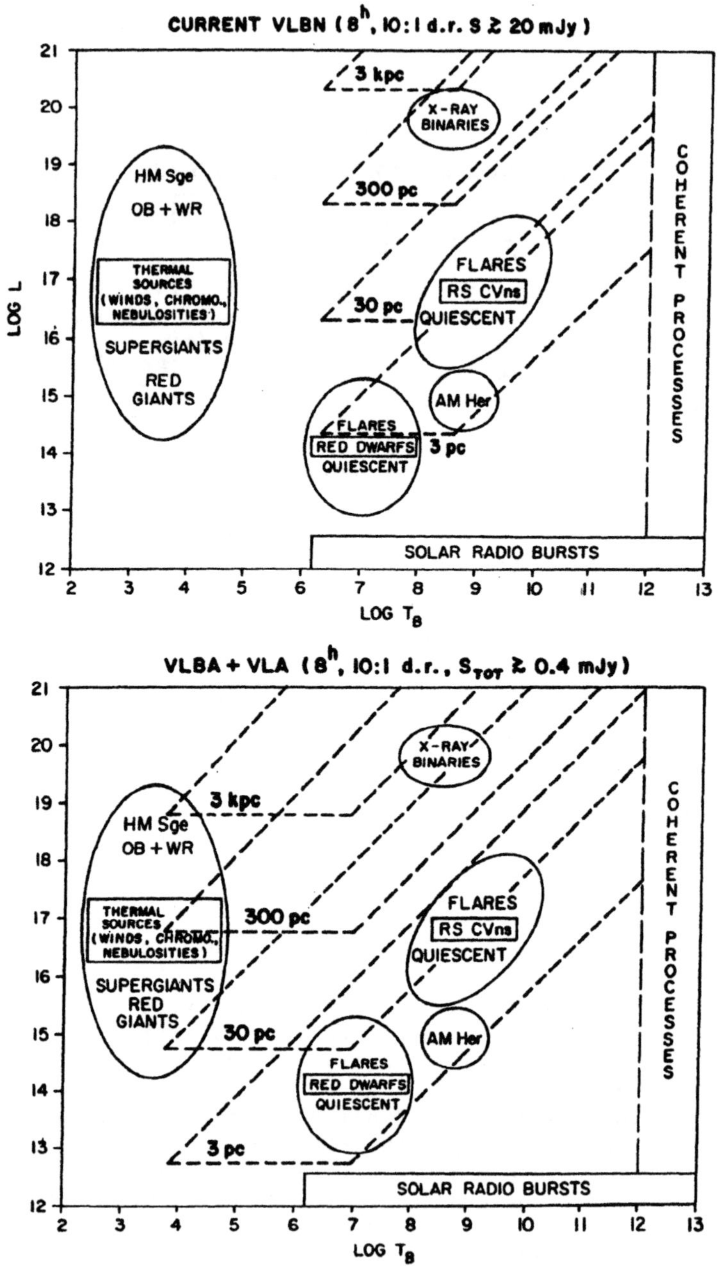

Figure 2. Luminosity-brightness temperature range for several classes of stellar radio emission. Figure 2(a) shows overlay of coverage for existing VLB Network in L-T_B coordinates. Dashed line indicates mappable regions at labelled distance: template moves vertically up for increasing distance. Figure 2(b) is the same, but for VLBA + phased VLA.

REFERENCES

Brown, R. L., and Crane, P. C., 1978, A.J., **83**, 1504.
Burke, B. F., 1984, in VLBI and Compact Radio Sources, IAU Symposium 110, ed. R. Fanti, K. Kellermann, and G. Setti (Reidel, Dordrecht), p. 397.
Chanmugan, G., and Dulk, G. A., 1982, Ap. J. (Letters), **255**, L107.
Clark, B. G., Kellermann, K. I., and Shaffer, D., 1975, Ap. J. Letters), **198**, L123.
Clark, T. A., et al., 1976, Ap. J. (Letters), **206**, L107.
Drake, S. A., and Linsky, J. L., 1983, Ap. J. (Letters), **274**, L81.
Gary, D. E., Linsky, J. L., and Dulk, G. A., 1982, Ap. J. (Letters), **263**, L79.
Geldzahler, B. J., Kellermann, K. I., and Shaffer, D. B., 1979, A.J., **84**, 186.
Hjellming, R. M., and Newell, R. R., 1984, Ap. J., **275**, 704.
Kellermann, K. I., 1984, in VLBI and Compact Radio Sources, IAU Symposium 110, ed. R. Fanti, K. Kellermann, and G. Setti (Reidel, Dordrecht), p. 377.
Kwok, S., Bignell, R. C., and Purton, C. R., 1984, Ap. J., **279**, 188.
Lang, K. R., Bookbinder, J., Golub, L., and Davis, M. M., 1983, Ap. J. (Letters), **272**, L15.
Lestrade, J.-F., Mutel, R. L., Preston, R. A., and Phillips, R. B., these proceedings.
Linsky, J. L., and Gary D. E., 1983, Ap. J., **274**, 776.
Lovell, B., Whipple, F. L., and Solomon, L. H., 1963, Nature, **198**, 228.
Melrose, D. B., and Dulk, G. A., 1982, Ap. J., **259**, 844.
Mutel, R. L., and Lestrade, J.-F., 1985, A.J. (in press).
Mutel, R. L., Lestrade, J.-F., Preston, R. A., Phillips, R. B., 1985, Ap. J. (in press).
Napier, P. J., Thompson, A R., and Ekers, R. D., 1983, Proc. IEEE, **71**, 1295.
Newell and Hjellming, 1982, Ap. J. (Letters), **263**, L85.
Rogers, A. E. E., et al., 1983, Science, **219**, 51.
Preston, R. A., et al., 1983, Ap. J. (Letters), **268**, L23.
Schilizzi, R. T., Burke, B. F., Booth, R. S., Preston, R. A., Wilkinson, P. N., Jordon, J. F., Preuss, E., Roberts, D., 1984, in VLBI and Compact Radio Sources, IAU Symposium 110, ed. R. Fanti, K. Kellermann, and G. Setti (Reidel, Dordrecht), p. 377.
Taylor, A. R., and Gregory, P. C., 1982, Ap. J., **255**, 210.
Underhill, A. B., 1984, Ap. J., **276**, 583.
White, R. L., and Becker, R. H., 1983, Ap. J. (Letters), **272**, L19.
Wright, A. E., and Barlow, M. J., 1975, M.N.R.A.S., **170**, 41.

The Time Resolution Domain of Stellar Radio Astronomy

Jay Bookbinder
Department of Astronomy
Harvard University
60 Garden Street
Cambridge, MA. 02138
U.S.A.

1. INTRODUCTION

While the origins of stellar radio astronomy have their roots in time-variable phenomena, high time resolution (HTR) radio observations of stellar sources is a very young technique. This is a somewhat surprising development in light of the historical development of stellar radio astronomy, which was exclusively concerned with the flare activity associated with dMe stars. One would expect that, having detected the flare emission, higher time resolution would have been employed to elucidate the time structure of the flare, thereby providing constraints on the emission and particle acceleration mechanisms. Rather, focus shifted from flare observations to detecting a variety of stellar sources at ever-lower flux levels, so that "quiescent" emission has been detected from several regions on the HR diagram. In fact, we shall see below that it may be possible that high time resolution studies can clarify the nature of the "quiescent" emission in some stellar sources.

A number of basic questions are appropriate for discussion at this point. What constitutes high time resolution in the context of stellar radio observations? For what sources are these observations likely to yield interesting results? And what current problems in understanding the emission can best be addressed by these observations? Can we use the solar analogue, and in particular the concept of the "solar-stellar connection", as a guide for our observations? Is it possible for these observations to provide us with a better understanding of related stellar properties, i.e. such as the nature of the stellar dynamo? Finally, are the facilities that are currently available or will be available in the near future suitable for making these observations?

For the purposes of our discussion, high time resolution refers to timescales that are short compared to either the hydrodynamic flow timescale or the thermal cooling timescale. We are thus selecting emission processes that are typically called "non-thermal", even if the underlying electron distribution is "thermal", i.e. Maxwellian. With this criterion, we eliminate from our discussion the free-free bremmstrahlung emission from, e.g., most O stars, which are variable, for example, on the timescales of months (see however, Abbott, Bieging and Churchwell 1985 and White 1985).

R. M. Hjellming and D. M. Gibson (eds.), Radio Stars, 371–378.

Higher time resolution makes it possible to obtain a better estimate for the brightness temperature, T_B, that is determined for the source. In general, one assumes that the timescale of variability represents some real propagation effect of the exciting phenomenon, and thus a characteristic velocity. This yields an upper limit on the true source size, which is simply related to the brightness temperature by:

$$T_B = 2.12 \times 10^7 (S_\lambda \ \lambda^2) (d/R)^2$$

where S_λ is the radio flux density at wavelength λ (in cm), d is the distance to the star in pc and R is the source size in units of solar radii. Obviously,in the absence of high angular resolution measurements, low time resolution observations only provide a lower limit on the source's true brightness temperature. If R is to be reasonably constrained for most stellar flare observations (using the rise-time of the flare), then the integration times must not exceed about one second for M dwarfs. By coincidence, these short timescales are also near the observational limits imposed by the current generation of radio telescopes. In general, observations on longer timescales have few instrumental restrictions since long integration times do not hamper variability studies on those timescales. A second effect of long integration times is to reduce the peak fluxes by averaging the instantaneous fluxes over the integration time. One example of this effect is found in the observations of Gary, Linsky and Dulk (1982) of a flare on L726-A (one of the components of UV Ceti). Their five minutes integrations gave a peak flux of under 12 mJy, while 10 second integrations had peak fluxes over 17 mJy and showed evidence that finer, but unresolved, time structure existed. The importance of a strong constraint on T_B cannot be overemphasized in terms of limiting the possible emission mechanisms: if $T_B > 10^8$, the emission is probably coherent; if $T_B > 10^{12}$, then there are no possible incoherent emission mechanisms.

Given the above restrictions on the emission mechanisms and timescales, what stellar sources lend themselves to HTR observations? Besides the sun, Gibson (1985) has compiled a list of 77 stellar sources that show evidence of "non-thermal" emission. Among these, nine are listed as sources of coherent radiation, and another nine have been observed with both coherent and incoherent emission. Each of these sources is therefore a good candidate for HTR observations. These sources cover a wide range of the HR diagram, with some significant gaps, but are mostly M giants, flare stars (M dwarfs), RS CVn systems and Wolf-Rayet and O stars and Algol-type binaries. We will restrict our discussion to the late-type stars and to the RS CVn systems, even though many of the points made in the following discussion are valid for the more exotic sources such as SS433, Cyg X-3 and Cir X-1. We will discuss as examples some of these sources and identify what information HTR observations can provide.

2. *SOLAR-TYPE STARS*

Flare stars have been the object of radio variability studies for over twenty years, since Lovell (1963) first observed stellar flares. Since then, observations have been made from below 20 MHz to about 15 GHz, with a variety of sensitivities and time resolutions. For these stars in particular, the use of the solar analogy may be quite apt, given that the overall structure of the stars is quite similar to the sun. One caveat, based upon solar analogy, is that in interpreting stellar observations we must expect short-lived radio phenomena to have vastly different properties, and may be sampling extremely different plasma conditions, depending on the wavelength of the observation. While the earliest detections of stellar flares occured at low frequencies (430 MHz), most recent work has been done in the 1.4 GHz and 5 GHz bands, especially since the advent of the VLA.

It is well accepted that solar and stellar activity at the chromospheric and coronal levels is a product of the magnetic field emerging from the stellar interior, and that this activity (at least in the solar case) is closely correlated with the active regions - i.e. spots and plages. Indirectly then, the observation and characterization of stellar flares may provide, via the constraints placed upon the magnetic fields involved with the flare, insights on the nature of stellar activity. We can then estimate how the magnetic field depends in detail on the two parameters most closely associated with stellar dynamos - the age and rotation rate of the star. Such studies of stellar activity will require large amounts of observing time dedicated to the task of establishing the statistical distribution of stellar flare energy (and perhaps other characteristics as well) as a function of event frequency from well chosen samples. Proper analysis of these distributions, to determine whether they can be represented by the solar distribution (and if not, determining which stellar parameters are responsible for the difference) will require the use of sophisticated statistical techniques. Thus, efforts should be made to extend detections of flares to all spectral types that are suspected of possessing solar-type closed magnetic structures (loops), since flares are typically associated with these features.

It is interesting to compare the highest time resolution observations yet reported for stellar radio flares with a solar example. Lang *et al.* (1983) observed an extremely strong flare (peak fluxes above 120 mJy at 1.4 GHz) on AD Leo that contained very fine structures that were unresolved on timescales of 0.2 seconds (see Figure 1a). Similar solar observations (Slottje 1978,1980) show complex structure down to the millisecond level (Figure 1b) that is quite similar in appearance to the stellar flare. The importance of the Lang *et al.* observation lies in their detection of high degrees of polarization and brightness temperatures in excess of 10^{13} K, which provides the first indication that coherent processes are responsible for (some of) the emission. Care must be taken in making comparisons to solar microwave bursts, however. While the typical solar burst consists of an impulsive phase followed by a gradual component, the gradual component that followed the extremely spiky emission on AD Leo could not have been the stellar analog of the solar gradual phase (cf. Holman, Bookbinder and Golub 1985).

Longer wavelength observations also have shown the existence of fine time structure in both the solar and stellar cases. Solar bursts with fine structure on the order of 1 - 100 msec are observed at a wide variety of frequencies: below 1 GHz (Slottje 1972), 1 to 1.4 GHz (Droge 1977), 2.6 GHz (Slottje 1978), 10 GHz (Hurford *et al.* 1979) and 22 and 44 GHz (Kaufman *et al.* 1980). Similarly, stellar bursts have been observed (usually with durations of less than a minute and variability on the order of a few seconds) at 196, 318 and 430 MHz (Spangler and Moffett 1976, Spangler, Shawhan and Rankin 1974) and 1.4 GHz (Lang *et al.* 1983). Since we are interested in the solar analogy, it may be worth noting from a solar example that variability may be best found by studying the polarized emission - in one example of a Type III burst, essentially no detailed structure was evident in the total intensity, but was present in the polarization (Pick *et al.* 1980, also Slottje 1974).

Furthermore, the timescale for the variability, if it does represent a real propagation effect, is interesting for another reason as well, since it then measures the size of the coronal structure giving rise to the emission. The source dimensions can be estimated by assuming a spatial growth rate between the assumed propagation speed (e.g. the Alfven speed) and c. These would be the first measurements (albeit indirect) of the sizes of coronal loops on stars other than the sun. Through the use of HTR obervations, even without detailed knowldege of the source spectra and polarization, one can learn something about the microphysics of the flare plasma - i.e., the electron density, the magnetic field, the particle energy, etc (cf. Holman, Bookbinder and Golub 1985). The

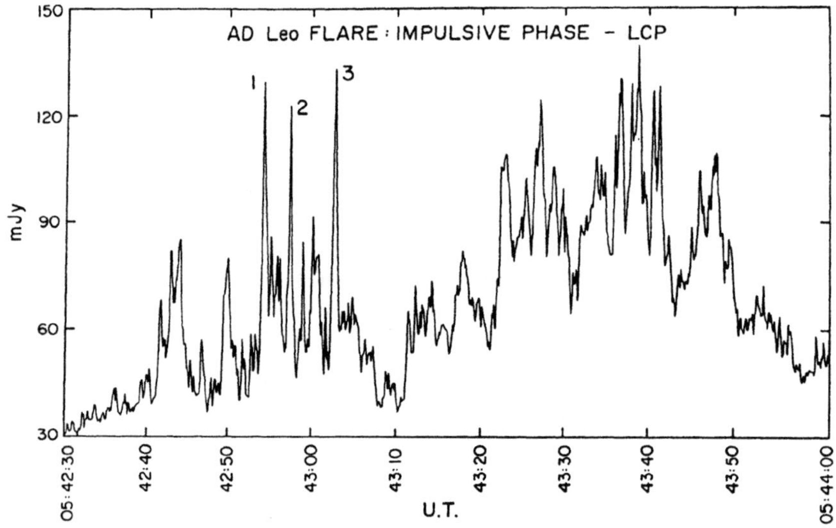

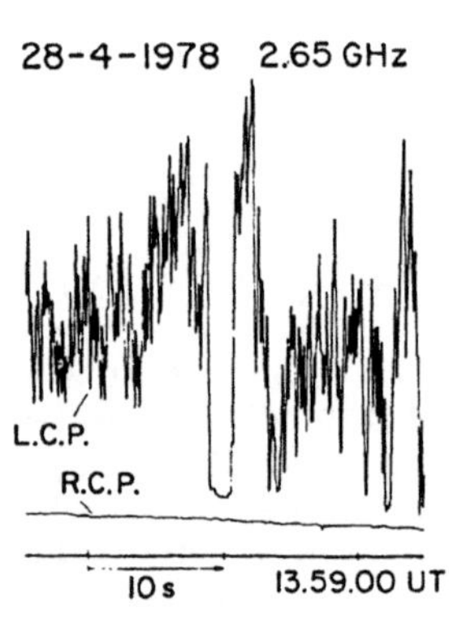

Fig. 1a. (above) A recording of the left circularly polarized (L.C.P) radiation from a large flare observed on AD Leo at 1.4 GHz with 200 msec integration times. Note the extensive (and unresolved) structure (from Lang *et al.* 1983). The right circularly polarized (R.C.P.) radiation showed no such variability, much like its counterpart in Fig. 1b.

Fig. 1b.(left) A solar observation at 2.65 GHz (after Slottje 1980) with a 50 msec time constant. The two figures are displayed with the same scales for their abcissas. The similarities are striking.

use of these diagnostics on a variety of active late-type stars, from dM through F (all of which are suspected to possess solar-like closed magnetic structures) can provide clues on the effects of the stellar parameters (such as efective temperature and surface gravity) on the stellar loops.

Another important consequence of HTR observations is the restriction on the particle acceleration mechanism. The rise and decay times of the emission and (in the case of spike) the interspike period provide information for this purpose. Most models prior to Slottje (1978) did not require acceleration mechanisms to be effective on timescales less than about a second. A number of theoretical mechanisms are available that can provide high brightness temperatures and spiky emission, for example, gyrosynchrotron masering (Melrose and Dulk 1982). If we consider an electron-cyclotron maser, for example, the rise time of the emission reflects either the increasing number of electrons in the trap (i.e. the duration of the electron pulse), the saturation time of the maser, or else simply variations in the electric field itself. The decay time reflects either the decreasing heating rate or the e-folding time for the electrons to precipitate, whichever is shorter. The interspike period is either a measure of the timescale required to accelerate the electrons, the loop transit time or, more likely, the time during which the rather stringent conditions required for maser operation are not satisfied. Moreover, accurate knowledge of the brightness temperature may clarify the operating mode of the maser. For a maser with emission at the second harmonic to operate in a steady state condition requires that $T_B \lesssim 10^{14}$ K, though an intermittent maser can generate spikes with T_b up to 10^{18} K (Melrose and Dulk 1982).

One of the principal difficulties in studying the characteristics of stellar flares is the relatively low rate of strong flares from these stars - and of course, as one moves to smaller flares, one runs into the additional problem of distinguishing a flare from noise. Fisher and Gibson (1981) provide rough estimates of flare frequencies (at 1.4 and 5 GHz) on the stars in the UV Ceti system: roughly one per 3.5 hours of observing. I have estimated the rate for AD Leo as one flare (greater than 20 mJy) per 8 hours at 1.4 GHz. Lovell (1963) has reported a flare rate of one flare per 35 hours at 240 MHz, though this result is for very strong flares. From Spangler and Moffett (1976) and Spangler, Shawhan and Rankin (1974) we can estimate stellar flare frequencies at lower frequencies for some of the more active M dwarfs. For Wolf 424 we obtain mean times between flares (greater than 0.5 Jy at 196 and 318 MHz and 1.5 Jy at 430 MHz) of 12, 2.4 and 1 hours at 196, 318 and 430 MHz respectively and for YZ CMi 2.8, 1.8 and 3 hours, respectively.

3. *RS CVn SYSTEMS*

Having looked at the information HTR can provide for us with respect to the solar-type stars, what can be said about the future of HTR observations for RS CVn systems? First, we note that HTR as applied to these objects involves timescales significantly longer than for the dM stars - and hence much easier to achieve in practice. Though the flare events generally occur on the timescales of hours to days, we note that observations of shorter-lived phemomenon are in the literature. Brown and Crane (1978) observed variations in the circularly polarized 2695 MHz flux density from V711 Tau on the order of a few minutes. Fix *et al.* (1980) have also observed variations of factors of two in the

1665 MHz flux on the order of several minutes. These observations provide us with an example of the different temporal behavior that can be found at different frequencies; Feldman (1983) reports that over 1000 hours of observing RS CVn systems at 3 cm has shown none of the rapidly changing (highly) circularly polarized fluxes that have been reported at lower frequencies by Feldman (1978) and Hjellming and Gibson (1980). The implication is that incoherent gyrosynchrotron emission seems adequate to explain the 3 cm emission, but that there is evidence for more exotic (coherent gyrosynchrotron or plasma emission) processes at the lower frequencies.

It is also possible that HTR can be used to establish the nature of the quiescent emission from these systems. Mullan (1984, see also Owen and Gibson 1978) has indicated that, unlike for the flare stars, the flare-frequency vs. flare-energy curve of RS CVn systems is consistent with the hypothesis that their "quiescent" emission is really the extrapolation of the outburst luminosity distribution to lower energies, i.e. the flares are just a manifestation of the coronal heating process. More high time resolution observations at high sensitivity could test this hypothesis more rigorously by searching for evidence of large numbers of small flares. Rosner and Vaiana (1978) have shown that it is possible to fit power laws to the flare-energy vs. flare-frequency curves for flares from many active stars under the assumption that the flaring is a stochastic relaxation phenomenon. Their model allows one to make inferences on the mode of energy storage and release. HTR observations of these systems on timescales much less than a second will likely be useful only to study the initial phase of the larger flares, when the primary energy release is occuring, unless the flares really are a manifestation of the overall coronal heating process. Along the same lines, but at the other end of the energy scale, are the so-called "superflares", such as reported by Simon *et al* (1980) Feldman (1983) and Hjellming and Gibson (1980), with flux densities on the order of 800 millijanskys (at 10.5 GHz), which are prime candidates for HTR observations. Fine structure, on the order of tens of milliseconds, should be resolvable. Although the smoothness of the rise and fall of the so-called superflares argues strongly against an intermittently operating mechanism, higher time resolution studies of these flares might provide interesting counterpoint to the current theory.

4. *INSTRUMENTATION*

As we have pointed out above, the high time resolution that is needed in stellar observations must compete against the need to have sufficient signal to noise (i.e. long integration times). In general, stellar observations should be made at the highest possible time resolution and then intergrated up to provide sufficient signal-to-noise. Some single dish instruments (i.e. the 305m dish at Arecibo) can succesfully reduce their integration times to the region of interest (on the order of a tenth of a second or less), provided that signal has a reasonably high peak flux density (cf. Lang *et al.* 1983, Boice *et al.* 1981). As usual with single dishes, confusion and terrestrial interference might be considered problems. However, in searching for variability at these timescales the problem of confusion is negligible, and interference monitors have proven quite effective in discriminating between local signals and those of stellar origin (Spangler and Moffett 1976). Recently, a proposal has been made (Stinebring 1984) that will permit extremely high time resolution observations using the VLA as a phased array, reaching down to about 20 microseconds from the current 3.3 second minimum integration time. Use of the VLA in this mode should be encouraged for all stellar observations, since it is run parallel to the aperture synthesis mode; i.e. maps can be made simultaneously with high time resolution flux measurements. It is worth stressing that the instrumental requirements for HTR radio observations do not demand major technological advances and can be easily met by existing technology (though perhaps at substantial costs).

5. SUMMARY

In conclusion we find that one of the main limitations on HTR observations is the refusal of stars to cooperate with the observer - i.e. low flare rates. Instrumental problems, obtaining the necessary sensitivity (i.e. low noise) on short integration times is also a major problem. Few instruments are likely to excel Arecibo or the VLA for this mode of observing. High time resolution observations are necessary to determine the nature of both the acceleration and emission mechanisms responsible for the short-lived radio phenomena that have already been observed. Further observations will undoubtedly provide us with novel events not familiar to us from the solar context. Besides the usual flare stars and RS CVn systems, it may be of interest to perform HTR observations on objects such as W UMa stars, PMS objects, OB stars, AM Her systems etc., all of which show evidence for non-thermal emission and/or short timescale variability.

Studies of coronal plasma processes at the Harvard-Smithsonian Center for Astrophysics are supported by NASA grant NAGW-112.

REFERENCES:

Abbott, D., Bieging, J. and Churchwell, E. 1985, these proceedings
Boice, D., Kuhn, J., Robinson, R. and Worden, S. 1981, *Ap. J. Lett.* 245, L71
Brown, R.L. and Crane, P. 1978, *A.J.* 83, 1504
Droge, F. 1977, *Astron. Astrophys.* 57, 255
Feldman, P. 1983, in *Activity in Red-Dwarf Stars* ed. P. Byrne and M. Rodono D. Reidel, p. 340
Feldman, P., Taylor, A., Gregory, P., Seaquist, E., Balonek, T. and Cohen, N. 1978, *Astron. J.* 83, 1471
Fisher, P. and Gibson, D. 1981, in *Second Cambridge Workshop on Cool Stars, Stellar Systems, and the Sun. Vol II.*, eds. M.S. Giampapa and L. Golub, (Cambridge: SAO) p. 109.
Fix, J. Claussen, M. and Doiron, D. 1980, *Astron. J.* 85, 1238
Gary, D., Linsky, J. and Dulk, G. 1982, *Ap. J. Lett.* 263, L79
Gibson, D. 1985, these proceedings
Holman, G. 1983, *Adv. Space Res.* 2, 181
Holman, G., Bookbinder, J. and Golub, L. 1985, these proceedings
Hjellming, R. and Gibson, D. 1980, in Proceedings of the IAU Symposium #86 *Radio Physics of the Sun*, eds. M. Kundu and T. Gergley, D. Reidel, p. 209
Hurford, G., Marsh, K., Zirin, H. and Kaufman, P. 1979, *BAAS* 11, 678
Kaufman, P., Strauss, F. and Opher, R. 1980, in IAU Symposium #86 *Radio Physics of the Sun*, eds. M. Kundu and T. Gergley, D. Reidel, p. 209
Lang, K., Bookbinder, J., Golub, L. and Davis, M. 1983, *Ap. J. Lett.* 272, L15
Lovell, B. 1963, *Nature* 198, 228
Lovell, B. 1971, *Quart. J. R. A. S.* 12, 98

Melrose, D. and Dulk, G. 1982, *Ap. J.* 259, 844
Mullan, D.J. 1985, these proceedings
Pick, M., Raoult, A. and Vilmer, N. 1980, in Proceedings of the IAU Symposium #86 *Radio Physics of the Sun*, eds. M. Kundu and T. Gergley, D. Reidel, p. 235.
Rosner, R. and Vaiana, G.S. 1978, *Ap. J.* 222, 104
Simon, T., Linsky, J. and Schiffer, F. 1980, *Ap. J.* 239, 911
Slottje, C. 1972, *Solar Phys.* 25, 210
Slottje, C. 1974, *Astron. Astophys.* 32, 107
Slottje, C. 1980, in Proceedings of the IAU Symposium #86 *Radio Physics of the Sun*, eds. M. Kundu and T. Gergley, D. Reidel, p. 209
Spangler, S. and Moffett, T. 1976, *Ap. J.* 203, 497
Spangler, S., Shawhan, S. and Rankin, J. 1974, *Ap. J. Lett.* 190, L129
Stinebring, D. 1984 *NRAO Research Expenditures Memo #* 12
White, R. 1985, These proceedings

DEDUCTION OF CORONAL MAGNETIC FIELDS USING MICROWAVE SPECTROSCOPY

G. J. Hurford[1], D. E. Gary[1] and H. B. Garrett[2]
[1] Solar Astronomy, Caltech, Pasadena, CA 91125
[2] Jet Propulsion Lab, Caltech, Pasadena, CA 91109

ABSTRACT. Gyroresonance opacity renders the solar corona optically thick at frequencies which are low integral multiples of the local gyrofrequency. This causes the microwave spectrum of sunspots to be sensitive to the strength of coronal magnetic fields. The concept is illustrated by high spectral resolution observations of a sunspot acquired with the Owens Valley frequency-agile interferometer. The observed spectrum is compared to the results of three-dimensional atmospheric model calculations in which the sunspot field is represented by the potential field of a dipole located beneath the photosphere. The comparison enables the depth, orientation and magnetic moment of the dipole that best fits the observations to be determined. Since such observations require that the microwave emission be resolved spectrally, not spatially, the technique may be applicable to the study of stellar coronal fields.

1. INTRODUCTION

In the absence of magnetic fields, microwave emission from the quiescent Sun is due to thermal bremsstrahlung. In the 10^6 K corona this free-free emission is optically thin so that observed brightness temperatures of a few times 10^4 K are determined by the plasma temperature of the underlying, optically thick transition zone and chromosphere. With the addition of magnetic fields, however, gyroresonance opacity can render the corona optically thick at frequencies which are low, integral harmonics of the local gyrofrequency, namely at ν [GHz] = 2.8 s B [kg] where s is a small integer. The resulting order-of-magnitude increase in brightness temperature when this condition is met suggests that the microwave spectrum is sensitive to coronal magnetic fields.

Qualitative features of the expected dependence of the microwave spectrum on coronal fields for an isolated sunspot are shown in Figure 1 by a sequence of cartoons for cases of progressively more realistic magnetic field geometries. In the simplest case (left), a uniform magnetic field gives rise to a sequence of optically thick cyclotron

R. M. Hjellming and D. M. Gibson (eds.), Radio Stars, 379–384.

lines. In the second case (centre) diverging fields fill in the spectrum below the optically thick lines. At these intermediate frequencies, the correspondingly weaker fields required to satisfy the gyroresonance condition are encountered along the line of sight higher up, but still in the 10^6 K corona. Discrete breaks in the spectrum remain, however, at frequencies corresponding the value of the field at the base of the corona. Permitting the field at the base of the corona to take on a range of values in the third case (right) still retains the basic spectral character: namely an increasing spectrum which is optically thick in both modes at low frequencies; a subsequent decrease in flux with further increase in frequency; a higher peak frequency in the extraordinary compared to the ordinary mode; and an optically thin flat spectrum at high frequencies.

2. MODEL CALCULATIONS

To put these considerations on a more quantitative basis, a numerical model was developed to evaluate solar microwave emission in the presence of strong coronal fields.

The model, illustrated in Figure 2, starts with a spherically symmetric Sun. For the results presented here, the variation of temperature and density with height was scaled from the tabulation by Papagiannis and Kogut (1975). For the magnetic field geometry, we used the potential field of a dipole located beneath the photosphere.

The microwave emission was evaluated by integrating down lines of sight and calculating the brightness temperature at the desired

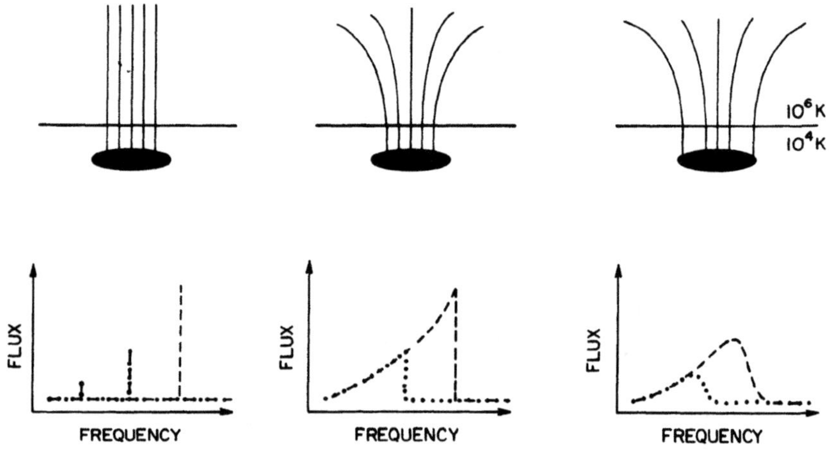

Figure 1. Sequence of schematic microwave spectra (below) corresponding to the sunspot magnetic field geometries (above) for a two-temperature solar atmosphere. In each case the corona is assumed to be optically thick up to the 2nd harmonic in the ordinary mode (dots) and up to the 3rd harmonic in the extraordinary mode (dashes).

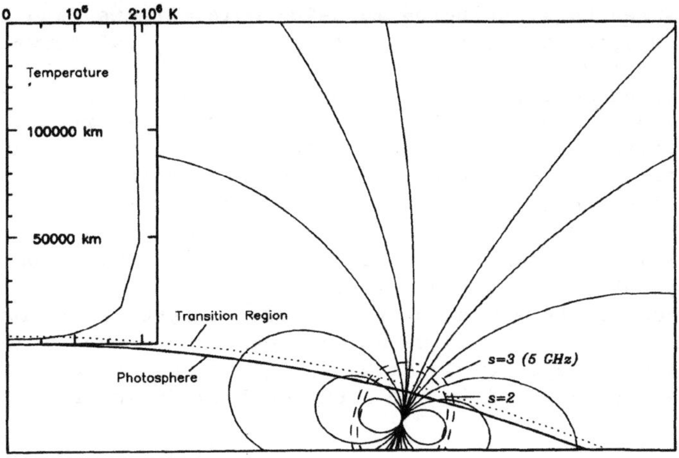

Figure 2. Geometry of the model used to calculate the microwave emission from a sunspot. The inset shows the run of temperature with height.

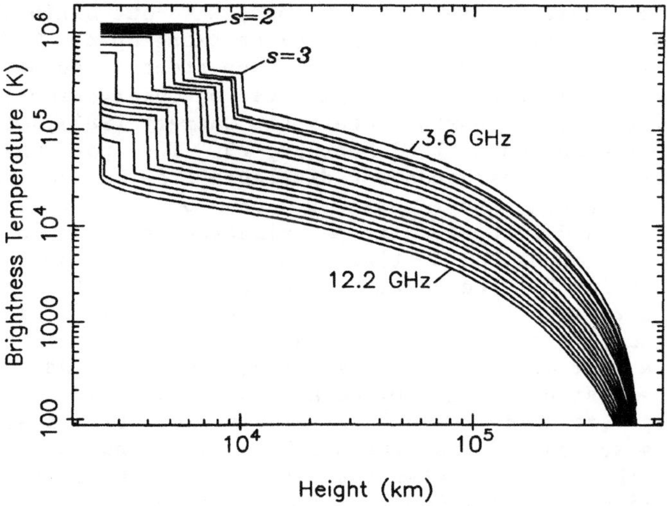

Figure 3. Curves of growth along a typical line of sight in the extraordinary mode for frequencies from 3.6 to 12.2 GHz. The steplike discontinuities are due to the contribution of gyroresonance opacity at the harmonics indicated. As the frequency is increased, gyroresonance opacity becomes effective at progressively lower heights in the corona until the transition zone is reached.

frequencies and polarizations. The calculations include the effects of thermal bremsstrahlung (including magneto-ionic effects) and gyroresonance opacity up to the 10th harmonic. Typical curves of growth are illustrated in Figure 3.

After evaluating such curves along $\sim 10^2$ lines of sight, the results can be displayed in a number of ways. For example, at each frequency/polarization, spatial maps can be generated. At intermediate frequencies, such maps show that the region of coronal brightness temperature has an inner core that is relatively cool, similar to the "ring" or "horseshoe" features that have been observed by Lang and Willson (1982) and Alissandrakis and Kundu (1982) in high spatial resolution maps of solar active regions.

An alternate representation of the model output can be made by spatially summing over the entire active region to generate an integrated microwave spectrum. As we shall see, such a spectrum is similar to that illustrated in the last panel in Figure 1.

3. OBSERVATIONS

High spectral resolution observations for comparison with these calculations were obtained with the frequency-agile interferometer of the Owens Valley Radio Observatory (Hurford, Read and Zirin 1984). This instrument has the ability to rapidly switch among up to 86 frequencies between 1 and 18 GHz, so that high resolution spectra can be determined every few seconds. In this case we make use of observations, discussed in more detail elsewhere (Hurford, Gary and Garrett 1984), of an isolated sunspot observed on 1983 February 25. The spot, which had an umbral diameter of 13 arcseconds was displaced from Sun centre by 24 degrees (heliocentric). The 17 point spectrum between 3.6 and 12.2 GHz obtained in right and left circular polarization is shown in Figure 4.

Figure 4 also shows the spatially integrated model spectrum discussed in Section 2. The parameters of the model (illustrated to scale in Figure 2) were chosen to reproduce as closely as possible the observed spectrum. Such a comparison is relevant since for these observations, a sufficiently short baseline (136 m) was used so that the active region emission was unresolved at each frequency. Correspondingly, the contribution of the quiet Sun brightness temperature was subtracted from the model calculations at each frequency so as to mimic the interferometer response. Thus the model and data represent the microwave emission in a comparable manner.

4. DISCUSSION

4.1. Comparison of Observations and Model

The agreement between the model calculations and observations in Figure 4 generally confirms the concept of the dependence of the microwave spectrum on coronal magnetic fields. Furthermore, such comparisons enable the salient parameters of the model to be determined.

For example in the case illustrated above, the maximum strength of the coronal field is found to be 1300 gauss, to a precision of ± 4%. (The maximum field strength determines the frequency at which the spectrum becomes flat.) The depth and orientation of the equivalent dipole moment are also well constrained. It is the depth ot the dipole (or characteristics of other more sophisticated magnetic models) that

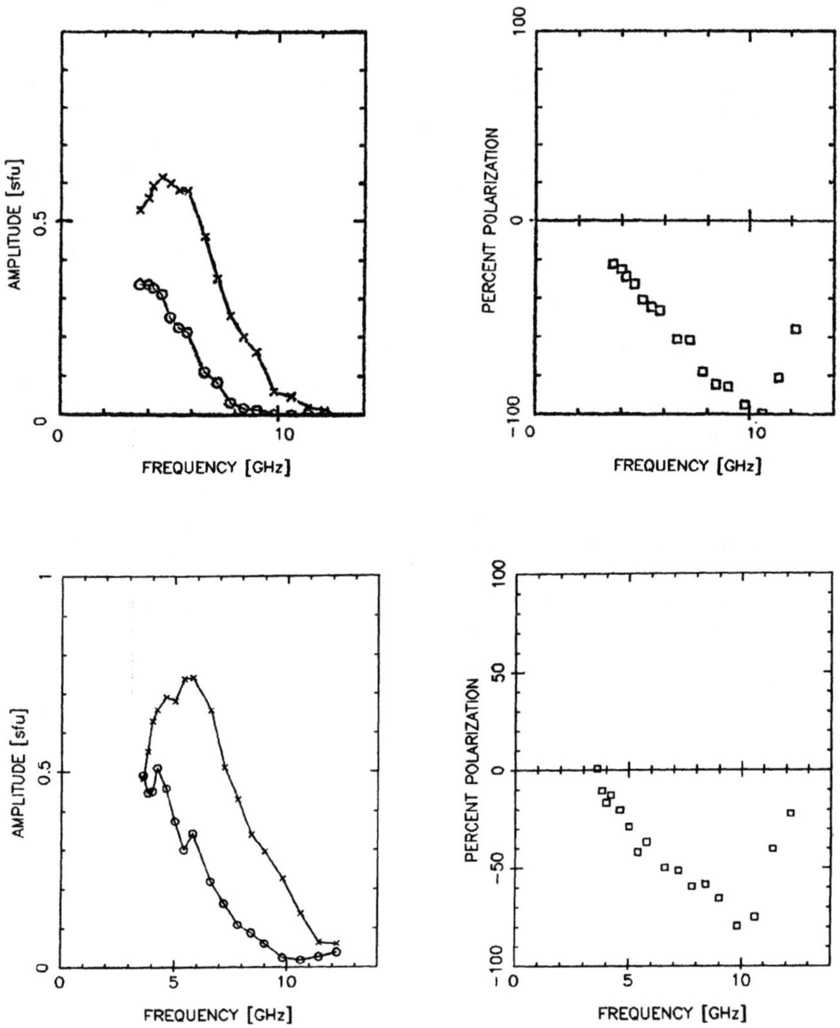

Figure 4. Comparison of the spatially-integrated model spectrum (top) with the observed spectrum (bottom). The curves in the left column represent the interferometric amplitudes in right (o) and left (x) circular polarization. The spectrum of percent polarization is shown at right.

influence the shape of the spectrum. The orientation of the dipole helps determine the polarization spectrum.

4.2. Application to Stellar Sources

As we have seen, the solar microwave spectrum is sensitive to coronal magnetic fields primarily because of the role of gyroresonance opacity. In stellar applications there may also be situations where gyroresonance opacity may be significant. (See the review by Gary in these proceedings.) This raises the possibility of using microwave spectroscopy to study magnetic fields in stellar coronae as well. In this connection, there is, of course, the task of evaluating the effects of solar/stellar differences in temperatures, densities, fields and geometry. Such an evaluation is beyond the scope of this paper. Nevertheless there are some insights to be gained from the solar case.

First, we note that the comparisons above were based on data which did NOT spatially resolve the source. (The role of the interferometer was to suppress the background quiet Sun.) In cases where the active region dominated the emission, as might be the case with star-spots, this would not be necessary.

Second, although we have not done so in this paper, the changes in the microwave spectrum that occur as the star rotates may also be effectively exploited. For example, for bipolar fields one would expect that the evolution of the polarization spectrum would depend on the orientation of the dipole with respect to the axis of rotation.

Third, we note that for this application, conventional frequency coverage of about one point per octave is not adequate. In the solar case, the spectral resolution should be significantly better than the frequency shift between the ordinary and extraordinary modes (ratio 3:2, 4:3 etc) so that spectral resolution, $d\nu/\nu$, of at least 10% is needed.

Finally, it should be recalled that microwave spectroscopy has played a pivotal role in identifying the diverse range of emission mechanisms (Dulk, 1985) which operate on the Sun. In this, as well as for more specific applications as illustrated here, microwave spectroscopy has the potential to play a corresponding role in stellar radio astronomy.

ACKNOWLEDGEMENTS

This work is supported by NSF grants ATM-8309955 and AST-8315217.

REFERENCES

Alissandrakis, C.E. and Kundu, M.R.: 1982, Astrophys. J. 253, L49.
Dulk, G. A.: 1985, Ann. Rev. Astron. Astrophys. 23, in press.
Hurford, G.J., Gary, D.E. and Garrett, H.B.: 1984, in preparation.
Hurford, G.J., Read, R.B. and Zirin, H.: 1984, Solar Physics, 94, 413.
Lang, K.R. and Willson, R.F.: 1982, Astrophys. J. 255, L111.
Papagiannis, M.D. and Kogut, J.A.: 1975, AFRCL-TR-75-0430.

A TECHNIQUE FOR REMOVING CONFUSION SOURCES FROM VLA DATA

D. E. Gary
Solar Astronomy, 264-33
Caltech
Pasadena, CA 91125
USA

ABSTRACT. A technique is described for removing strong confusion sources from VLA visibilities. The technique uses the source components obtained from the CLEAN algorithm to specify the source structure, and subtracts these contributions from the visibilities. Point source fluxes can then be obtained from the resultant "clean" database in the same way as is done for unconfused areas of the sky.

1. INTRODUCTION

The observer of point radio sources uses the u-v coverage of the VLA not to map source structure, but simply to distinguish between the flux of the source of interest and flux due to confusing sources that may be within the primary beam. For some stars (at some frequencies), there may be no confusing sources nearby, and the making of large maps is unnecessary. In these cases, the source flux can be determined simply by shifting the phase center of the data to the position of the source and performing a vector average. (This is equivalent to making a "one cell map" using a discrete Fourier transform. Better results can be obtained by making a small discrete transform map of several cells around the source location and comparing the source shape to the beam shape.) For many stellar objects, however, confusing sources are present, especially at 20 cm, and there has been no recourse but to laboriously make large maps and clean them of confusing source sidelobes in order to measure the flux of the stellar source. In the case of a flaring object, whose flux may vary on timescales less than the 3 s time resolution of the VLA and last for more than an hour, this technique is prohibitive--imagine the prospect of making and cleaning 1200 maps of 512 cells each. A procedure for removing confusing sources from visibility data is illustrated using the field near the flare star AD Leo.

R. M. Hjellming and D. M. Gibson (eds.), Radio Stars, 385–390.

2. PROCEDURE

Step 1. A map containing the offending confusing sources is made, using the entire time range of the observation to maximize signal to noise. (See Figure 1 and Figure 3.)

Step 2. The map is cleaned for all sources <u>except</u> the source of interest, and the positions (x_i, y_i) and fluxes f_i of the cleaned components are saved in a file for input into Step 3.

Step 3. The "dirty" visibilities ($\underline{V}_j$) are then "cleaned" by subtracting the contribution of each cleaned component obtained in Step 2, e.g.,

$$V_j = \underline{V}_j - \sum_i f_i \, e^{2\pi i(x_i u_j + y_i v_j)} \quad .$$

The resulting cleaned visibilities now contain no contribution from the confusing sources, and maps made with the "clean" database will have only the remaining (uncleaned) source of interest. (See Figure 2 and Figure 4.)

3. CONCLUSION

We have shown that by making and cleaning a single map and subtracting the visibilities defined by the clean components thus obtained, confusion sources may be removed from visibility data. This allows efficient flux determination techniques to be used that previously were not allowed due to the presence of sidelobes from the confusing sources. Figure 5 shows that flux measurements from shorter timescale maps can also be made from the "clean" database without the need to clean each map. After "cleaning" the 6 cm database, we were able to make the flux versus time plots of Figure 6. Figure 6a shows that two flares occurred on AD Leo during the 8 hour observations. The second flare is shown at 3 s time integration in Figure 6b (from 0900 to 0910 IAT). The flare is nearly 100% LH circularly polarized, and shows significant structure on the 3 s timescale. The plot in Figure 6b would have been prohibitively expensive to obtain without the technique described here.

It was brought to the author's attention at the workshop that a program, UVSUB, already exists in the AIPS package (written for another purpose) that performs the operation described here. Those with access to the AIPS package apparently can run APCLN and use the component file in UVSUB exactly as described here.

Acknowledgement: This work was supported in part by NSF grant ATM-8309955 to the California Institute of Technology.

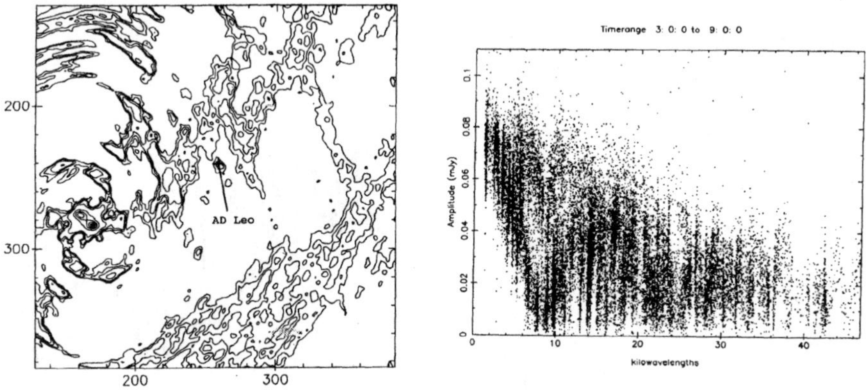

Figure 1.
(a) A 6 cm map of the field near the flare star AD Leo, showing the confusing source at left center, and its side lobes. The position of AD Leo is marked. The contours are at 1, 3, 5, 10, 30, 50, 70, 90, and 99% of the 24.72 mJy peak. (b) The visibility amplitude versus u-v distance for the 6 cm data used to make the map in (a).

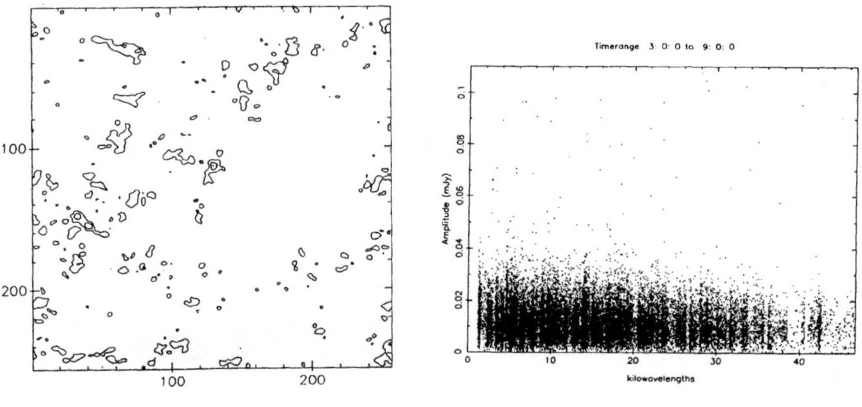

Figure 2.
(a) A dirty 6 cm map made from the "clean" database. The contours are the same as in Figure 1. The confusing source and its side lobes have been eliminated, and AD Leo is now the strongest source in the map. (b) The corresponding visibility amplitude versus u-v distance plot, showing the nearly constant amplitude with u-v distance expected for a point source.

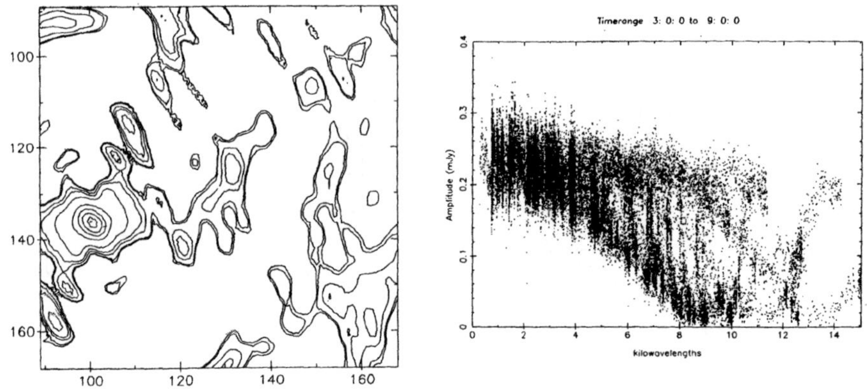

Figure 3.
(a) A dirty map of the same area of the sky as Figure 1, but for 20 cm. The contours are at 0.5, 1, 3, 5, 10, 30, 50, 70, 90, and 99% of the 163.5 mJy peak. (b) The corresponding visibility amplitude versus u-v distance plot.

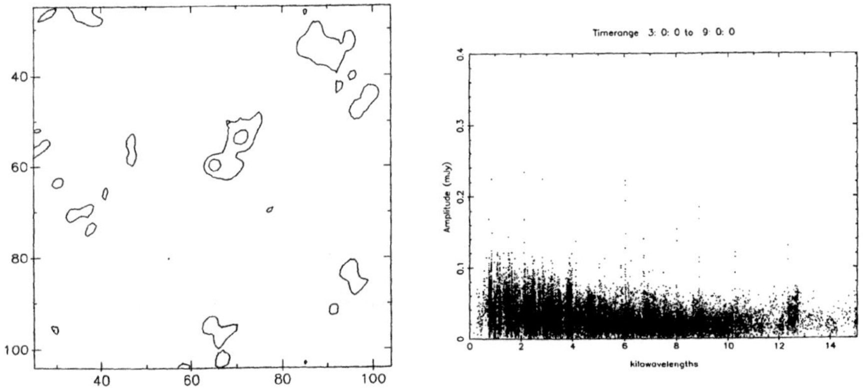

Figure 4.
(a) A dirty 20 cm map made from the "clean" database. The contours are the same as in Figure 3. AD Leo and another weak source appear near the map center. (b) The corresponding visibility amplitude versus u-v distance plot, showing the nearly constant amplitude with u-v distance.

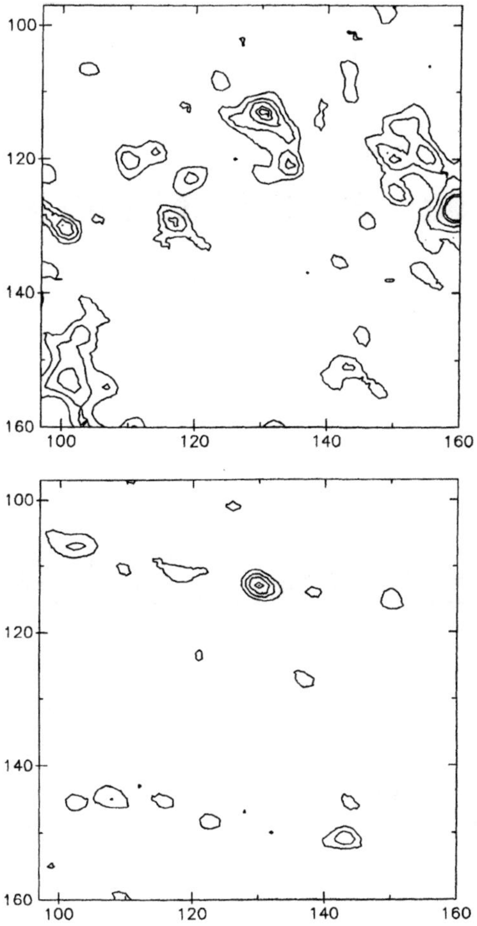

Figure 5.

The comparison between 1 minute integration maps made from (a) the "dirty" database and (b) the "clean" database. The maps correspond to the inner quarter of the maps shown in Figs. 1 and 2, and the contours are at 30, 50, 70, 70, and 99% of the 10.73 mJy peak.

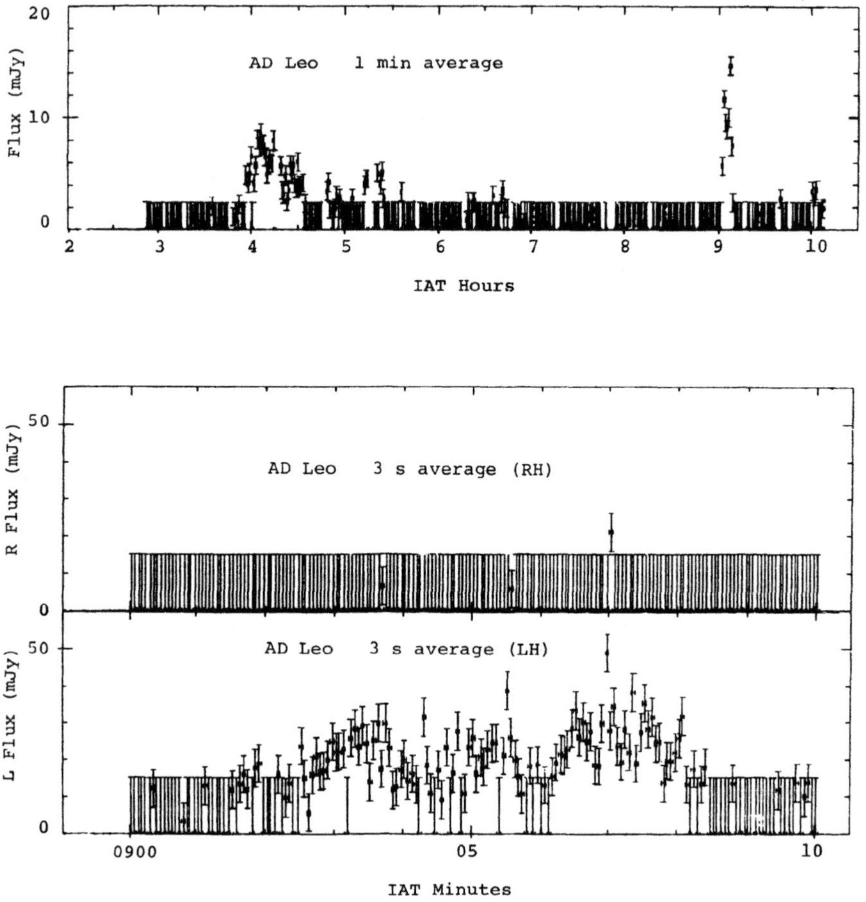

Figure 6.
Plots of the 6 cm flux profile of AD Leo, made using a discrete transform technique after "cleaning" the database. These plots would have been impossible to obtain without the cleaning technique described here. To obtain the plot in (b) would have required 362 maps to be made and cleaned.

DISCUSSION AFTER MELROSE'S TALK IN PART V

LINSKY: You stated that the quiescent and flaring radio emission from RS CVn systems is not a direct consequence of binarity, but only an indirect consequence of the rapid rotation of the active star induced by tidal forces of the companion star. I wish to disagree with your statement. While I cannot prove that the interaction of flux tubes from the two stars is the source of free energy (through field annihilation), energetic particles (produced by electric fields induced by reorganization of the fields), or the radio emission itself (synchrotron emission presumably), I can cite some compelling circumstantial evidence:

1) the component stars in RS CVn systems are separated by only a few stellar radii;
2) x-ray observations of AR Lac during eclipse (Walter, Gibson, and Basri, Ap.J., **267**, 665, 1983) indicate an extended corona about the K-star;
3) VLBI observations of UX Ari (see paper by Lestrade et al., this volume) indicate an extended component to the radio emission;
4) UV observations of UX Ari during a flare (Simon, Linsky, and Schiffer, Ap.J., **239**, 911, 1980) suggest flow of matter from one star to the other; and
5) scaling laws for the flux tubes (ibid.) imply flux tube lengths larger than a stellar radius.

In my opinion, these circumstantial arguments, particularly when you put them all together, provide a strong case for the interacting flux tube or interacting magnetosphere model.

MELROSE: Taking a broad view, there have been flares observed on both single and binary star systems. There is a danger in explaining flares in flare stars with such an electrodynamic coupling.

HOLMAN: I have three comments. I agree that free-free ebsorption is a big problem for plasma emission at the plasma frequency. At 1400 MHz, however, I found that it may be reasonable for emission at twice the plasma frequency.

Emission at twice the upper hybrid frequency should also be included as a possible emission mechanism.

I don't believe that anyone is saying that incoherent synchrotron emission must be self-absorbed. The fact that the source must self-absorb places constraints (in many cases very strong constraints) upon the source size and magnetic field strength. Existing calculations such as those by Ramaty (1969), give a good idea of the degree of polarization that is possible near the self-absorption frequency.

MELROSE: Commenting on your second point, I also would include upper-hybrid emission as another type of plasma emission.

With regard to your third point, I think you are trying to have your cake and eat it too! If you push the brightness temperature to 10^{10}°K, the observed value, then you must have more relativistic particles, which makes the absorption problem worse.

R.M. Hjellming and D.M. Gibson (eds.), Radio Stars, 391–396.

MUTEL: I want to emphasize that, whenever VLBI experiments have directly measured $T_B \sim 10^{10}$°K, the observed circular polarization has been quite low. This is positive evidence for $\tau \gtrsim 1$ in these sources.

KUIJPERS: Could you summarize the present status of the relative importance of the x-mode versus the z-mode in the electron-cyclotron masers?

MELROSE: When the plasma frequency is much greater than the cyclotron frequency, one expects the bulk of the radiation to be in the x-mode. When the two frequencies become comparable, the cyclotron maser will favor pumping energy into the z-mode. It can, in principle, get out by coupling the two together through the second harmonic. There will be a gradual variation in the drift as each mode comes out. We may be able to observe this in some other source such as a terrestrial source or possibly the Sun.

DISCUSSION AFTER MUTEL'S TALK IN PART V

DRAKE: A comment concerning VLBI observations. In order to get high resolution images of most stellar wind sources one needs to increase the VLBI sensitivity, perhaps by increasing the bandwidth.

KUNDU: With regard to QUASAT observations, I believe one needs 24-48 hrs of observation to obtain complete uv-coverage. You listed 50 mJy for the sensitivity in a 10 hour observation. Is it possible to map some stellar wind sources which are steady?

MUTEL: The general answer is no. It may be possible for objects like α Ori.

UNDERHILL: The optical disks of the apparently bright Be stars have sizes ~ 0.5 milliarcsec. These stars are suspected to be immersed in shells of 10^4 K plasma extending to 10-100 R_*. However, I think their radio fluxes are small. If you could obtain sensitivities down to 0.1 mJy, you might be able to resolve the structure around Be stars. It would be valuable to determine whether these structures are spherical or disk-like.

MUTEL: I think you will find it hard to detect sources that are fainter than a few mJy. We now have receivers operating close to the theoretical noise limit. I don't see any large increases in aperture in the plans for the next ten years.

LINSKY: I would like to request that a study be made of the feasibility and cost of increasing the bandwidth of the VLA from 100 MHz to 500 MHz or even 1 GHz. In the last several decades optical astronomers have been able to make their existing telescopes orders of magnitude more powerful in sensitivity and in other categories by attaching the newest focal plane instruments. In this way they have, in effect, obtained much larger apertures at relatively low cost. The VLA was designed with the technology of 10 years ago. While it is true that the receivers are being upgraded, the array is being used at lower frequencies, and the computing power is being increased, the bandwidth is the same as originally designed. If one can increase the signal-to-noise by another factor of 2 - 3 by increasing the bandwidth another factor of 4 - 9, then the possibility arises to observe a large number of sources that may be on the threshold of detectability. I have in mind the detection and imaging of many more wind sources (both hot and cold) and nonthermal stellar sources. A fundamental question is whether this project could be done at moderate cost or not.

MUTEL: Increasing the bandwidth to 1 GHz would only buy a factor of three in sensitivity, but would cost a factor of ten more than the present receivers. But that's the only real opportunity we have unless the VLA wants to install 15 K masers. The problem is also not quite as simple as you have stated it because one would also have to develop GHz wide correlators, and that is really at the cutting edge of today's technology.

ABBOTT: I agree with Jeff. Among the hot stars there is only one, γ^2 Vel, which is brighter than 10 mJy. What we need most of all is sensitivity.

MUTEL: The full VLBA (if it works right) will be able to map anything you can now map with the VLA.

BEIGING: There is probably another factor of 2 improvement to be gained in receiver sensitivity from a new type of transistor now being developed, namely the HEMT (high electron-motility transistor). Hence, the VLA may, in ~3-5 yrs, be able to detect and map a substantially larger number of stars of all types, when HEMT receivers can be installed on the whole array.

HJELLMING: Let me mention another instrument that may be available in the distant future. During the last couple of years both the millimeter astronomy community and the NRAO have been getting organized to support and build an array for aperture synthesis at millimeter wavelengths. The part of this array that would be most useful for stellar problems is currently conceived of as 21 antennas, each with 10 meter diameter, 1 GHz bandwidth, and 100 K system temperatures operating at all useful wavelengths above 1 mm. This array would operate in configurations ranging from 90 m to 2000 m on an ordinary basis, but would have the potential of operating in even larger configurations (like up to 35 km) using VLBI techniques. The utility of such an instrument for mapping stellar winds, bipolar flows, and other types of mass outflow objects is obvious. If things go well, NRAO may submit a proposal to build such an instrument in late 1985 to 1986, with completion planned for the 1990's.

MUTEL: I mentioned this in my talk but let me repeat that if you had an ideal 1 mm array, you could actually map stellar disks. I find this exciting. For example, Sirius would have a flux of 16 mJy. It would be resolvable with relatively modest baselines (~ 100 km). You could do Capella but not RS CVn binaries or M-dwarfs. The situation would be best for M-supergiants like Betelguese where you would have 32 mJy/beam.

FLORKOWSKI: With regard to Bob Mutel's point on astrometry with VLBA, the large improvement in the precision of optical positions from the use of Hipparchus and the Space Telescope, besides relating the FK4/FK5 to the quasar reference frame, will also provide values for the offsets between the centroids of optical and radio emission. This may help determine the location of radio source sites. How does the binary nature influence the location of radio emission, e.g. as in RS CVn binaries and Algols? Optical space observations may help us with this question.

GIBSON: I'd like to mention another "array" which will be important for stellar work. The 327 MHz system at the VLA should be complete in about two years. It will provide relatively high sensitivity at a frequency where little reliable data has been obtained previously.

HJELLMING: That's right. There are presently three antennas with 327 MHz receivers. The plan is to continue installing them at a rate of about one every 2-3 months. By January 1985 there will be five operational antennas at 327 MHz. The entire array may be operational at 327 MHz by the end of 1986.

Continuing in this line, three VLA antennas are planned to be equipped with 75 MHz capabilities. Initially one needs to test ionospheric stability and the interference environment. If it pans out, there may eventually be a 75 MHz capability at every VLA antenna.

DISCUSSION AFTER HURFORD'S TALK IN PART V

GARY: I would like to point out that radio spectra are only relevant when the source size is the same at all frequencies. When source size varies with frequency, thermal emission may appear nonthermal, etc. High spectral resolution instruments of the future are essential for distinguishing these possibilities.

MULLAN: If dynamic spectra could be obtained for stellar work, what frequency resolution would be useful? To answer that, I'd like to ask a historical question. What spectral resolution did Paul Wild have when he first obtained the dynamic spectra of solar bursts that allowed him to arrive at the useful classification of bursts into types I - V? If that classification could be repeated for stellar flares, it would be a major improvement in the study of stellar flares.

KUNDU: I think Wild did his work between 40 and 240 MHz. The only working instrument that could be used at these frequencies is the Clark Lake array. However, it currently works only at one frequency at a time and with a relatively small bandwidth. It does have the advantage of high time resolution - a fraction of a millisecond. We can change frequencies but the flare must be large enough so that we can detect it and vary slowly enough so that the spectra don't change significantly over a cycle of observations.

DISCUSSION RELATED TO PART V

KUNDU: We have come to the end of formal panel presentations and discussions. In the remaining time we may wish to discuss issues like VLA observing time, coordinated campaigns, etc. Does anyone have any comments on this?

LINSKY: I have organized or participated in a number of coordinated observations of dMe flare stars and RS CVn binary stars during the last several years involving x-ray, UV, optical, and radio observations. In general, we have been quite successful in acquiring observing time and in detecting the transient events that were goals of the programs. However, it has been difficult to acquire the time, and in the Space Telescope era it is likely become more difficult.

The major difficulty is that each instrument has a different set of rules, different deadlines for proposals, and different observing constraints. Also, observing time allocation committees are often reluctant to grant time, for a program that is truly excellent science only with the coordinated observations, prior to the approval of time by the other observatories. This is a Catch-22 situation. To get out of this impossible problem, I urge that those who allocate observing time on major facilities (VLA, IUE, KPNO, CTIO, EXOSAT, Space Telescope, etc) recognize that coordinated observing should be a modest portion of the observing time, provided the proposals are worthy. To accomodate such observations, they should retain some scheduling flexibility, perhaps as Director's discretionary time, and communicate with schedulers at other observatories to coordinate time allocations. Also, those who propose such programs should help by adding to their proposals a statement that time has been requested at other observatories, and include the name and telephone numbers of the contact persons at the other observatories - as well as their deadlines.

WADE: Coordinated stellar observing should probably be approached by an organizational structure similar to the VLBI network, whereby individual observatories turn over part of their scheduling to an outside committee. However, occasional quick response to unexpected events must be allowed for.

UNDERHILL: Coordinated observing at several wavelengths is a moderately new procedure. What is necessary is to develop a new philosophy of doing science which recognizes that the value of the research is not solely a product of the sum of several observatories, but is a positive power-law function. The persons who act on boards to select observing programs are usually concerned with maximizing the outcome of the use of the instrument. A useful first step might be to broaden the philosophy of how one does science to emphasize the power-law effect in putting together simultaneous observations in several wavelengths regions. Observing boards are sensitive to the philosophy of doing science and how to maximize outcome.

KUNDU: Ron Ekers has told me that the VLA ends up using about 25% of its time for stellar observing and 4% for solar observing. By those standards we seem to be doing rather well.

HJELLMING: That has been the approximate figure for VLA stellar proposals for the last few years. I have heard Ron Ekers grumble mildly about the amount of VLA time that goes to radio stars, with the comment that "they would use all of it if we let them"; however, it is very friendly grumbling because the priority is to schedule interesting science, and a considerable fraction of the proposals to observe stars meet that criterion.

I would like to emphasize that the primary reason for the high rate of success is that the proposals have been predominately interesting ones that referees and schedulers find to be well worth scheduling. I believe that the current system is much better for getting VLA observing time than any attempt to argue for a "quota" for stellar radio astronomy.

MOLNAR: Some flexibility in scheduling coordinated observations does currently exist, judging from my experience with a recent campaign on Cyg X-3. On the negative side, the idea of a network analogous to the VLBI network does not seem promising because the needs and the facilities are much more varied. If the situation becomes unsatisfactory, a solution more radical than a loose network may become necessary.

LINSKY: I think the needs are most uniform in the case of the spotted flaring stars where observations of quiescent structures and flares require the full range of the spectrum and the best sensitivities available. In particular, the observation of time delays between the onset of flaring in the different bands is critical to determining the flare process.

SUMMARY OF THE CURRENT AND FUTURE PROBLEMS IN RADIO STARS

M. R. Kundu
Astronomy Program, University of Maryland
College Park, MD 20742 U.S.A.

R. M. Hjellming
National Radio Astronomy Observatory*
Socorro, NM 87801 U.S.A.

ABSTRACT. We summarize the current state and future problems of research on stars that exhibit continuum radio emission. This summary is based upon comments presented by the authors during the meeting together with material suggested by the papers and discussions.

1. KNOWN PHENOMENA AND PROBLEMS

In the absence of flares, there are two components of radio emission in the Sun: the quiet Sun component of free-free emission from the parts of the atmosphere in which there are no active regions (sunspots, plages, etc.), and the slowly varying component of radio emission from active regions. The slowly varying component is due to either free-free emission from high-density regions or thermal gyroresonant emission caused by magnetic fields in regions associated with sunspots. In other stars the quiescent stellar emission, which would be a counterpart of the quiet Sun component, probably exists but has not been identified conclusively. Long-lasting variable quiescent emission, probably associated with starspots, has been identified in late-type dwarf stars; it has been interpreted in terms of either thermal gyroresonant emission or nonthermal gyrosynchrotron radiation from energetic electrons having a power-law energy distribution.

In the Sun, flares last from minutes to hours. There are several ways in which flare energy is released: 1) simple loops contain tearing-mode instabilities that are the causes of the flares; 2) two oppositely polarized interacting flux tubes produce a current sheet and trigger flares; and 3) large-scale current sheets between open magnetic-field lines, which are believed to exist at the base of "helmet" streamers, interact to produce flaring.

Flares in stellar systems, such as RS CVn binaries, last from hours

* The National Radio Astronomy Observatory is operated by Associated Universities, Inc., under contract with the National Science Foundation.

R. M. Hjellming and D. M. Gibson (eds.), Radio Stars, 397–402.

to weeks. One of the ideas discussed extensively during this workshop is that flares in RS CVn stars are powered by magnetic-field annihilation of interacting flux tubes that are analogous to those in solar flares but are much larger in size scale because the flux tubes of two separate stars are responsible. The long time scale of RS CVn flares is possibly a consequence of the large size scales. High degrees of circular polarization, found in microwave flare emission (e.g. in HR 1099) indicate that the emission process is magnetic in character. Uchida and Sakurai (1983) proposed that magnetic flux tubes of active regions on both stars will interact and thus interconnect the two stars. Coronal heating and flares could result from magnetic reconnection as individual starspots move to locations such that pairs of active regions can interact across the distance between the stars. The low-temperature plasma is interpreted as confined to small loops, whereas high-temperature plasma is confined to large loops interconnecting the two stars. Magnetic coupling between the two stars in a binary system is one of the most poorly understood problems in stellar radio physics (Melrose, 1985). In addition to the approach of Uchida and Sakurai, there have been discussions of two other approaches. One idea is that, as the stars rotate and move around their orbits, the system finds a magnetic configuration in which dissipation in a magnetic region allows the build-up in magnetic stresses to be continuously relaxed. The second idea is based upon an assumed analogy with the Io-Jupiter system in that one star is regarded as a conductor moving through the magnetic field of the other star. At the present time no detailed models based upon these concepts have been worked out.

Fast millisecond spikes in microwave bursts are commonly seen in the Sun. These spike bursts, with brightness temperatures $\sim 10^{14}$–10^{16} K, have been interpreted in terms of the electron-cyclotron-maser process. Similar bursts on time scales of ~200 milliseconds have been observed in the flare star AD Leo, and possibly also in the RS CVn binary HR 1099. The 200-millisecond bursts have high brightness temperatures, and electron-cyclotron masering has been suggested as their emission mechanism, analogous to solar spike bursts. Observations at high spatial resolution are necessary to determine whether coherent processes are responsible for some components of stellar-flare radio emission.

The observed radio emission from extended chromospheres, winds from late- and early-type stars, nova shells, and interacting winds has been interpreted as free-free emission, allowing determination of mass-loss rates, radial dependence of electron density and temperature, and fractional ionization. These determinations have been based typically on wind models with spherical symmetry and constant velocity. Direct spatial resolution of such objects, made with the VLA in a few cases, but with VLBI (and the new VLBA in particular) in most cases, will provide the following information: 1) the angular size of stellar winds at several frequencies, permitting more straightforward computation of mass-loss rates, 2) the two-dimensional structure, checking the validity of assumptions about symmetry, 3) the effective angular size and hence the actual brightness temperature, allowing us to distinguish between thermal and nonthermal radiation processes, and 4) the temporal variation of angular size, and therefore of expansion velocity, ionization fraction, and morphology.

The bipolar and other types of outflows observed in star-forming regions are phenomena for which multi-frequency mapping will be critical. The VLA, VLBA, and proposed Millimeter Array will be major tools for such objects.

Much information has been obtained, using the VLA, on long-lasting quiescent variable emission from M-type dwarf stars at centimeter wavelengths. However, very few real flares have been observed with the VLA, although in the past non-imaging observations have reported many flares at meter wavelengths. On the other hand, observations made over many hundreds of hours with imaging instruments such as the Clark Lake Teepee Tee array at ~50 MHz, and with the Ooty aperture-synthesis radio telescope at 327 MHz have reported no flares above 1 and 0.1 Jy, respectively, for known flare stars such as UV Ceti, YZ CMi, and AD Leo. It is important to establish the reality of so-called meter-wavelength radio emission from flare stars.

The nearby RS CVn stars appear to be suitable for high spatial-resolution studies with the existing VLBI systems at levels of ~10 mJy, and with the future VLBA at levels of ~0.1 mJy. Such observations will provide information on the characteristic size and shape of radio-emission regions during quiescent and active periods. Even more important will be the information they provide on the mode of energy release during flarings. These measurements will lead to direct determination of the brightness temperatures in burst components, thus permitting us to discover whether coherent-emission processes are involved in stellar flares.

In identifying the radio-emission mechanisms of stars, one of the important parameters is the brightness temperature T_B. It permits us to estimate whether the source is thermal ($T_B \leqq 10^6$ K for stellar coronae and winds), incoherent nonthermal ($kT_B \leqq \varepsilon$, where k = Boltzmann's constant, ε = energy of radiating particles), or coherent nonthermal ($kT_B >> \varepsilon$). The radio emission from stellar winds is usually attributed to thermal emission, although some of the results discussed in this workshop raise doubts in particular cases. The emission from most flare stars has $T_B \leqq 10^{10}$ K, so it is usually interpreted as incoherent gyrosynchrotron emission from electrons with $\varepsilon \leqq 1$ MeV. However, the emission with $T_B \approx 10^{10}$ K also can be coherent rather than incoherent. Polarization data can distinguish between these two possibilities. A gyrosynchrotron source with $T_B \approx 10^{10}$ K would be optically thick and therefore weakly polarized, whereas a cyclotron-maser source would be strongly polarized. The brightest emission ($T_B >> 10^{10}$ K) from some M-dwarf flare stars requires a coherent-emission process.

Future problems concerning stellar winds involve necessarily greater detail in both observations and theory, particularly when comparing the observed spectra with predicted spectra. Drawing from our experience with solar- flare radio emission, we understand the gyrosynchrotron mechanism reasonably well. The interpretation of nonthermal stellar radio emission is concerned, at the present time, with detailed models to explain the magnetic structure and particle spectra. Among the pertinent questions in this process are 1) the methods by which the electrons are accelerated and 2) the source of free energy. For some systems, like the RS CVn and AM Her binaries, it is plausible that the ultimate source of free energy arises from some kind of magnetic coupling between two stars. As mentioned earlier, this coupling is not well understood.

Coherent radio-emission mechanisms, such as the electron-cyclotron-maser process, have been applied to the high-brightness-temperature spike bursts observed in the Sun. This process has been invoked to explain the 200-millisecond fine-structure spikes in the microwave flare emission from AD Leo found by Lang et al. (1983). It has been argued by Holman et al. (1985) that the underlying smooth component of this stellar flare emission could also be due to a coherent process, the electron-cyclotron maser, unlike the gradual component of solar microwave spike bursts that is attributed to incoherent gyrosynchrotron emission. One should note that, at the moment, the only convincing case of stellar maser emission is the AD Leo flare observed by Lang et al. (1983) at 1400 MHz with the Arecibo telescope, using a time resolution of 200 milliseconds. The importance of this observation lies in their detection of a high degree of circular polarization together with an inferred (from rise-time arguments) brightness temperature in excess of 10^{13} K. This provides the first indication that coherent processes may be responsible for the radio emission from some stellar flares. It is possible that, in this particular case, the underlying smooth component could consist of many mini-spikes if it were observed with a time resolution of ~ 1 millisecond.

It is obvious that one must use time resolution of the order of 1 millisec to properly understand the time scale of the spike emission from stellar flares. At present this is not possible with the normal mode of operation of either the VLA or the Arecibo telescope. It is possible for the time resolution at Arecibo to be as small as ~ 50 millisec; however, the current VLA can normally provide only 3.3-second time resolution. Under discussion at NRAO is a proposal that may permit high time-resolution observations using the VLA as a phased array with up to 20-μsec-time resolution. Use of the VLA in this mode of operation is very important for both solar and stellar observations because of the unique capability to make good aperture-synthesis maps with high time-resolution.

2. SUMMARY OF IMPORTANT OBSERVATIONS

We conclude that the following measurements will be very important for understanding the basic processes involved in stellar radio emission.

2.1 Multi-wavelength Observations

Multi-wavelength observations of late-type dwarf stars should be carried out at centimeter, meter, and decameter wavelengths to constrain the existing models of quiescent and slowly varying stellar emission based upon nonthermal gyrosynchrotron radiation. This will permit us to estimate the radio-source size, temperature, density, magnetic field, and electron-energy distribution. The VLA appears to be an ideal instrument for this purpose, particularly with the current implementation of 327-MHz capabilities and the possibility of adding a 75-MHz capability. Further attempts should be made to detect radio flares from the dMe flare stars at meter-decameter wavelengths with imaging instruments such as the Clark Lake Teepee Tee (300-100 MHz) and the Ooty aperture-synthesis radio telescope (327 MHz). Failure to detect any meter-wavelength flares from the well known flare stars, such as UV Ceti and YZ CMi, raises fundamental questions.

The importance of coordinated observations of flares at several observatories must be emphasized. A larger range in frequencies and higher time resolution can be more easily obtained with single-dish telescopes, but these must be coordinated with an imaging system similar to the VLA in order to obtain confidence in the single-dish measurements and to obtain simultaneous measurements of the centimeter-meter radio spectrum.

2.2 Millimeter-Wavelength Observations of Stars

The available continuum observations of some cool stars (α Ori, Newell and Hjellming 1982; α Sco, Hjellming and Newell 1983; α Her, Drake and Linsky 1983) indicate that high-frequency continuum observations will be critical in estimating the physical properties of the photospheres and chromospheres of such stars.

High-resolution imaging of stellar winds and other outflow phenomena such as bipolar outflows are greatly needed at millimeter wavelengths because of the predominantly thermal nature of these phenomena.

The Millimeter Array, currently being discussed as an instrument to be built by NRAO, would be very important in these and other areas. This workshop has not discussed radio spectral-line observations of stars, but they are important for cool stars and will be among the major areas of research for a high-resolution spectral-line imaging array at millimeter wavelengths.

2.3 VLBA Observations of Stars

As discussed in detail by Mutel (1985), high spatial-resolution observations with current VLBI networks and the VLBA represent a fundamental frontier for radio-star observations. Resolution at the milliarcsec level will be very important in studying the regions of energy release in close binary systems, such as RS CVn systems, and consequently the nature of the magnetic coupling between stars in binary systems. The VLBA will be a nearly ideal instrument for this purpose, permitting us to determine the core-halo structure during RS CVn flares, and how this structure changes with time. Imaging information coupled with high time resolution will be a very important capability of the VLBA, but exploratory observations should continue with the currently available VLBI networks.

2.4 High Time-Resolution Observations of Stars

As discussed by Bookbinder (1985), high time-resolution observations (~1-10 millisec) will be necessary to understand coherent radio emission from stellar flares in both dMe flare stars and active binaries. This can be accomplished with the Arecibo telescope, as demonstrated by Lang et al. (1983), but it is very important that this capability be implemented for a phased-array mode of the VLA.

3. THEORETICAL PROBLEMS

The lessons learned from solar phenomena emphasize the complexity of the interpretation of stellar radio emission. As discussed by Kuijpers, Mullan, and Melrose during their talks in this workshop, there is a wide range of complex theoretical possibilities involved in stellar radio physics. Improved theoretical insight is very important if we are to improve our understanding of stellar radio-source phenomena.

REFERENCES

Bookbinder, J. 1985, Radio Stars (Reidel:Dordrecht), ed. R.M. Hjellming and D.M. Gibson, 371.

Drake, S.D. and Linsky, J.L. 1983, Ap.J.(Letters), **274**, L77.

Hjellming, R.M. and Newell, R.T. 1983, Ap.J., **275**. 704.

Holman, G., Bookbinder, J., and Golub, L. 1985, Radio Stars (Reidel:Dordrecht), ed. R.M. Hjellming and D.M. Gibson, 35.

Lang, K., Bookbinder, J., Golub, L, and Davis, M. 1983, Ap.J.(Letters), **272**, L15.

Melrose, D.B. 1985, Radio Stars (Reidel:Dordrecht), ed. R.M. Hjellming and D.M. Gibson, 351.

Mutel, R. L. 1985, Radio Stars (Reidel:Dordrecht), ed. R.M. Hjellming and D.M. Gibson, 359.

Newell, R.T. and Hjellming, R.M. 1982, Ap.J.(Letters), **263**, L85.

Uchida, Y. and Sakurai, T. 1983, Activity in Red Dwarf Stars (Reidel:Dordrecht), ed. P.B. Byrne and M. Rodono, 629.

OBJECT INDEX

SUBJECT INDEX

ASTROPHYSICS AND SPACE SCIENCE LIBRARY

1. C. de Jager (ed.), *The Solar Spectrum, Proceedings of the Symposium held at the University of Utrecht, 26–31 August, 1963.* 1965, XIV + 417 pp.
2. J. Orthner and H. Maseland (eds.), *Introduction to Solar Terrestrial Relations, Proceedings of the Summer School in Space Physics held in Alpbach, Austria, July 15–August 10, 1963 and Organized by the European Preparatory Commission for Space Research.* 1965, IX + 506 pp.
3. C. C. Chang and S. S. Huang (eds.), *Proceedings of the Plasma Space Science Symposium, held at the Catholic University of America, Washington, D.C., June 11–14, 1963.* 1965, IX + 377 pp.
4. Zdeněk Kopal, *An Introduction to the Study of the Moon.* 1966, XII + 464 pp.
5. B. M. McCormac (ed.), *Radiation Trapped in the Earth's Magnetic Field. Proceedings of the Advanced Study Institude, held at the Chr. Michelsen Institute, Bergen, Norway, August 16–September 3, 1965.* 1966, XII + 901 pp.
6. A. B. Underhill, *The Early Type Stars.* 1966, XII + 282 pp.
7. Jean Kovalevsky, *Introduction to Celestial Mechanics.* 1967, VIII + 427 pp.
8. Zdeněk Kopal and Constantine L. Goudas (eds.), *Measure of the Moon. Proceedings of the 2nd International Conference on Selenodesy and Lunar Topography, held in the University of Manchester, England, May 30–June 4, 1966.* 1967, XVIII + 479 pp.
9. J. G. Emming (ed.), *Electromagnetic Radiation in Space. Proceedings of the 3rd ESRO Summer School in Space Physics, held in Alpbach, Austria, from 19 July to 13 August, 1965.* 1968, VIII + 307 pp.
10. R. L. Carovillano, John F. McClay, and Henry R. Radoski (eds.), *Physics of the Magnetosphere, Based upon the Proceedings of the Conference held at Boston College, June 19–28, 1967.* 1968, X + 686 pp.
11. Syun-Ichi Akasofu, *Polar and Magnetospheric Substorms.* 1968, XVIII + 280 pp.
12. Peter M. Millman (ed.), *Meteorite Research. Proceedings of a Symposium on Meteorite Research, held in Vienna, Austria, 7–13 August, 1968.* 1969, XV + 941 pp.
13. Margherita Hack (ed.), *Mass Loss from Stars. Proceedings of the 2nd Trieste Colloquium on Astrophysics, 12–17 September, 1968.* 1969, XII + 345 pp.
14. N. D'Angelo (ed.), *Low-Frequency Waves and Irregularities in the Ionosphere. Proceedings of the 2nd ESRIN-ESLAB Symposium, held in Frascati, Italy, 23–27 September, 1968.* 1969, VII + 218 pp.
15. G. A. Partel (ed.), *Space Engineering. Proceedings of the 2nd International Conference on Space Engineering, held at the Fondazione Giorgio Cini, Isola di San Giorgio, Venice, Italy, May 7–10, 1969.* 1970, XI + 728 pp.
16. S. Fred Singer (ed.), *Manned Laboratories in Space. Second International Orbital Laboratory Symposium.* 1969, XIII + 133 pp.
17. B. M. McCormac (ed.), *Particles and Fields in the Magnetosphere. Symposium Organized by the Summer Advanced Study Institute, held at the University of California, Santa Barbara, Calif., August 4–15, 1969.* 1970, XI + 450 pp.
18. Jean-Claude Pecker, *Experimental Astronomy.* 1970, X + 105 pp.
19. V. Manno and D. E. Page (eds.), *Intercorrelated Satellite Observations related to Solar Events. Proceedings of the 3rd ESLAB/ESRIN Symposium held in Noordwijk, The Netherlands, September 16–19, 1969.* 1970, XVI + 627 pp.
20. L. Mansinha, D. E. Smylie, and A. E. Beck, *Earthquake Displacement Fields and the Rotation of the Earth, A NATO Advanced Study Institute Conference Organized by the Department of Geophysics, University of Western Ontario, London, Canada, June 22–28, 1969.* 1970, XI + 308 pp.
21. Jean-Claude Pecker, *Space Observatories.* 1970, XI + 120 pp.
22. L. N. Mavridis (ed.), *Structure and Evolution of the Galaxy. Proceedings of the NATO Advanced Study Institute, held in Athens, September 8–19, 1969.* 1971, VII + 312 pp.

23. A. Muller (ed.), *The Magellanic Clouds. A European Southern Observatory Presentation: Principal Prospects, Current Observational and Theoretical Approaches, and Prospects for Future Research, Based on the Symposium on the Magellanic Clouds, held in Santiago de Chile, March 1969, on the Occasion of the Dedication of the European Southern Observatory.* 1971, XII + 189 pp.
24. B. M. McCormac (ed.), *The Radiating Atmosphere. Proceedings of a Symposium Organized by the Summer Advanced Study Institute, held at Queen's University, Kingston, Ontario, August 3–14, 1970.* 1971, XI + 455 pp.
25. G. Fiocco (ed.), *Mesospheric Models and Related Experiments. Proceedings of the 4th ESRIN-ESLAB Symposium, held at Frascati, Italy, July 6–10, 1970.* 1971, VIII + 298 pp.
26. I. Atanasijević, *Selected Exercises in Galactic Astronomy.* 1971, XII + 144 pp.
27. C. J. Macris (ed.), *Physics of the Solar Corona. Proceedings of the NATO Advanced Study Institute on Physics of the Solar Corona, held at Cavouri-Vouliagmeni, Athens, Greece, 6–17 September 1970.* 1971, XII + 345 pp.
28. F. Delobeau, *The Environment of the Earth.* 1971, IX + 113 pp.
29. E. R. Dyer (general ed.), *Solar-Terrestrial Physics/1970. Proceedings of the International Symposium on Solar-Terrestrial Physics, held in Leningrad, U.S.S.R., 12–19 May 1970.* 1972, VIII + 938 pp.
30. V. Manno and J. Ring (eds.), *Infrared Detection Techniques for Space Research. Proceedings of the 5th ESLAB-ESRIN Symposium, held in Noordwijk, The Netherlands, June 8–11, 1971.* 1972, XII + 344 pp.
31. M. Lecar (ed.), *Gravitational N-Body Problem. Proceedings of IAU Colloquium No. 10, held in Cambridge, England, August 12–15, 1970.* 1972, XI + 441 pp.
32. B. M. McCormac (ed.), *Earth's Magnetospheric Processes. Proceedings of a Symposium Organized by the Summer Advanced Study Institute and Ninth ESRO Summer School, held in Cortina, Italy, August 30–September 10, 1971.* 1972, VIII + 417 pp.
33. Antonin Rükl, *Maps of Lunar Hemispheres.* 1972, V + 24 pp.
34. V. Kourganoff, *Introduction to the Physics of Stellar Interiors.* 1973, XI + 115 pp.
35. B. M. McCormac (ed.), *Physics and Chemistry of Upper Atmospheres. Proceedings of a Symposium Organized by the Summer Advanced Study Institute, held at the University of Orléans, France, July 31–August 11, 1972.* 1973, VIII + 389 pp.
36. J. D. Fernie (ed.), *Variable Stars in Globular Clusters and in Related Systems. Proceedings of the IAU Colloquium No. 21, held at the University of Toronto, Toronto, Canada, August 29–31, 1972.* 1973, IX + 234 pp.
37. R. J. L. Grard (ed.), *Photon and Particle Interaction with Surfaces in Space. Proceedings of the 6th ESLAB Symposium, held at Noordwijk, The Netherlands, 26–29 September, 1972.* 1973, XV + 577 pp.
38. Werner Israel (ed.), *Relativity, Astrophysics and Cosmology. Proceedings of the Summer School, held 14–26 August 1972, at the BANFF Centre, BANFF, Alberta, Canada.* 1973, IX + 323 pp.
39. B. D. Tapley and V. Szebehely (eds.), *Recent Advances in Dynamical Astronomy. Proceedings of the NATO Advanced Study Institute in Dynamical Astronomy, held in Cortina d'Ampezzo, Italy, August 9–12, 1972.* 1973, XIII + 468 pp.
40. A. G. W. Cameron (ed.), *Cosmochemistry. Proceedings of the Symposium on Cosmochemistry, held at the Smithsonian Astrophysical Observatory, Cambridge, Mass., August 14–16, 1972.* 1973, X + 173 pp.
41. M. Golay, *Introduction to Astronomical Photometry.* 1974, IX + 364 pp.
42. D. E. Page (ed.), *Correlated Interplanetary and Magnetospheric Observations. Proceedings of the 7th ESLAB Symposium, held at Saulgau, W. Germany, 22–25 May, 1973.* 1974, XIV + 662 pp.
43. Riccardo Giacconi and Herbert Gursky (eds.), *X-Ray Astronomy.* 1974, X + 450 pp.
44. B. M. McCormac (ed.), *Magnetospheric Physics. Proceedings of the Advanced Summer Institute, held in Sheffield, U.K., August 1973.* 1974, VII + 399 pp.
45. C. B. Cosmovici (ed.), *Supernovae and Supernova Remnants. Proceedings of the International Conference on Supernovae, held in Lecce, Italy, May 7–11, 1973.* 1974, XVII + 387 pp.
46. A. P. Mitra, *Ionospheric Effects of Solar Flares.* 1974, XI + 294 pp.
47. S.-I. Akasofu, *Physics of Magnetospheric Substorms.* 1977, XVIII + 599 pp.

48. H. Gursky and R. Ruffini (eds.), *Neutron Stars, Black Holes and Binary X-Ray Sources.* 1975, XII + 441 pp.
49. Z. Švestka and P. Simon (eds.), *Catalog of Solar Particle Events 1955–1969. Prepared under the Auspices of Working Group 2 of the Inter-Union Commission on Solar-Terrestrial Physics.* 1975, IX + 428 pp.
50. Zdeněk Kopal and Robert W. Carder, *Mapping of the Moon.* 1974, VIII + 237 pp.
51. B. M. McCormac (ed.), *Atmospheres of Earth and the Planets. Proceedings of the Summer Advanced Study Institute, held at the University of Liège, Belgium, July 29–August 8, 1974.* 1975, VII + 454 pp.
52. V. Formisano (ed.), *The Magnetospheres of the Earth and Jupiter. Proceedings of the Neil Brice Memorial Symposium, held in Frascati, May 28–June 1, 1974.* 1975, XI + 485 pp.
53. R. Grant Athay, *The Solar Chromosphere and Corona: Quiet Sun.* 1976, XI + 504 pp.
54. C. de Jager and H. Nieuwenhuijzen (eds.), *Image Processing Techniques in Astronomy. Proceedings of a Conference, held in Utrecht on March 25–27, 1975.* 1976, XI + 418 pp.
55. N. C. Wickramasinghe and D. J. Morgan (eds.), *Solid State Astrophysics. Proceedings of a Symposium, held at the University College, Cardiff, Wales, 9–12 July, 1974.* 1976, XII + 314 pp.
56. John Meaburn, *Detection and Spectrometry of Faint Light.* 1976, IX + 270 pp.
57. K. Knott and B. Battrick (eds.), *The Scientific Satellite Programme during the International Magnetospheric Study. Proceedings of the 10th ESLAB Symposium, held at Vienna, Austria, 10–13 June 1975.* 1976, XV + 464 pp.
58. B. M. McCormac (ed.), *Magnetospheric Particles and Fields. Proceedings of the Summer Advanced Study School, held in Graz, Austria, August 4–15, 1975.* 1976, VII + 331 pp.
59. B. S. P. Shen and M. Merker (eds.), *Spallation Nuclear Reactions and Their Applications.* 1976, VIII + 235 pp.
60. Walter S. Fitch (ed.), *Multiple Periodic Variable Stars. Proceedings of the International Astronomical Union Colloquium No. 29, held at Budapest, Hungary, 1–5 September 1976.* 1976, XIV + 348 pp.
61. J. J. Burger, A. Pedersen, and B. Battrick (eds.), *Atmospheric Physics from Spacelab. Proceedings of the 11th ESLAB Symposium, Organized by the Space Science Department of the European Space Agency, held at Frascati, Italy, 11–14 May 1976.* 1976, XX + 409 pp.
62. J. Derral Mulholland (ed.), *Scientific Applications of Lunar Laser Ranging. Proceedings of a Symposium held in Austin, Tex., U.S.A., 8–10 June, 1976.* 1977, XVII + 302 pp.
63. Giovanni G. Fazio (ed.), *Infrared and Submillimeter Astronomy. Proceedings of a Symposium held in Philadelphia, Penn., U.S.A., 8–10 June, 1976.* 1977, X + 226 pp.
64. C. Jaschek and G. A. Wilkins (eds.), *Compilation, Critical Evaluation and Distribution of Stellar Data. Proceedings of the International Astronomical Union Colloquium No. 35, held at Strasbourg, France, 19–21 August, 1976.* 1977, XIV + 316 pp.
65. M. Friedjung (ed.), *Novae and Related Stars. Proceedings of an International Conference held by the Institut d'Astrophysique, Paris, France, 7–9 September, 1976.* 1977, XIV + 228 pp.
66. David N. Schramm (ed.), *Supernovae. Proceedings of a Special IAU-Session on Supernovae held in Grenoble, France, 1 September, 1976.* 1977, X + 192 pp.
67. Jean Audouze (ed.), *CNO Isotopes in Astrophysics. Proceedings of a Special IAU Session held in Grenoble, France, 30 August, 1976.* 1977, XIII + 195 pp.
68. Z. Kopal, *Dynamics of Close Binary Systems.* XIII + 510 pp.
69. A. Bruzek and C. J. Durrant (eds.), *Illustrated Glossary for Solar and Solar-Terrestrial Physics.* 1977, XVIII + 204 pp.
70. H. van Woerden (ed.), *Topics in Interstellar Matter.* 1977, VIII + 295 pp.
71. M. A. Shea, D. F. Smart, and T. S. Wu (eds.), *Study of Travelling Interplanetary Phenomena.* 1977, XII + 439 pp.
72. V. Szebehely (ed.), *Dynamics of Planets and Satellites and Theories of Their Motion. Proceedings of IAU Colloquium No. 41, held in Cambridge, England, 17–19 August 1976.* 1978, XII + 375 pp.
73. James R. Wertz (ed.), *Spacecraft Attitude Determination and Control.* 1978, XVI + 858 pp.

74. Peter J. Palmadesso and K. Papadopoulos (eds.), *Wave Instabilities in Space Plasmas. Proceedings of a Symposium Organized Within the XIX URSI General Assembly held in Helsinki, Finland, July 31–August 8, 1978.* 1979, VII + 309 pp.
75. Bengt E. Westerlund (ed.), *Stars and Star Systems. Proceedings of the Fourth European Regional Meeting in Astronomy held in Uppsala, Sweden, 7–12 August, 1978.* 1979, XVIII + 264 pp.
76. Cornelis van Schooneveld (ed.), *Image Formation from Coherence Functions in Astronomy. Proceedings of IAU Colloquium No. 49 on the Formation of Images from Spatial Coherence Functions in Astronomy, held at Groningen, The Netherlands, 10–12 August 1978.* 1979, XII + 338 pp.
77. Zdeněk Kopal, *Language of the Stars. A Discourse on the Theory of the Light Changes of Eclipsing Variables.* 1979, VIII + 280 pp.
78. S.-I. Akasofu (ed.), *Dynamics of the Magnetosphere. Proceedings of the A.G.U. Chapman Conference 'Magnetospheric Substorms and Related Plasma Processes' held at Los Alamos Scientific Laboratory, N.M., U.S.A., October 9–13, 1978.* 1980, XII + 658 pp.
79. Paul S. Wesson, *Gravity, Particles, and Astrophysics. A Review of Modern Theories of Gravity and G-variability, and their Relation to Elementary Particle Physics and Astrophysics.* 1980, VIII + 188 pp.
80. Peter A. Shaver (ed.), *Radio Recombination Lines. Proceedings of a Workshop held in Ottawa, Ontario, Canada, August 24–25, 1979.* 1980, X + 284 pp.
81. Pier Luigi Bernacca and Remo Ruffini (eds.), *Astrophysics from Spacelab.* 1980, XI + 664 pp.
82. Hannes Alfvén, *Cosmic Plasma,* 1981, X + 160 pp.
83. Michael D. Papagiannis (ed.), *Strategies for the Search for Life in the Universe,* 1980, XVI + 254 pp.
84. H. Kikuchi (ed.), *Relation between Laboratory and Space Plasmas,* 1981, XII + 386 pp.
85. Peter van der Kamp, *Stellar Paths,* 1981, XXII + 155 pp.
86. E. M. Gaposchkin and B. Kołaczek (eds.), *Reference Coordinate Systems for Earth Dynamics,* 1981, XIV + 396 pp.
87. R. Giacconi (ed.), *X-Ray Astronomy with the Einstein Satellite. Proceedings of the High Energy Astrophysics Division of the American Astronomical Society Meeting on X-Ray Astronomy held at the Harvard-Smithsonian Center for Astrophysics, Cambridge, Mass., U.S.A., January 28–30, 1980.* 1981, VII + 330 pp.
88. Icko Iben Jr. and Alvio Renzini (eds.), *Physical Processes in Red Giants. Proceedings of the Second Workshop, helt at the Ettore Majorana Centre for Scientific Culture, Advanced School of Agronomy, in Erice, Sicily, Italy, September 3–13, 1980.* 1981, XV + 488 pp.
89. C. Chiosi and R. Stalio (eds.), *Effect of Mass Loss on Stellar Evolution. IAU Colloquium No. 59 held in Miramare, Trieste, Italy, September 15–19, 1980.* 1981, XXII + 532 pp.
90. C. Goudis, *The Orion Complex: A Case Study of Interstellar Matter.* 1982, XIV + 306 pp.
91. F. D. Kahn (ed.), *Investigating the Universe. Papers Presented to Zdenek Kopal on the Occasion of his retirement, September 1981.* 1981, X + 458 pp.
92. C. M. Humphries (ed.), *Instrumentation for Astronomy with Large Optical Telescopes, Proceedings of IAU Colloquium No. 67.* 1981, XVII + 321 pp.
93. R. S. Roger and P. E. Dewdney (eds.), *Regions of Recent Star Formation, Proceedings of the Symposium on "Neutral Clouds Near HII Regions - Dynamics and Photochemistry", held in Penticton, B.C., June 24–26, 1981.* 1982, XVI + 496 pp.
94. O. Calame (ed.), *High-Precision Earth Rotation and Earth-Moon Dynamics. Lunar Distances and Related Observations.* 1982, XX + 354 pp.
95. M. Friedjung and R. Viotti (eds.), *The Nature of Symbiotic Stars,* 1982, XX + 310 pp.
96. W. Fricke and G. Teleki (eds.), *Sun and Planetary System,* 1982, XIV + 538 pp.
97. C. Jaschek and W. Heintz (eds.), *Automated Data Retrieval in Astronomy,* 1982, XX + 324 pp.
98. Z. Kopal and J. Rahe (eds.), *Binary and Multiple Stars as Tracers of Stellar Evolution,* 1982, XXX + 503 pp.
99. A. W. Wolfendale (ed.), *Progress in Cosmology,* 1982, VIII + 360 pp.
100. W. L. H. Shuter (ed.), *Kinematics, Dynamics and Structure of the Milky Way,* 1983, XII + 392 pp.

101. M. Livio and G. Shaviv (eds.), *Cataclysmic Variables and Related Objects,* 1983, XII + 351 pp.
102. P. B. Byrne and M. Rodonò (eds.), *Activity in Red-Dwarf Stars,* 1983, XXVI + 670 pp
103. A. Ferrari and A. G. Pacholczyk (eds.), *Astrophysical Jets,* 1983, XVI + 328 pp.
104. R. L. Carovillano and J. M. Forbes (eds.), *Solar-Terrestrial Physics,* 1983, XVIII + 860 pp.
105. W. B. Burton and F. P. Israel (eds.), *Surveys of the Southern Galaxy,* 1983, XIV + 310 pp.
106. V. V. Markellos and Y. Kozai (eds.), *Dynamical Trapping and Evolution on the Solar System,* 1983, XVI + 424 pp.
107. S. R. Pottasch, *Planetary Nebulae,* 1984, X + 322 pp.
108. M. F. Kessler and J. P. Phillips (eds.), *Galactic and Extragalactic Infrared Spectroscopy,* 1984, XII + 472 pp.
109. C. Chiosi and A. Renzini (eds.), *Stellar Nucleosynthesis,* 1984, XIV + 398 pp.
110. M. Capaccioli (ed.), *Astronomy with Schmidt-type Telescopes,* 1984, XXII + 620 pp.
111. F. Mardirossian, G. Giuricin, and M. Mezzetti (eds.), *Clusters and Groups of Galaxies,* 1984, XXII + 659 pp.
112. L. H. Aller, *Physics of Thermal Gaseous Nebulae,* 1984, X + 350 pp.
113. D. Q. Lamb and J. Patterson (eds.), *Cataclysmic Variables and Low-Mass X-Ray Binaries,* 1985, XII + 452 pp.
114. M. Jaschek and P. C. Keenan (eds.), *Cool Stars with Excesses of Heavy Elements,* 1985, XVI + 398 pp.
115. A. Carusi and G. B. Valsecchi (eds.), *Dynamics of Comets: Their Origin and Evolution,* 1985, XII + 442 pp.
116. R. M. Hjellming and D. M. Gibson (eds.), *Radio Stars,* 1985, XI + 411 pp.

GPSR Compliance

The European Union's (EU) General Product Safety Regulation (GPSR) is a set of rules that requires consumer products to be safe and our obligations to ensure this.

If you have any concerns about our products, you can contact us on ProductSafety@springernature.com

In case Publisher is established outside the EU, the EU authorized representative is:

Springer Nature Customer Service Center GmbH
Europaplatz 3
69115 Heidelberg, Germany

Batch number: 10815974

Printed by Printforce, the Netherlands